Aging Men's Health
A Case-Based Approach

Aging Men's Health
A Case-Based Approach

Robert S. Tan, M.D.

Clinical Director
Extended Care Line
Michael E. DeBakey Veterans Affairs Medical Center
Associate Professor of Medicine (Geriatrics)
Baylor College of Medicine
Clinical Associate Professor of Family Medicine
University of Texas
Houston, Texas
Advisory Board
Men's Health Network
Washington, D.C.

Thieme
New York • Stuttgart

Thieme Medical Publishers, Inc.
333 Seventh Ave.
New York, NY 10001

Associate Editor: Owen Zurhellen
Consulting Editor: Esther Gumpert
Vice President, Production and
    Electronic Products: Anne Vinnicombe
Production Editor: Becky Dille
Marketing Director: Phyllis Gold
Sales Director: Ross Lumpkin
Chief Financial Officer: Peter van Woerden
President: Brian D. Scanlan

Compositor: Techset Composition, Ltd.
Printer: Maple-Vail Book Manufacturing Group

Library of Congress Cataloging-in-Publication Data

Tan, Robert S.
Aging men's health : a case-based approach / Robert S. Tan
    p. ; cm.
Includes bibliographical references and index.
ISBN 1-58890-296-X (TMP : alk. paper) - - ISBN 3-13-139391-2 (GTV : alk. paper)1. Older
men- -Diseases. 2. Older men- -Medical care.
[DNLM: 1. Aging- -Case Reports. 2. Androgens- -Case Reports. 3. Geriatrics- -methods- -Case
Reports. 4. Men- -Aged- -Case Reports. 5. Men- -Middle Aged- -Case Reports.] I. Title.
RC952.5.T35 2005
618.97'0081- -dc22
                                                           2005002486

Important note: Medical knowledge is ever-changing. As new research and clinical experience
broaden our knowledge, changes in treatment and drug therapy may be required. The authors and
editors of the material herein have consulted sources believed to be reliable in their efforts to
provide information that is complete and in accord with the standards accepted at the time of publication. However, in the view of the possibility of human error by the authors, editors, or publisher of
the work herein or changes in medical knowledge neither the authors, editors, nor publisher, nor any
other party who has been involved in the preparation of this work, warrants that the information
contained herein is in every respect accurate or complete, and they are not responsible for any errors
or omissions or for the results obtained from use of such information. Readers are encouraged to
confirm the information contained herein with other sources. For example, readers are advised to
check the product information sheet included in the package of each drug they plan to administer
to be certain that the information contained in this publication is accurate and that changes have not
been made in the recommended dose or in the contraindications for administration. This recommendation is of particular importance in connection with new or infrequently used drugs.

Some of the product names, patents, and registered designs referred to in this book are in fact registered trademarks or proprietary names even though specific reference to this fact is not always made
in the text. Therefore, the appearance of a name without designation as proprietary is not to be con-
strued as a representation by the publisher that it is in the public domain.

Printed in The United States of America.

5 4 3 2 1

TMP ISBN 1-58890-296-X
GTV ISBN 3-13-139391-2

In Thanks

*I wish to thank Solvay Pharmaceuticals, University of Texas,
Baylor College of Medicine, Veterans Administration,
and the many contributors for their kind support of this work.
Special mention must be made of the patience of the editorial staff at Thieme,
in particular Owen Zurhellen and Rebecca Dille.
This book is dedicated to my aging parents and to
the many patients whose quality of life we hope to improve.*

Contents

Foreword

"While a gender-specific approach is often used to identify persistent inequalities in the status of women, the specific situation of men, particularly of older men, also requires investigation and further studies, especially with regard to the determinants of health."

Men, Aging and Health; WHO 2001

Despite many well-recognized facts, for example, that life expectancy differs between men and women, and that osteoporosis and depression are no longer "women's diseases," there has been near silence about men's health in general among politicians, the medical community, and the media. It is impossible to understand aging and health without a gender-based approach. From a physiological and psychosocial point of view, the determinants of health as we age are related to gender. It has only been a few years that aging men's health has received increased attention in medical research, in publication of evidence-based scientific articles, and in scientific meetings such as the World Congress on Men's Health (www.wcmh.info). Gender-sensitive and gender-specific medicine is needed to provide longer and healthier lives for men in general.

Aging Men's Health presents the state of the art in aging men's health. It contains knowledge that the medical community needs in order to do its work professionally. Twenty chapters written by leaders in the field encompass the information needed by practicing colleagues and academic institutions interested in gender medicine, preventive gerontology, geriatrics, and men's health. This information is needed today more than ever. The book provides the most recent information on longevity and cardiovascular health. It gives straight answers on the controversies and effects of androgen and growth hormone replacement. It includes chapters on andropause, erectile dysfunction, the depression–erectile dysfunction–coronary heart disease (D.E.C.) syndrome, and prostate cancer, which all are of great interest to practicing colleagues.

This book broadens the context in which we think about aging men's health and tries to fill the gap created by the paucity of existing research data on men. This, in general, is a difficult issue that needs to be addressed immediately. It will be essential to stimulate appropriate research for men of all ages and to develop comprehensive gender-specific databases and suitable evaluation methods for the gender differences between men and women. Our definition of gender-specific medicine and men's health will inevitably expand in direct proportion to research and to the emergence of new concepts concerning the importance of sex and gender in health.

It is a privilege to write this foreword as the president of the International Society for Men's Health and Gender (ISMH, www.ismh.org) and Editor-in-Chief of the *Journal of Men's Health and Gender* (JMHG). The ISMH was established in 2000 as an international independent organization to provide a platform for all interested organizations and to promote concepts concerning men's health and gender medicine in order to contribute to the development of good practice. The JMHG was launched in 2004. It is an international interdisciplinary journal offering updates on practice-based issues, current research, and policy matters covering all aspects of men's health and gender medicine.

This book is an invaluable research-oriented source of knowledge that addresses, and promises to influence, the medical care and management of aging men.

SIEGFRIED MERYN, M.D.
Professor of Medicine
President of the ISMH
Editor-in-Chief, JMHG
Center for Advanced Medical Education
Medical University of Vienna, Austria

Overview

We are entering an age of global graying of men as well as women, and physicians in the majority of specialties, not just geriatricians, need the latest information and ideas on ways of preventing and treating diseases of later life. This is just what Professor Robert Tan and his co-authors provide for men in this very practical, case-based book.

In most countries, modern medicine has added years to life, but now we need to be able to add life to years. Prevention is certainly better than treatment of disease, particularly in the elderly where disease processes are more difficult to halt and reverse. Slowing older men's progress toward the 3D's, debility, dependency, and death requires action in midlife as well as later life. To help the individual limit the time spent in the first two of these stages, and live life as far as possible like an alkaline battery, going full charge to the end, involves skilled input from a variety of health professionals.

Those at the frontline in this battle will find throughout this book a wealth of advice and guidance. In an early chapter on delaying aging in men, it is argued that because on average in most of the Western world women live seven years longer than men, there are lessons that can be learned from women. While there is little that men can do about the alleged genetic limit put on longevity by the Y chromosome, lifestyle changes even late in life can make a large difference in the rate at which aging occurs and the diseases suffered. It seems the well-explained theories of why we age, especially as relates to endocrine function, free radicals, and calorie restriction, become ever more complex. Though the medical reasons for trying to get people to adopt an optimistic, healthy, food-limiting lifestyle, rich in antioxidants and physical activity are clearly put forth, they still have to be applied against a worsening background of commercial pressures toward stress, sloth, and gluttony.

Gerontologists will appreciate the chapter on the real role of prevention to achieve optimal aging, even though if successfully applied in middle age to reduce premature aging, it would keep many people away from their clinics and in-patient beds. Many new endocrine aspects of osteoporosis and arthritis are also explored and practical advice given. There is also interesting coverage on the role nutraceuticals can play in preventive medicine in relation to benign enlargement, and possibly cancer, of the prostate, sex hormone formation and action, and the role of the compounding pharmacist in advising clinicians and patients alike on these issues. This seems to offer ways in which large sectors of the population can be encouraged to adopt positive measures to maintain their own health.

Urologists will find some of the latest research on many topics making up a large part of their specialist practice. The controversies and developments in both screening and treating prostate cancer are covered from the point of view of unmodifiable risk factors such age, race, and familial background, and modifiable ones such as diet and lifestyle. As elsewhere in the book, there is emphasis on the active participation of the patient in all stages of the diagnosis and treatment of his cancer, and, if it has become incurable, converting it to a controllable chronic illness.

Erectile dysfunction is dealt with in great detail, particularly highlighting the interlinking contributory factors of depression and coronary heart disease. This again emphasizes the need for a multidisciplinary approach to promoting health in the aging male. The overview of erectile dysfunction explores the extent to which it is related to age, diabetes, and iatrogenic causes such as medications and prostate surgery. The various treatments for erection problems described underline the fact that this is one of the most rapidly advancing, and widely appreciated, fields of pharmacotherapy in the whole of medicine.

Cardiologists are likely to enjoy the excellent description of the important role of androgen deficiency in coronary heart disease, which may well be the underlying thread in these three interlinked conditions of depression, erectile dysfunction, and coronary heart disease. It presents exciting new epidemiological evidence of the inverse relationship between bio-available testosterone levels and atherosclerosis, blood clotting factors, and the incidence of coronary thrombosis. It describes the important part that lack of testosterone may play in the impaired carbohydrate tolerance, hypertension, hyperlipidemia, obesity, and "metabolic syndrome." Also, this section further highlights the effects of androgens in dilating both coronary and peripheral arteries, and in improving arterial compliance. This impressive array of androgen actions in relation to cardiovascular disease emphasizes the need for further research in this area, leading hopefully to improvements in this key area of men's health.

Psychiatrists will be equally fascinated by the many insights given into two of the major conditions they have to treat, depression and dementia. The disturbing statistics that 8 to 16% of community-dwelling seniors, and up to 20% of institutionalized ones, may have significant depression, combined with the fact that the suicide rate for men is four times higher than that for women, lend urgency to the debate on the origins and treatment of this comment on the quality of a person's life.

Both depression and memory loss are prominent in the highly characteristic pattern of symptoms of androgen deficiency comprising the increasingly recognized and treated condition of the andropause. Professor Tan has particular interest and expertise in this area and provides impressive evidence on the links between low endogenous testosterone levels and depression, and the importance of considering the differential diagnosis, as well as the effect that stress can have in lowering androgen levels, both before starting treatment, and in considering alternatives in patients resistant to antidepressant drugs.

The authors report the abundance of human evidence linking sex hormones and dementia. When testosterone is suppressed in men undergoing hormonal treatment for prostate cancer, beta-amyloid levels in the blood rise, and fall again when the treatment is stopped. There have been limited studies showing both enhanced cerebral glucose metabolism in men with low androgen levels given testosterone, and improved cognition as well as reduced frailty in men with Alzheimer's disease.

Throughout this stimulating book there are many thought-provoking ideas to encourage doctors in every field of medicine to consider how they can apply the knowledge it provides to their sphere of medicine for the benefit of men's health in reaping the mixed blessings of living to a ripe old age.

MALCOLM CARRUTHERS, M.D., F.R.C.Path., M.R.C.G.P.
Chairman, Andropause Society
London, United Kingdom

Preface

Aging Men's Health was conceived to fill the need for a greater understanding of the specific aging issues of men. It is targeted at busy practitioners treating aging men in primary care, geriatrics and gerontology, endocrinology, urology, and other disciplines, and is case-based to suit the clinician. The book defines "aging" as a continuous process but targets the male in midlife and beyond.

It is often difficult to reverse disease states, especially in old age. Chronic disease tends to plague older individuals. The prevention of aging, also known as preventive gerontology, is therefore key to helping men (and women) to enjoy a long, healthy life. Exercise, the maintenance of ideal body weight, and taking of selected nutraceuticals can indeed modify the aging process, and we examine these measures for men in this book.

There is a sustained difference in longevity between men and women, and possible reasons for this are explored.

Aging is accompanied by many changes, which may have endocrine, cardiovascular, or neurologic bases. Often these physiological and pathological changes are interrelated and overlapping. A common theme in many of the chapters is the link between these changes and the declining production of androgens in aging men. Androgens are found in relative abundance in men compared with women, and these hormones are responsible for much of gender differentiation. In addition to affecting sexuality, decline in androgens can affect several physiological systems such as the brain, heart, and joints. The impact of hormone replacement in men to mediate these effects is explored in detail in several chapters.

Illnesses specific to the aging male such as prostate cancer, erectile dysfunction, and premature ejaculation are also covered.

Aging Men's Health is the collaborative effort of experts in different clinical disciplines across the continents. It brings into perspective the challenges every practitioner faces in treating the older male. Newer and novel therapeutics have spearheaded the interest in men's health in recent years but there is no separate discipline of men's health at the moment. It falls within the domain of primary care or the specialties. Many groups do exist to advance the health of men, including the International Society for Men's Health, International Society for the Study of the Aging Male, Andropause Society, and Men's Health Network in the United States.

The aim of this book is to educate and stimulate interest in this emerging medical field. Its ultimate goal is to give the practitioner better tools to maintain the health of aging men. My sincere hope is that this book will be useful and enlightening.

ROBERT S. TAN, M.D.
Clinical Director
Extended Care Line
Michael E. DeBakey Veterans Affairs
Medical Center
Associate Professor of Medicine (Geriatrics)
Baylor College of Medicine
Clinical Associate Professor of Family Medicine
University of Texas
Houston, Texas
Advisory Board
Men's Health Network
Washington D.C.

Contributors

Bruce Biundo, R.Ph.
Adjunct Clinical Professor
University of Houston College of Pharmacy
Houston, Texas
Professional Compounding Centers of America
   (PCCA)
Houston, Texas

**Kew-Kim Chew, M.D., F.R.C.P.(Edin.),
   F.R.C.P.(Glasg.)**
Senior Clinical Fellow
Keogh Institute for Medical Research
Queen Elizabeth II Medical Centre
Nedlands, Perth, Western Australia

Emma Cid, M.D.
Department of Cardiology
University of Texas Medical School
Houston, Texas

Melanie G. Cree, B.A.
Medical Student
Department of Preventive Medicine and
   Community Health
Metabolism Unit
University of Texas Medical Branch
Galveston, Texas

Christopher B. Cutter, M.D.
Private Practice
Mamou, Louisiana
Active Staff
Savoy Medical Center
Mamou, Louisiana

Grant C. Fowler, M.D.
Professor and Vice Chairman
Family and Community Medicine
University of Texas
Houston, Texas

Ginny Fullerton, B.S.
Graduate Research Assistant
Department of Psychology
University of Houston
Houston, Texas

Christina Ho, B.S.
Research Assistant
Houston, Texas

Michelle Iannuzzi-Sucich, M.D.
Modena Family Practice
Modena, New York

Meenu Jacob, M.D.
Staff Physician
Extended Care Line
Michael E. DeBakey Veterans Affairs
   Medical Center
Houston, Texas

Darren W. Lackan, M.D.
Assistant Professor of Medicine
Department of Internal Medicine
Metabolism Unit
University of Texas Medical Branch
Galveston, Texas

Peter Huat-Chye Lim, M.D.,
 M.Med. (Surg.)., Dip.Urol. (Lond.)
Adjunct Professor and Senior Consultant
Edith Cowan University
Perth, Western Australia
Professor of Andrology and Men's Health
Andrology, Urology, and Continence Centre
Gleneagles Hospital
Singapore, Singapore

Ralph N. Martins, Ph.D.
The Sir James McCusker Alzheimer's Disease
 Research Unit
Hollywood Private Hospital
Nedlands, Perth, Western Australia

Michael Mistric, R.N., M.N.Sc., A.P.R.N., B.C.,
 F.N.P.-C.
Extended Care Line
Michael E. DeBakey Veterans Affairs Medical
 Center
Houston, Texas
Adjunct Clinical Professor
School of Nursing
University of Texas Medical Branch
Galveston, Texas

Perianan Moorthy, Ph.D.
Andrology, Urology, and Continence Centre
Gleneagles Hospital
Singapore, Singapore

Thomas Curtis Namey, M.D., F.A.C.P.,
 F.A.C.R.
Professor of Medicine, Nutrition,
 and Exercise Science
Director
Applied Physiology Laboratory
Associate Director
The Nutrition Institute
University of Tennessee
Knoxville, Tennessee

Akira Nishikawa, M.D.
Clinical Professor
Department of Internal Medicine
Division of Cardiology
University of Texas Health Science Center
Houston, Texas

Owner and President
Advanced Cardiac Care Association
Houston, Texas

Claudia A. Orengo, M.D., Ph.D.
Associate Professor of Psychiatry
Department of Psychiatry
Baylor College of Medicine
Houston, Texas
Psychiatry Service Line
Michael E. DeBakey Veterans Affairs Medical
 Center
Houston, Texas

Douglas Paddon-Jones, Dip.App. Sc. (Diag.
 Rad.), B.Sc. (Hons.), M.S., Ph.D.
Assistant Professor
Department of Surgery
Metabolism Unit
University of Texas Medical Branch
Galveston, Texas

Carlos A. Plata, M.D.
Staff Rheumatologist
Amarillo Veterans Affairs Medical Center
Amarillo, Texas
Associate Clinical Professor
Texas Tech Medical School
Lubbock, Texas

Shou-Jin Pu, M.D.
Department of Internal Medicine
Chang Gung Memorial Hospital
Taipei, Taiwan, Republic of China

Melinda Sheffield-Moore, Ph.D.
Associate Professor
Department of Internal Medicine
Division of Endocrinology and Metabolism
University of Texas Medical Branch
Galveston, Texas

Robert S. Tan, M.D.
Clinical Director
Extended Care Line
Michael E. DeBakey Veterans Affairs
 Medical Center
Associate Professor of Medicine (Geriatrics)
Baylor College of Medicine

Clinical Associate Professor of Family Medicine
University of Texas
Houston, Texas
Advisory Board
Men's Health Network
Washington, D.C.

Pamela Taxel, M.D.
Assistant Professor of Medicine
Department of Endocrinology
University of Connecticut Health Center
Farmington, Connecticut

**Seng-Hin Teoh, M.D., M.R.C.O.G. (London),
  M.Med. (S'pore)**
Private Practice
Singapore, Singapore

Randall J. Urban, M.D.
Interim Administrative Chair of Internal
  Medicine
Department of Internal Medicine
Metabolism Unit
University of Texas Medical Branch
Galveston, Texas

Robert Yong, M.D., F.R.C.S.C., F.A.C.S.
Division of Urology
Department of Surgery
Surrey Merial Hospital
Surrey, British Columbia
Canada

Preventive Gerontology in Men's Health: Optimal Aging Concepts for Midlife and Beyond

ROBERT S. TAN AND AKIRA NISHIKAWA

Mortality, Morbidity, and Longevity in Men

Preventive gerontology is a concept applied to all stages of life whereby modifying risk factors that can promote aging can, in turn, improve longevity. Obvious examples include smoking cessation, diet, and exercise. This is to be differentiated from *preventive geriatrics*, which focuses on health maintenance and disease prevention when old age, with its associated frailty, has been reached.[1,2]

In the United States today, the average life span of men is about 7 years short of that of women. Arguably, women may have a long period of chronic illness, as life span is greater. The longevity of women is due to a variety of factors including genetics, endocrine, and perhaps lifestyle, along with arguably better preventive screening programs. Women, by and large, visit doctors more frequently, partly because of the need for help with reproductive issues. Screening for comorbid states such as hypertension, hypercholesterolemia, and diabetes occurs more frequently in women. There are data supporting excess mortality and morbidity in men. In general, the four most important causes of death in men are cardiovascular disease, cancer, accidents, and suicide.

Coronary heart disease (CHD) was responsible for one of every five deaths in the United States in 1997. Although there is currently evidence that CHD may be as common in women as in men, symptoms of presentation may *differ* between men and women. In general, mortality rates of myocardial infarction in men are believed to be lower than those in woman.[3] For example, 42% of women who have heart attacks die within 1 year compared with 24% of men. The reasons for this are not well understood. The explanation accepted by many is that women tend to get heart disease *later* in life than do men and are more likely to have other coexisting, chronic conditions. Studies also indicate that

men and women react to drugs prescribed for heart disease and other conditions quite differently.

Because the United States is a heterogeneous society, cancer rates can differ based on ethnic backgrounds. The leading cancer in men, regardless of race, is prostate cancer, followed by lung/bronchus and colon/rectal. Prostate cancer rates are 1.5 times higher in black men than white men. Although it is important to realize the frequency of cancers, it is not quite the same with the way cancer affects morbidity and mortality. Prostate cancer and breast cancers are reported to be more common partly because of the availability of screening tools such as prostate-specific antigen (PSA) and mammograms. Autopsy results reveal that most men die *with* prostate cancer but not *of* prostate cancer. This contrasts with lung cancer, for instance, because lung cancer, when diagnosed, often is the terminal event. The leading cancer in women, regardless of race, is breast cancer, followed by lung/bronchus and colon/rectal in white women, and colon/rectal and lung/bronchus in black women. Breast cancer rates are ~20% higher in white women than in black women. Melanomas of the skin and cancer of the testis are among the top 15 cancers for white men but not black men. Melanomas of the skin and cancer of the brain or other nervous systems are among the top 15 cancers for white women, but not black women. Multiple myeloma and cancer of the stomach are among the top 15 cancers for black women, but not white women. Multiple myeloma and cancer of the liver are among the top 15 cancers for black men, but not white men.[4]

As a measure of *morbidity* trends in men, a community-based British study revealed that the ailments that plagued men more than women include gout, duodenal ulcer, venereal disease, coronary heart disease, bladder cancer, and alcoholism.[5] Knowledge of morbidity and mortality trends is useful, as it helps us plan for preventive strategies. Gout, duodenal ulcer, venereal disease, coronary heart disease, bladder cancer, and alcoholism are usually the result of lifestyle and environment issues and are modifiable.

It is also interesting to study the *life expectancy* of men in different countries, as this may give us insight as to lifestyle factors such as diet and exercise. Life expectancy from birth, representing the average life span of a newborn, is an indicator of the overall health of a country. By and large, better life expectancy is the result of improvements in public health nutrition and prevention. The overall health of a country, however, is not a direct correlation of the wealth of the country (Fig. **1–1**). For example, Saudi Arabia, which has a very high gross national product (GNP) per capita, has an average life expectancy of ~68 years. This is actually less than that in El Salvador by 2 years and about the same as that in the Philippines. The countries with the highest average life expectancies are Andorra (84 years), San Marino (81 years), Japan (81 years), and Singapore (80 years). The average life expectancy in the United States in comparison is 77 years.[6]

In most countries, there is a *longevity differential* between men and women of ~4 to 6 years. Andorra is a small country sandwiched between France and Spain that boasts of a Mediterranean climate. The foods are typically seafood and oils, including olive. Wine is a characteristic of the region. San Marino is also a small

European country to the north of Italy, with similar Mediterranean attributes. It is also hilly, which implies that much walking has to be done. The Asian countries of Japan and Singapore are both modern and have good public health systems in place. The Japanese are known for their fondness for seafood and green tea, which may be responsible factors for their longevity.

Recent research has suggested that there may be physiological markers for longevity. Three *physiological measures* associated with long-term caloric restriction in monkeys have been linked to longevity in men and include skin temperature, insulin, and dehdroepiandrosterone sulfate (DHEAS) levels.[7] Roth et al[7] compared more than 700 healthy men, ages 19 to 95, who participated in the Baltimore Longitudinal Study of Aging (BLSA), with 60 rhesus monkeys, ages 5 to 25. The men were divided into two groups, based on whether they were in the upper or lower halves of the population for each of the three biomarkers: body temperature, blood insulin levels, and blood levels of DHEAS. The monkeys also were divided into two groups: one group was allowed to feed freely, typically consuming between 500 to 1000 calories daily; the other group was fed a diet composed of at least 30% fewer calories than consumed by the unrestricted monkeys. After analyzing the age-adjusted data, the investigators concluded that among men who participated in the BLSA, those who had *lower* body temperatures, *lower* blood insulin levels, or *higher* blood levels of DHEAS as they aged tended to live longer (Fig. **1–2**). The calorie-restricted

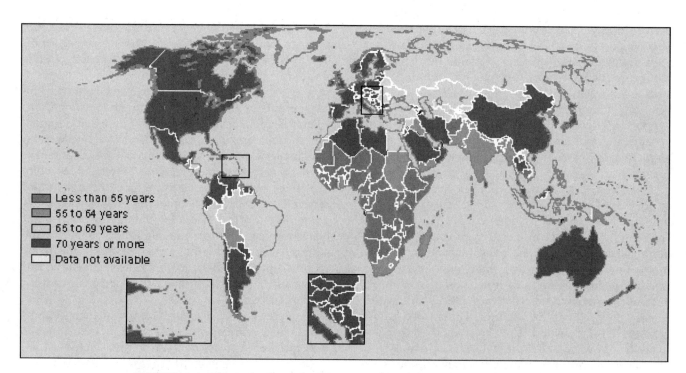

FIGURE 1–1. **This map shows the lifespans in different countries of the world.**

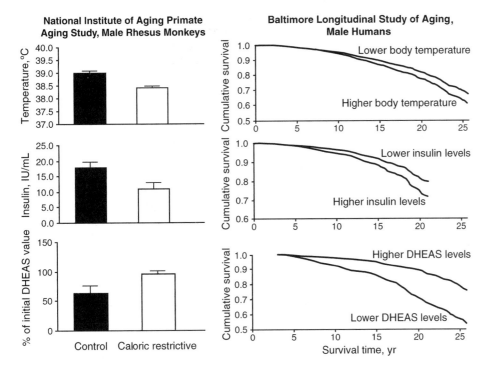

FIGURE 1–2. Biomarkers of caloric restriction such as low temperature, low insulin, and high DHEAS levels may predict longevity. (Adapted from Roth GS, Lane MA, Ingram DK, et al. Biomarkers of caloric restriction may predict longevity in humans. Science 2002;297:811, with permission.)

monkeys showed a similar trend and had half the death rate of monkeys allowed to feed freely. This study is obviously not conclusive, but we know that obesity is associated with higher levels of insulin, and obesity certainly is one of the modifiable risk factors for premature death.

The association of high insulin with diminished life span suggests that insulin resistance may play a part in the mortality of men. However, the association of low DHEAS with diminished life span does not necessarily imply that consuming dehydroepiandrosterone (DHEA) would improve longevity. That has yet to be determined, and there are no prospective controlled trials as yet.

Prevention and Implications for Longevity

Prevention has different components. In primary prevention, we educate our patients about risk factors and lifestyle changes that reduce risk. We also identify and alter risk factors to prevent the onset of disease. In contrast, secondary prevention refers to education and treatment after the onset of the disease, to limit its progress. It is intuitive to think that preventive strategies result in early detection of disease and as a result early intervention. Does it in the end lead to increase

life expectancy, and better quality of life? There is ample scientific evidence that diet, exercise, alcohol moderation, and cessation of smoking are the major modifiable factors that lead to increased life expectancy. Interventions with hormonal replacement such as estrogen in women and testosterone in men have demonstrated improved quality of life, but the effects on longevity are not clear. The Women's Health Initiative demonstrated that in women, replacement with an estrogen/progesterone preparation improved quality of life such as by decreasing hot flashes and improving osteoporosis.[8] Unfortunately, this combination of hormones leads to an increase in thrombotic events. Testosterone replacement in physiological doses in men has definitely improved quality of life, but the long-term effects, though promising, remain unclear. This is not unlike many aspects of medicine, and as such testosterone replacement in men should be carefully monitored and individualized. *Lifestyle changes should accompany any preventive medicine program.* Preventive strategies altering cardiovascular risk factors such as treating high blood pressure, hypercholesterolemia, and arguably inflammatory processes in the cardiovascular system and elsewhere can certainly decrease morbidity and mortality, and in turn result in enhanced longevity. Nutraceuticals is a large area of controversy in terms of whether it actually prevents disease. Chapter 20 discusses nutraceuticals and their impact on men.

Limitations of Current Guidelines for Preventive Medical Care

A very large interface with preventive care is cost and managed care. It is prudent to maintain health, and thereby decrease the long-term cost of health care. Indeed, health maintenance organizations (HMOs) are set up with that purpose in mind, as prevention decreases risk. Unfortunately, it would *not* be possible to be comprehensive with preventive programs under managed care programs because of cost. As such, only the basic preventive programs are recommended, such as a physical examination, rectal examinations, Pap smears, mammograms, blood pressure, cholesterol levels, etc. These procedures, although essential, must not be mistaken for a complete passage of a good bill of health. There are several recommendations from different bodies for preventive care. Interest groups such as the American Cancer Society have very tight guidelines for preventive care in the area of cancer. These contrast with those of the U.S. Preventive Services Task Force (USPTF), which relies much on evidence-based medicine in community-based practice. Though the USPTF recommendations are widely used, they may lack the following:

1. Gender-specific recommendations for men above 40 years of age. For instance, there is no mention of screening for hypogonadism, especially with the symptomatic. There are screening tools such as the Androgen Decline in Aging Males (ADAM) questionnaire. Erectile dysfunction may be a quality-of-life issue, but it also may be a harbinger for cardiovascular disease, and screening tools such as the International Index of Erectile Function (IIEF) exist. Despite the fact that prostate cancer is the most common cancer in men, USPTF does not recommend routine screening. This may be because of the limitations of the current methods for screening for prostate cancer.

2. Gender-specific recommendations for prevention in men above 65 years of age. Osteoporosis also occurs in men, albeit later in life than in women. In fact, the mortality from hip fractures in men in later life is higher than that in women. Unfortunately, few men are screened for osteoporosis, or educated about smoking and alcohol as contributory factors (Table 1–1). Dementia is a function of age, and no routine screening is mentioned such as the Mini–Mental State Examination (MMSE). Perhaps the reason is that the MMSE is not sensitive or specific enough as a tool to predict early dementia.[9] There is also no screening for benign prostate hyperplasia (BPH) and lower urinary tract symptoms (LUTSs), which are common ailments affecting men in this age group.

Emerging Concepts for Specific Areas of Prevention in Men

Prevention for our patients should take into consideration the guidelines, which at times should be individualized especially as new concepts are raised. But it takes time for a consensus among physicians to be built. Some of these emerging concepts are the following:

- Obesity is a universal health problem and is usually the result of an inactive lifestyle and paradoxically malnutrition. Obesity is often associated with the metabolic syndrome. Researchers at the Centers for Disease Control and Prevention (CDC) estimated that as many as 47 million Americans may exhibit a cluster of medical conditions often termed the metabolic syndrome, which is characterized by insulin resistance and the presence of obesity, abdominal fat, high blood sugar and triglycerides, high blood cholesterol, and high blood pressure.[10] Loss of weight is crucial to maintaining health and decreasing morbidity and mortality. One of the emerging concepts that has been debated is the use of the appropriate diet to lose weight and hence decrease the possibility of

TABLE 1–1. Osteoporosis and Risk of Hip Fracture in Men and Women: A Comparison

Factor	Men	Women
Peak bone mass[5,7]	10% to 12% greater than in women	
Lifetime risk of hip fracture at age 50[3]	6%	17.5%
Sex distribution of hip fractures worldwide[3,7]	30%	70.0%
U.S. incidence of hip fracture at age 65[3]	4 to 5 per 1000	8 to 10 per 1000
Mortality from hip fracture[7]	31%	17.0%

Adapted from Campion JM, Maricic MJ. Osteoporosis in men. Am Fam Physician 2003;67:1521–1526, with permission.

the metabolic syndrome. Proponents of a low-carbohydrate diet, including Atkins, Zone, Ornish, Wadden, and others have challenged the traditional Food and Drug Administration (FDA) food pyramid. To validate the Atkins diet, researchers recently completed a 1-year, multicenter, controlled trial involving 63 obese men and women who were randomly assigned to either a low-carbohydrate, high-protein, high-fat diet or a low-calorie, high-carbohydrate, low-fat (conventional) diet. Professional contact was minimal to replicate the approach used by most dieters. The low-carbohydrate diet produced a 4% greater weight loss than did the conventional diet for the first 6 months, but the differences were not significant at 1 year. The low-carbohydrate diet was associated with a greater improvement in some risk factors for coronary heart disease. Unfortunately, adherence was poor and attrition was high in both groups.[11] Men seen in practice often ask which diet is best for them to lose weight, but exercise is important to keep the weight off in the long term. There is also ongoing research as to whether androgen replacement can decrease fat mass in men, and preliminary data suggest that it does.[12] Chapter 11 discusses the impact of androgens on obesity in men.

- There have been many inroads in cardiovascular health prevention in recent years. High low-density lipoprotein (LDL) cholesterol and triglycerides, low high-density lipoprotein (HDL), hypertension, and diabetes are no longer seen to be the *only* risk factors for coronary heart disease in men. LDL cholesterol is a serum lipoprotein that contains apolipoprotein B (ApoB), cholesterol, and triglycerides. LDL is the most atherogenic of the lipoproteins. Recent evidence suggests that the oxidized form of LDL may play a key role in the initiation and progression of atherosclerosis, and like lipoprotein(a), oxidized LDL forms foam cells, which are associated with the formation of atherosclerotic plaques. Smoking and family history are other obvious risk factors. The following factors are other emerging concepts that can contribute to coronary heart disease:

  ○ *Elevated homocysteine*: Homocysteine is a sulfur-containing amino acid, and when it becomes elevated, it can damage coronary arteries, cell structures, blood lipids, and artery walls, eventually leading to the development of atherosclerosis and other forms of heart disease. Vitamins B_6, B_{12}, and folate, involved in homocysteine metabolism, act to regulate and reduce homocysteine. The assessment of homocysteine status and B vitamins, particularly B_6, B_{12}, and folate, is useful because heart disease is the leading cause of fatality in the United States. In addition, current medical research reports that an elevated homocysteine status and/or deficiency of B_6, B_{12}, or folate increases the risk for heart disease. Approximately two thirds of cases with elevations in homocysteine are related to *deficiencies* of one or more of these B vitamins, and can be reversed potentially by supplementation of these micronutrients.

  ○ *Elevated lipoprotein(a)* [LP(a)] is a complex of ApoA and LDL, and an elevated status is associated with an increased risk for atherosclerosis and cardiovascular disease. The pathogenic role of Lp(a) is similar to that of LDL in the development of atherosclerosis; it is localized in the blood vessel walls, and then oxidized. When oxidized, it forms the foam cells associated with atherosclerotic plaques. Diet and exercise can potentially decrease levels of Lp(a).

  ○ *Elevated C-reactive protein* (CRP) is a nonspecific indicator of systemic inflammation and infection. One of the emerging concepts of heart disease is that it is inflammatory in nature, and results in endothelial dysfunction. CRP levels rise rapidly in response to tissue injury and inflammation. Exercise, aspirin, and a healthy diet can decrease CRP levels.

  ○ *Elevated fibrinogen status*: Fibrinogen is an important coagulation protein that is involved in the mesh-like network of the common blood clot. Studies have shown that elevated fibrinogen status is associated with early coronary heart disease.

  ○ *Low apolipoprotein A-I and apolipoprotein B status*: Apolipoprotein A-I (ApoAI) is the major protein constituent of HDL. This molecule is responsible for the activation of two enzymes that are necessary for the formation of HDL, and this process may be a key factor in the relationship between HDL levels and the incidence of atherosclerosis. On the other hand, apolipoprotein B (ApoB) is the primary protein found in LDL. Studies suggest that ApoB plays a major role in targeting the selective uptake of LDL by the liver, and it has been identified as one component of the syndrome known as atherogenic lipoprotein phenotype, which is a common disorder in persons at risk for atherosclerosis.

- Osteoporosis does not occur only in women. For instance, ~30% of hip fractures occur in men, and one in eight men older than 50 years will have an osteoporotic fracture.[13] As such, prevention of osteoporosis in men is also becoming an emerging health concept. Men's bone integrity differs from that of women as they generally have greater peak bone mass. In addition, men usually present with hip, vertebral body, or distal wrist fractures 10 years later than women. Hip fractures in men, however, result in a much higher mortality rate at 1 year after fracture as compared with women (31% versus a rate of 17% in women). The major risk factors for osteoporosis in men are not only age related but also are associated with steroid use for longer than 6 months (e.g., chronic obstructive pulmonary disease). Hypogonadism remains a risk factor in men that is potentially reversible, and as such there has been recent interest in the area of testosterone replacement.[14] The FDA has recently approved bisphosphonates and teriparatide (recombinant parathyroid hormone) for use in men, which should be considered along with supplemental calcium and vitamin D. If physicians are aware of modifiable risk factors for male osteoporosis such as alcoholism and smoking, the prevention of osteoporosis takes on another dimension. Preventing osteoporosis in the long term leads to the decrease in morbidity and mortality events that occur with falls and fractures. Chapter 9 discusses osteoporosis in men.

- Mental health in men is an often-neglected aspect of preventive medicine, despite the staggering fact that elderly men are more likely to commit suicide than elderly women. Epidemiological studies based on subjective reporting by patients suggest that depression may be less common in men than in women. However, men are less likely to report depression, and may manifest other problems such as anger, alcoholism, slacking work performance, and domestic violence. Prevention includes early detection and intervention with supportive therapy, psychotherapy, and medications. Gender differences and gender-specific treatments are discussed in a later chapter. Dementia, too, is perceived to be less common in men than in women. The difference can be explained by the longer life span of women because dementia is a function of age. Hormones such as estrogens and testosterone may influence cognitive function (see Chapter 7). Preventive strategies for dementia such as aspirin, gingko biloba, "brain exercises," vitamin E, antiinflammatory drugs, etc., are exciting areas, and their efficacies are being examined in clinical trials at present.

- The most common cancers in men are *skin, prostate, lung,* and *colorectal.* These four cancers alone

are expected to kill more than 150,000 men in the United States each year. Fortunately, with lifestyle changes and proper screening, physicians can help prevent these cancers or successfully treat them when detected in their early stages. Another common cancer that is found only in men is testicular cancer. The incidence of testicular cancer is ~7000 per year. Fortunately, testicular cancer is treatable when detected early. The world-champion cyclist Lance Armstrong has been one of the advocates in this area of preventive health, having survived this cancer himself. As the disease tends to develop at *younger ages* than other cancers, young men should be taught to regularly practice self-exams, not unlike women performing breast self-examinations. As in heart disease, research has shown that diet and exercise can be pivotal in the prevention of cancers in men. Genetics is also an important factor.

- Prostate-specific antigen (PSA) has had a profound impact on the early diagnosis, treatment, and follow-up of prostate cancer, the most common malignancy in men. However, it is not only a marker for prostate cancer but is also often expressed in benign conditions such as benign prostatic hyperplasia, prostatitis, and other inflammatory disorders. We have been trying to make the PSA test even more useful for several years. Most of these attempts, such as PSA density, PSA velocity, and age-dependent PSA ranges, were impossible to verify and thus not very useful. One of the newer tests developed, and an emerging concept in preventive health in prostate cancer, is the *free PSA test.* Free PSA is both reproducible and useful. Free PSA is the percentage of the total PSA that circulates in the blood without a carrier protein. Most patients with prostate cancer have a free PSA *less* than 15%. The ratio of free to total PSA improves specificity while maintaining a high sensitivity for prostate cancer detection for men with a total PSA of 2.5 to 10 ng/mL who also have a normal digital rectal examination. Complex PSA seems to be a reliable tool and equivalent alternative to total PSA to improve specificity at high sensitivity levels in men with suspected prostate cancer, mainly for PSA levels below 4 ng/mL. Several newly discovered *isoforms* of free PSA (bPSA, [-2]pPSA, and inactive intact PSA) may also impact the early detection of prostate cancer, with encouraging preliminary results that warrant further clinical investigation.[15]

  ○ *Human glandular kallikrein 2* (hK2) also has the potential to be a valuable tool in combination with both total and free PSA for the early diagnosis of prostate cancer. Patients with high hK2 measurements have a five- to eightfold increase in risk for prostate cancer, adjusting for PSA level and other established risk factors. As such, hK2 measurements may be a useful adjunct to PSA in improving patient selection for prostate biopsy. However, the optimal clinical use of human glandular kallikrein 2 still remains to be clarified.[16]

○ The *education* of both men and women is one of the first important steps in prevention of *sexually transmitted disease* (STD). Other preventive strategies include administering the hepatitis B vaccine series to unimmunized men who present for STD evaluation and administering hepatitis A vaccine to illegal drug users and men who have sex with men. Nonoxynol 9 is a chemical that has been introduced recently for STD prevention, but the CDC recommends against prescribing it at present because the results from clinical trials have been conflicting. New treatment strategies include avoiding the overuse of quinolone therapy in patients who contract gonorrhea because of resistance. Testing for herpes simplex virus serotype is advised in patients with genital infection because recurrent infection is less likely with the type 1 serotype than with the type 2 serotype. HIV screening remains important in men at risk, such as homosexuals, intravenous drug abusers, and those who had transfusions in the past. Informed consent is necessary for testing.

• Many men are now appearing for health care because they are requesting help with *erectile dysfunction* (ED). This gives an excellent opportunity for screening and preventive medicine. A British study suggests that ED shares many of the same risk factors as coronary artery disease, so the level of *occult cardiovascular risk* in men with ED was assessed. A total of 174 men presenting with ED underwent cardiac risk stratification according to the Princeton Consensus Guidelines. Thirty percent were stratified as intermediate/high cardiovascular risk and had ED treatment deferred until further cardiological assessment; 37% had abnormal lipid profiles; 24% had elevated hemoglobin HbA_{1C}/glucose levels; 17% had uncontrolled hypertension; and 6% were suspected of having significant angina.[17] The role of stress testing for detection of occult cardiovascular disease in men is currently being explored in clinical trials, in light of the finding that many of these patients presenting with ED have underlying untreated occult heart disease. ED is a window to many illnesses, including not only coronary heart disease but also depression.[18] Chapter 15 discusses the interlinked syndrome of depression, ED, and CHD, known as the DEC syndrome.

• Total-body computed tomography (CT) screening has been becoming very popular with patients partly because of direct marketing by radiological practices. However, many organizations, including the FDA and the American College of Radiology (ACR), do not endorse total-body CT screening because of the lack of evidence that it is useful in prevention. This is especially so in patients with no symptoms or a family history suggesting disease. In addition, there is no evidence that total-body CT screening is cost-efficient or -effective in prolonging life. In addition, the ACR is concerned that this procedure will lead to the discovery of numerous findings that will not ultimately affect patients' health but will result in unnecessary follow-up examinations and treatments and significant wasted expense. However, the ACR stated recently that CT screening targeted at specific diseases might be useful. Early data suggest that these targeted examinations may be clinically valid. Large, prospective, multicenter trials are currently under way or in the planning phase to evaluate whether these screening exams reduce the rate of mortality. As such, these exams may again represent yet another area of emerging concepts in prevention in men.

• Lung scanning for cancer in current and former smokers remains controversial but promising, if cost is not a factor. This has potential significance as chest x-rays lack sensitivity and specificity, and lesions are only detected late. Moreover, lung cancer is one of the most common cancers in men and is associated with the highest mortality when diagnosed. In a recent study by Mahadeivia et al,[19] using a computer-simulated model, annual helical CT screening was compared with no screening for hypothetical cohorts of 100,000 current, quitting, and former heavy smokers, aged 60 years, 55% of whom were men. The authors simulated efficacy by changing the clinical stage distribution of lung cancers so that the screened group would have fewer advanced-stage cancers and more localized-stage cancers than the nonscreened group (i.e., a stage shift). The model incorporated known biases in screening programs such as lead time, length, and overdiagnosis bias. Over a 20-year period, assuming a 50% stage shift, the current heavy smoker cohort had 553 fewer lung cancer deaths (13% lung cancer–specific mortality reduction) and 1186 false-positive invasive procedures per 100,000 persons. The incremental cost-effectiveness for current smokers was $116,300 per quality-adjusted life year (QALY) gained. For quitting and former smokers, the incremental cost-effectiveness was $558,600 and $2,322,700 per QALY gained, respectively. The authors concluded that even if efficacy is eventually proven, screening must overcome multiple additional barriers to be highly cost-effective. They also said that given the current uncertainty of benefits, the harms from invasive testing, and the high costs associated with screening, direct-to-consumer marketing of helical CT is *not advisable* at this point, and further research is needed.[19]

- Coronary artery calcium scoring as a predictor of cardiac events is also becoming more common and is fairly well studied. There are reports that very high calcium scores, in particular those >1000, indicate a significantly increased cardiovascular risk. The advent of cardiac spiral CT has made coronary calcium scanning more widely available. In a recent study in Germany, Pohle et al[20] compared the presence and extent of coronary calcifications in young patients with first, unheralded acute myocardial infarction with matched controls without a history of coronary artery disease. Calcifications were present in 95.1% of patients with acute myocardial infarction but in only 59.1% of controls ($p = .008$). Using calcium scoring at this point remains promising but is seen to be an additional aid to screening for heart disease and *does not* replace existing standards for screening.

- CT colonography (virtual colonoscopy) for colon cancer is another area of controversy. Preliminary small studies have suggested that in good hands, the accuracy may supersede that of barium enema and approaches that of colonoscopy. It offers convenience to the patient and is less uncomfortable as a procedure. The accuracy of the technique is continually being challenged. At this point, it is probably the preferred screening test for patients with an incomplete colonoscopy, or for those patients who cannot undergo colonoscopy because of advanced age, for instance. Its precise role in screening average-risk patients for colon cancer remains to be defined by ongoing research and clinical trials.

- Hormonal decline with aging is inevitable, and there are generally two schools of thought about whether to intervene with hormones. There is a general consensus that thyroxine should be substituted when the patient is hypothyroid. However, if DHEA levels were low such as in the so-called "adrenopause," or if growth hormone were low in the "somatopause," then the role of substitution is less clear, especially with the lack of symptoms. There are no large clinical trials that are sufficiently long term to suggest

the benefits of hormonal replacement as a means of *prevention*. However, clinical experience does suggest that these compounds can be extremely potent, and can improve quality of life substantially. The main concern about the long-term use of these hormones is the possibility of carcinogenesis. There is one school of thought that suggests that the low hormonal milieu protects one from cancer, and that replacement to youthful physiological levels may do harm. On the other hand, there is another school of thought that believes that the low hormonal state in itself is associated with cancer. One argument for the use of testosterone has been that it has been observed that prostate cancer tends to occur in older men, and that older men tend to be more likely to be hypogonadal as compared with younger men. Aging is a complex event that involves genetic expression, environmental interactions, and lifestyle habits. Aging and death are inevitable. Hormones, when replaced to youthful physiological levels, can improve substantially the quality of life for the patient. At present, there is no evidence to suggest that hormones have an anti-aging effect, but rather that they can restore function and improve biological and physiological markers. Measurements of biological age against chronological age remains a novel approach, and if used appropriately, may motivate patients to improve their health through lifestyle changes. The area of hormonal replacement for men is evolving; at present, replacement should be reserved for those patients who are symptomatic, and patients should be carefully monitored. The role of hormonal replacement for men for prevention remains promising, and only time will tell of its efficacy.

The previous issues are discussed throughout this book, and are illustrated through the use of case histories.

Conclusion: Optimal Aging

Rather than pursue the reversal of age or anti-aging, perhaps the real role of prevention in aging is to achieve the concept of *optimal aging*. In the hospital, we often attend to patients who are in their 40s or 50s but who look as if they are in their 60s or 70s because of their chronic illnesses. In fact, Medicare recognizes this and allows the enrollment of patients before age 65 if they have multiple chronic illnesses and are disabled at the same time. Hence we see patients that are *chronologically young, but biologically old*. Optimal aging intends that we practice the best form of preventive medicine so that patients will remain healthy as they age gracefully. As such, optimal aging as applied to a 40- or

50-year-old man may mean drastic attention to weight and controlling blood pressure, risk factors for heart disease, and diabetes. It also means attention to lifestyle factors such as smoking and excessive drinking. Patients who consult for problems of hypogonadism or erectile dysfunction should have a thorough evaluation of their associated risk factors. Hormones can be prescribed if symptomatic and biological measurements support a low level. On the other hand, optimal aging as applied to a 75-year-old man may mean maintaining his functionality, such as the prevention of falls. This could be achieved through muscle strengthening exercises and balance training, and in some instances hormonal replacement. Bone integrity can be improved and maintained with drugs such as Fosamax. A 75-year-old man may also face memory loss, and prevention may entail the prescription of acetylcholinesterase inhibitors such as Aricept, Exelon, or Reminyl. These groups of drugs do not improve memory but they should be seen as preventing further decline. Maintaining use of cognitive functions is important to prevent loss of function. Family and social support can also be seen as a form of preventive medicine.

There is an emerging concept that chronological age is not quite the same as biological age, and one of the principles of preventive gerontology is to decrease biological age, as chronological age will not change. There are many so-called biomarkers of aging, which are really tests of physiological functionality as compared with the same age cohorts. Normative age-matched values are used for comparison. For example, grip strength has normative values and can be used for age comparison (Fig. **1–3**).

Another commonly used test includes taking measurements of subjects and then comparing them with normative values for neuropsychological batteries (e.g., Brief Visual Spatial Memory Test), cognitive reflexes, and pulmonary function. These tests could be used as a motivation for patients to achieve a healthier lifestyle. Normative values can vary based on factors such as body size, activity, genetics, and ethnicity. As such, these physiological measures are useful only for *intraindividual comparison* over time, rather than interindividual comparisons.

Genetics also plays a major role in longevity; as such, prevention even to the highest level may have limitations. There is a discussion on genetics and why we age in Chapter 2. However, aging is also a function of how genes express themselves, and some people are just more fortunate than others. Longevity should be accompanied by improvement of quality of life. It should be appreciated that human life expectancy has improved dramatically through achievements in public health, therapy, nutrition, and general living standards. But as humans reach their possible age limits, it is always important to remember the old axiom, "Death and taxes are inevitable."

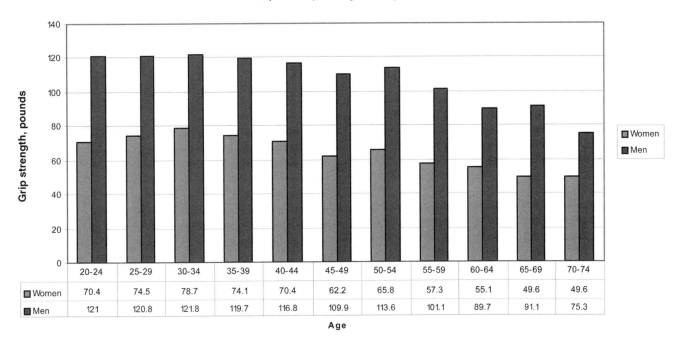

Graph of Grip Strength vs. Age

	20-24	25-29	30-34	35-39	40-44	45-49	50-54	55-59	60-64	65-69	70-74
Women	70.4	74.5	78.7	74.1	70.4	62.2	65.8	57.3	55.1	49.6	49.6
Men	121	120.8	121.8	119.7	116.8	109.9	113.6	101.1	89.7	91.1	75.3

FIGURE 1–3. Normative values of dominant hand grip strength versus age. (Adapted from Mathiowetz V, et al. Grip and pinch strength: normative data for adults. Arch Phys Med Rehabil 1985;66:69–74, with permission.)

REFERENCES

1. Hazzard WR. Preventive gerontology: optimizing health and longevity for men and women across the lifespan. J Gend Specif Med 2000;3:28–34

2. Hazzard WR. The gender differential in longevity. In: Hazzard, et al. *Principles of Geriatric Medicine*. 4th ed. New York: McGraw-Hill; 1999:69–80

3. Ettinger SM. Myocardial infarction and unstable angina: gender differences in therapy and outcomes. Curr Womens Health Rep 2003;3:140–148

4. CDC. United States Cancer Statistics, 1999. Atlanta: CDC; 1999

5. McCormick A, Charlton J, Fleming D. Assessing health needs in primary care: morbidity study from general practice provides another source of information. BMJ 1995;310:1534

6. 2000 United States Census Bureau International Data Base

7. Roth GS, Lane MA, Ingram DK, et al. Biomarkers of caloric restriction may predict longevity in humans. Science 2002;297:811–815

8. Wassertheil-Smoller S, Hendrix SL, Limacher M, et al. Effect of estrogen plus progestin on stroke in postmenopausal women: the Women's Health Initiative: a randomized trial. JAMA 2003;289:2673–2684

9. Anthony JC, LeResche L, Niaz U, et al. Limits of the "Mini-Mental State" as a screening test for dementia and delirium among hospital patients. Psychol Med 1982;12:397–408

10. Lakka HM, Laaksonen DE, Lakka TA, et al. The metabolic syndrome and total and cardiovascular disease mortality in middle-aged men. JAMA 2002;288:2709–2716

11. Foster GD, Wyatt HR, Hill JO, et al. A randomized trial of a low-carbohydrate diet for obesity. N Engl J Med 2003;348:2082–2090

12. Tan RS, Pu SJ. Impact of obesity on hypogonadism in the andropause. Int J Androl 2002;25:195–201

13. Campion JM, Maricic MJ. Osteoporosis in men. Am Fam Physician 2003;67:1521–1526

14. Tan RS, Culberson JW. An integrative review on current evidence of testosterone replacement therapy for the andropause. Maturitas 2003;45:15–27

15. Haese A, Partin AW. New serum tests for the diagnosis of prostate cancer. Drugs Today (Barc) 2001;37:607–616

16. Nam RK, Diamandis EP, Toi A, et al. Serum human glandular kallikrein-2 protease levels predict the presence of prostate cancer among men with elevated prostate-specific antigen. J Clin Oncol 2000;18:1036–1042

17. Solomon H, Man J, Wierzbicki AS, et al. Erectile dysfunction: cardiovascular risk and the role of the cardiologist. Int J Clin Pract 2003;57:96–99

18. Tan RS, Pu SJ. The interlinked depression, erectile dysfunction, and coronary heart disease syndrome in older men: a triad often underdiagnosed. J Gend Specif Med 2003;6: 31–36

19. Mahadevia PJ, Fleisher LA, Frick KD, et al. Lung cancer screening with helical computed tomography in older adult smokers: a decision and cost-effectiveness analysis. JAMA 2003;289:313–322

20. Pohle K, Ropers D, Maffert R, et al. Coronary calcifications in young patients with first, unheralded myocardial infarction: a risk factor matched analysis by electron beam tomography. Heart 2003;89:625–628

2

Delaying Aging for Men: Longevity Lessons from Women

ROBERT S. TAN AND MEENU JACOB

Case History

A 58-year-old college professor, who was considering retirement. He was happily married to his wife of 30 years, who currently was postmenopausal and on estrogens. He had heard that good preventive medicine could be one of the keys for life extension. He had had annual physicals since he was 45 years old and was generally well. He did not smoke but drank a glass of wine every night. He admitted to exercising infrequently. His main aim was to remain healthy through his "golden years," and he was inquiring about possible interventions including diet, exercise, and hormones. He jokingly said that he wanted to live as long as his wife, who was 5 years younger than he. He requested to have blood drawn for evaluation. He realized that most of these tests would not be covered by insurance, but he wanted to proceed nonetheless. Physical examination revealed that he was 5 feet 7 inches, weighed 151 pounds, and had a body fat content of 23% and blood pressure (BP) of 120/85 mm Hg. He also had a mild degree of gynecomastia and normal testicular size. He did have some hair loss over his scalp and thinning of his skin, relative to his age. He was asked to catch a falling ruler with his fist to test his reflexes, and he did well on that test.

Significant laboratory results included: fasting glucose 96 mg/dL, cholesterol 183 mg/dL, triglyceride 180 mg/dL, hemoglobin HbA_{1C} 5.1, insulin-like growth factor I (IGF-I; somatomedin C) 83 ng/mL (range 90–360), total testosterone 273 ng/dL (range 260–1000), free testosterone 47.5 pg/mL (range 50–210), prolactin 5 ng/mL, dehydroepiandrosterone sulfate (DHEAS) 429 μg/dL (range 20–413), thyroid-stimulating hormone (TSH) 2.05 μIU/mL, free triiodothyronine (T_3) 3.2 pg/mL (range 2.3–4.2), free thyroxine (T_4) 1.0 ng/dL (range 0.8–1.8), luteinizing hormone (LH) 2.2 mIU/mL, estradiol 8 pg/mL (range 10–50), insulin 21 μU/mL (range <20), homocysteine 12.9 μmol/L (range 5.4–11.4), prostate-specific antigen (PSA) 0.4 ng/dL.

The results were discussed with him, as it was noted that he had marginally low free testosterone and IGF-I. He also had marginally high homocysteine levels. He inquired about hormone replacement with testosterone and human growth hormone. He also asked for intervention to bring his homocysteine levels down, which he had heard was a risk factor for cardiac disease. He inquired about whether he should change his diet and if exercise would alter some of his biomarkers.

An Overview of Aging

In the United States, average life expectancy at birth is ~79 years for women and ~72 years for men. The oldest person for whom reliable records exist was a woman who recently died in France at the age of 123. The odds of someone living this long are about one in 6 billion. From a practical point of view, we can consider a century as the average maximum of human life. We are not there yet, of course. At present, average life expectancy for those born after 1960 is ~85 years.

Before learning about delaying aging, it is imperative that one understands the processes of aging and the limitations of human life expectancy. The process of aging is universal for all forms of life. It begins at birth, or arguably on conception. Since time began, men have tried to go against aging, and to search for immortality. In reality, at the microscopic level, the accumulation of the diverse deleterious changes produced by aging in cells and tissues progressively impair function and can eventually cause death. Aging, in general, can be attributed broadly to the following five categories:

- Genetic defects
- Development
- Disease states
- The environment
- An inborn process—the aging process

The probability of death at a specific age can be said to be a measure of the average number of aging changes accumulated by persons of the physiological age and the rate of change of this measure as the rate of aging. Overall, the probability of death is decreased by improvements in general living conditions, preventive medicine, immunizations, and good habits like abstinence from smoking and drinking alcohol in moderation. During the past 2000 years, it is amazing that the average life expectancy at birth (ALE-B) of the human race has risen from 30 years in the Roman Empire to almost 80 years today in the developed countries. This is unprecedented in history. Most of the increase in ALE-B is attributed not so much to advances in medicine but to the general improvement of economies and perhaps to immunizations. Probabilities of death in the developed countries are now near limiting values, and ALE-B is approaching plateau values. Overall, one can argue that there is no limit, but biologically as a species we are limited to ~85 years in general. In the United States, we are thus on average 6 to 9 years less than the potential maximum at this point. Obviously, there are variations from person to person.

In general, the inherent aging process now largely determines the chances of death after the age of 28 years. It is interesting to note that, in Sweden, only 1.1% of female cohorts die before this age; the remainder die off at an exponentially increasing rate with advancing age. The inherent aging process limits ALE-B to around 85 years, and the maximum life span (MLS) to ~122 years.

Past efforts to increase ALE-B did not require an understanding of the biological processes of aging. Such knowledge will be necessary in the future to significantly increase ALE-B and MLS. This knowledge is also required to satisfactorily plan for the medical, economic, and social problems associated with advancing age. For instance, many developed countries are increasingly burdened by social problems and increasing taxation as the population pyramid changes from one with a larger base of young people to one with a larger base of older people.[1]

There are many theories that have been proposed to account for aging, and they should be used to the extent they are feasible. The difficulty of the science of delaying aging or anti-aging principles is that often these theories are built on animal models, and they may or may not apply to human beings. Overall, the previous measures evolved by societies to ensure adequate care for older individuals are rapidly becoming inadequate because of changes in lifestyle, the growing percentage of older people, declining fertility rates, and the diminishing size of the workforce to provide for the elderly. Measures are being advanced to help with this problem but they largely address the preventive level, where lifestyle changes can significantly impact longevity. Prospects are bright for further increases in the span of functional life and improvements in the quality of lives of older individuals.

Limits to Human Longevity

As mentioned previously, the human life span at this point is limited to ~122 years and average life expectancy to 85 years in developed countries, although there is a predilection for longevity in women. However, demographic approaches to modeling and forecasting mortality are often based on the observation of short-term trends in death statistics and the assumption that future mortality will exhibit patterns similar to those of the recent past. This extrapolation method has led some demographers to conclude that ALE-B in the near future may reach 100 years.

Similar predictions follow from other demographic models that establish a hypothetical link between risk factor modification and changes in death rates. Risk factor modification would include attention to diet, smoking cessation, and exercise. These predictions are examined within the context of the observed mortality records and their biological plausibility assessed based on the current theories of aging. Results indicate that these demographic models lead to mortality schedules that do not follow from the observed mortality record and that are inconsistent with predictions of biologically based limits to longevity. Although there is probably not a genetic program for death, the biology of our species places inherent limits on human longevity. As a human race, we were not built to last forever, but attention to some lifestyle habits may increase our chances of reaching the biological limit for our life span.[2]

Sex Differential in Longevity

There is a sex differential in longevity, with women outliving men. There may be biogenetic, environmental, and psychosocial perspectives that may be responsible for this sex differential[3]:

- Wellness and preventive health are practiced more by women than by men, and may explain some of the discrepancies. Women are more apt to go for Pap smears, mammograms, and cholesterol and

BP checks than men are apt to go for cholesterol and BP checks.

- In general, men abuse their physical health more than women, and are more likely to indulge in alcohol and smoking.

- Women arguably have better coping mechanisms with respect to mental health as they may have better support and social systems.

- Women's occupational role, by and large, focuses on family, and hence the stressor events are much different from those of men.

- Spiritual health in women on the average is better than in men, and hence their coping mechanisms are much better.

- Poor environmental health can decrease longevity, and in some men who are employed in vocations such as mining, fishing, the military, etc., life expectancy can be shortened.

- Last, but not least, the genetic makeup of men and women is different. Perhaps the Y chromosome puts a limit on longevity.

Over the years, there have been many theories put forth as to the longevity differential. Most of these theories have been based on the animal kingdom, but much can be learned. For instance, higher death rates among male animals have been linked to the more risky behaviors that males exhibit, including fighting each other over females. A Scottish study from the University of Stirling shows that male wild animals have more parasitic infections than do females at the time of death.[4] The researchers studied parasites in all kinds of mammals and determined that males have significantly more infections. The researchers postulate that males are generally larger than females, thus providing larger targets for parasites and animals that carry them. Another reason may be testosterone, which can suppress the immune system, making the body more prone to infection. The accompanying editorial in *Science* states that the parasite explanation might apply to humans, too, because men are more susceptible to parasitic and infectious diseases than are women. The editorialists point out that men in the United States, the United Kingdom, and Japan are twice as vulnerable as women to parasite-induced death. And in Kazakhstan and Azerbaijan, where parasite-related death rates are high, men are more than four times more vulnerable than women.

In another theory, based on studying people who live 100 years, researchers conclude that menopause may be a major determinant of the life spans of both women and men. Women's life span depends on the balance of two forces: the evolutionary drive to pass on one's genes, and the need to stay healthy enough to rear as many children as possible. Menopause draws the line between the two. It protects older women from the risks of bearing children late in life, and lets them live long enough to take care of their children and grandchildren. It is interesting to note that most animals do not undergo menopause. It seems that menopause evolved in part as a response to the amount of time that the young remain dependent on adults to ensure their survival. Pilot whales, for example, suckle their young until the age of 14 years, and they, along with humans, are two of the few species that menstruate. The theory assumes that longevity is linked to a later menopause. Hence, there is the argument for hormone replacement therapy for postmenopausal women. This may extend perhaps to testosterone replacement in postandropausal men.

Interestingly, in their studies of centenarians, Perls and Fretts[5] found that a surprising number of women who lived to be 100 or more gave birth in their 40s. These 100-year-old women were four times as likely to have given birth in their 40s as women born in the same year who died at age 73. A study of centenarians in Europe by the Max Planck Institute of Demography in Germany found the same relationship between longevity and fecundity. Factors that allow certain older women to bear children, including a slow rate of aging and decreased susceptibility to disease, improve a woman's chances of living a long time. Extending that idea, it is possible that the driving force of human life span is maximizing the time during which woman can bear children. The age at which menopause eliminates the threat of female survival by ending further reproduction may therefore be the determinant of subsequent life span.

The menopause theory, however, does not fully explain why women live so much longer than men. In all developed countries and most undeveloped ones, women outlive men, sometimes by a margin of 10 years. The gender gap for the longevity differential is most pronounced in those who live 100 years or more. Among centenarians worldwide, women outnumber men nine to one. The mortality gap even varies during other stages of life. For example, between the ages of 15 and 24 years, men are four to five times more likely to die than women. This time frame coincides with the onset of puberty and an increase in reckless and violent behavior in males. Researchers sometimes refer to it as a "testosterone storm" as most deaths in this male group come from motor vehicle accidents, followed by homicide, suicide, cancer, and drowning. After the age of 24 years, the difference between male and female mortality narrows until late middle age. In the 55- to 64-year-old range, more men than women die, due mainly to heart disease, suicide, car accidents, and illnesses related to smoking and alcohol use. Heart disease kills 5 of every 1000 men in this age group.

It does seem that women have been outliving men for centuries and perhaps longer. Even with the sizable risk conferred by childbirth, women have outlived men since at least the 16th century. It is interesting to note that, in the United States between 1900 and the 1930s, the death risk for women of childbearing age was as high as that for men. Since then, improved health care, particularly in childbirth, has put women ahead of men again in the survival struggle, as well as raising life expectancy for both sexes. Sadly, almost like an equalizer, longevity doesn't equate with quality of life. Although men may die of fatal illnesses like heart disease, stroke, and cancer, women tend to live on with nonfatal conditions such as arthritis, osteoporosis, diabetes, and dementia.

One contributor to the gender difference in life span may be the influence of sex hormones, in particular estrogens. The impact of testosterone on longevity is yet to be established. Estrogen lowers low-density lipoprotein (LDL) cholesterol and raises high-density lipoprotein (HDL) cholesterol. Another theory of longevity is based on survival. The fittest live longer, and arguably the longer a woman lives, the more slowly she ages, and the more offspring she can produce and rear to adulthood. Therefore, evolution would naturally select the genes of such women over those who die young. Long-lived men would also have an evolutionary advantage. Studies of chimps, gorillas, and other species closely related to humans suggest that a male's reproductive capacity is actually limited more by access to females than by life span. And because men have not been involved in child care as much as women, survival of a man's offspring, and thus his genes, depended not so much on how long he lived, but on how long the mother of his children lived.

The hypothesis that the Y chromosome may be a limiter to life span could be challenged. The longevity differential can be closed with lifestyle modification. Whether the average person drinks and smokes, on the one hand, or exercises and eats vegetables, on the other, subtracts 5 to 10 years from one's life or adds 5 to 10 years to one's life. But to live an additional 30 years requires the kind of genes that slow down aging and reduce susceptibility to conditions such as Alzheimer's disease, stroke, heart disease, and cancer. Clues about what those genes are and how they work could come from studying those who survive 100 years or more. The New England Centenarian Study is the only scientific investigation of the oldest people performed in the United States.[6] Centenarians are a tremendous resource for the discovery of genes responsible for aging and the ways in which aging occurs. Discovering these genes could lead to testing people

and determining who might be disposed to accelerated aging via diseases such as Alzheimer's, cancer, heart disease, and stroke. Such individuals might eventually be treated to extend the prospect of their living longer.

Although women can expect to live longer than men, the gap is closing. Death rates have begun to converge in the past 20 years. Some researchers attribute the convergence to women taking on the behaviors and stresses formerly considered the domain of men, including smoking, drinking, and working outside the home. Death rates from lung cancer have almost tripled in women in the past 20 years. On average, middle-aged female smokers live no longer than male smokers.

Mood and Longevity

It has been assumed that people with better moods end up living longer. Depressed individuals, especially men, are more likely to take their own lives. In a recent study, however, it was found that mildly depressed older women tend to live longer than those who are not depressed at all.[7] The findings are contrary to most other studies on the link between depression and mortality.

The finding that women with *mild* depression live longer suggests a survival mechanism. The Duke study[7] was based on a group that started with 2401 women and 1269 men, all older than 65. They were interviewed about their health at roughly 3-year intervals from 1986 to 1997 and were separated into three categories: depressed, mildly depressed, and not depressed. Of the women, 10.5% were considered mildly depressed. The women with mild depression were, on average, 60% were less likely than other women to die during any 3-year period. Researchers took into account age, chronic illness, and other factors in calculating the mortality rate.

Interestingly, the researchers found that *depression had no influence on the mortality of men*. This study may support a theory that says mild depression may allow people to cope more easily with their problems and remove themselves from dangerous or harmful situations.

Summary of Existing Theories

There are several theories that explain longevity in different species.[8] Most, if not all, of the theories are derived from animal models. As such, it is important for the clinician not to go overboard in recommending a particular strategy for life extension based on theory alone. However, these theories should be used in conjunction with conventional medical practice, and that would include lowering cardiovascular and cerebrovascular risk factors including blood pressure and cholesterol levels. These simple measures can realistically increase the life span. There is a lot of hype about life extension, but the clinician should stress lifestyle changes, exercise, and nutrition. It is important not to give false hope to patients or to rely on the "pseudoscience" of some anti-aging products.[9]

Gene Regulation Theory

This theory focuses on the genetic programming encoded within our DNA. We are born with a unique genetic code and a predetermined tendency to certain types of physical and mental functioning. That genetic inheritance has a great deal to say about how quickly we age and how long we live. Each of us has a biological clock ticking away, set to go off at a particular time, give or take a few years. When that clock goes off it signals our bodies first to age and then to die. However, this genetic clock is subject to enormous variations, depending on what happens to us as we grow and on how we actually live. This is essentially the nature versus nurture debate. The differential life span of Japanese in Hawaii versus those in Japan typifies this debate about whether nature or nurture is more important.

CLINICAL IMPLICATIONS

The life span seems to be the longest in certain areas of the world including Okinawa (Japan), Sardinia (Italy), Kerala (India), and Scandinavia. It could be the genetics of these groups of people that result in longevity. However, the nurture debate may confound the true longevity rates. For instance, Okinawans eat mainly seafood, which is rich in omega oils that may protect the heart. Sardinians likewise have a Mediterranean diet that includes seafood, olives, and red wine, which have antioxidant properties. Keralans have one the highest educational levels in India, and education could facilitate taking care of oneself. Scandinavians, on the other hand, have a good socialized health care system that provides good access to care. All these factors can contribute to longevity, other than genes alone.

It has been hypothesized that genes control longevity. In one such study, the frequencies of 80 human leukocyte antigen (HLA) phenotypes in 82 centenarians and 20 nonagenarians in Okinawa, Japan, were compared with those in other healthy adults in various age brackets.[10] Subjects over age 90 had an *extremely low frequency of HLA-DRw9 and an increased frequency of HLA-DR1.* In this age group the corrected relative risk (for number of antigens) and the *p* value for HLA-DRw9 were 5.2 and .0001, respectively; those for HLA-DR1 were 13.3 and .0367, respectively. Because a high frequency of DRw9 and a low frequency of DR1 are associated with autoimmune or immune deficiency diseases, the genetic protection against these disorders may contribute to longevity. A review by Ivanova et al[11] suggested links with other alleles as well, and this is demonstrated in Table **2–1.**

DNA Repair Capability

In 1963, Dr. Leslie Orgel of the Salk Institute suggested that because the "machinery for making protein in cells is so essential, . . . an error in that machinery could be catastrophic to human survival." The body's DNA is so vital that natural repair processes kick in when an error occurs. However, the body's system is unable to do this all the time, and thus accumulation of these flawed molecules can cause disease and other age changes to occur. If the DNA repair process did not exist, scientists estimate that enough damage would accumulate in cells in 1 year to make them nonfunctional. Others believe that aging is caused by DNA damage through intrinsic mutations or other

TABLE 2–1. Human Leukocyte Antigen HLA-DRB1 Allele Frequencies among Controls and Long-Lived Groups

Allele	Controls $n = 2950$	Female Centenarians $n = 259$	Male Centenarians $n = 66$	Female Siblings $n = 77$	Male Siblings $n = 86$
DR1	11.5	9.1	14.4	9.1	8.0
DR3	11.0	10.8	8.3	11.0	13.4
DR4	13.3	13.1	12.9	10.7	10.4
DR7	14.0	15.0	21.2	15.6	23.1
DR8	2.8	3.3	1.5	0.6	4.1
DR9	0.7	1.0	0.8	0.0	1.7
DR10	0.8	0.8	0.8	1.6	1.2
DR11	13.1	11.6	6.8	23.4	11.7
DR12	1.6	1.3	0.0	0.6	0.6
DR13	12.8	17.4	17.4	13.6	11.0
DR14	3.8	3.7	1.5	1.0	3.2
DR15	12.4	9.5	10.6	9.2	8.4
DR16	2.3	3.5	3.8	3.5	3.2

Adapted from Ivanova R, Henon N, Lepage V, et al. HLA-DR alleles display sex-dependent effects on survival and discriminate between individual and familial longevity. Hum Mol Genet 1998;7:187–194, with permission.

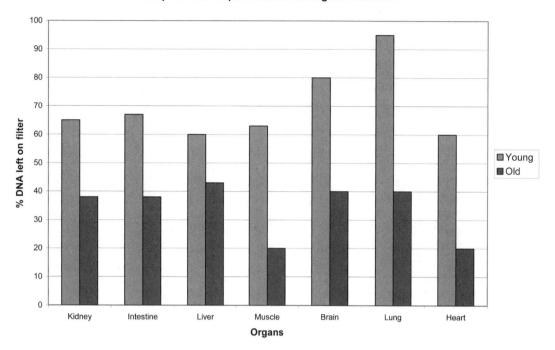

FIGURE 2–1. DNA status of different organs in young and old mice. (Adapted from Zahn RK. DNA status in brain and heart as prominent co-determinants for life span? Mech Ageing Dev 1996;89:79–94, with permission.)

age-related nonmutational or epigenetic changes such as altered methylation.

In a laboratory experiment, Zahn[12] demonstrated that the DNA repair capacities of young and old mice were indeed different (Fig. **2–1**). Alkaline filter elution has been modified by a freeze-grinding step that allows the evaluation of the DNA status of whole tissue, including mouse tail cross sections, with only small additional artifacts. Four to seven different organs from individually coordinated female mice, rated as young as 2 to 3 months of age and as old as 24 to 27 months of age, have been used. Tissues of individual mice differ significantly in their DNA status. Alkali-labile sites are relatively rare and differ in amount in the different organs in the young. They show significant increases in the old, reaching the highest values in the brain and the heart. Proteinase K–dependent DNA–protein cross-links are not prominent, nor are they increased with age in some organs, except in the brain and the heart. DNA damage susceptibility was measured after application of 3.5 μM nitroquinoline-*N*-oxide to 15-mg fresh tissue pieces for 90 minutes. The susceptibility is large and varies in wide ranges in the different organs. Upon 3-hour postexposure incubation in full medium, all samples showed DNA repair; the young reach nearly complete repair in the lung, whereas repair in the old is generally significantly decreased. In old brain and heart it is even near zero. This together with high values in

alkali-labile sites and DNA–protein cross-linking suggests that these two organs may act as pacemakers and play a role as prominent codeterminants for the life span of the species.

CLINICAL IMPLICATIONS

Some patients may be tempted to purchase and use nutraceuticals or cosmetics that claim to be able to restore vitality by "repairing DNA." The clinician should tell the patient that although the theory sounds interesting, at least in fruit flies and mice, there is no evidence that these substances actually work in human beings.

Telomeres and Telomerase

Telomeres are bits of DNA on the ends of our chromosomes; the parallel would be the hard ends of shoelaces. Fig. **2–2** is a graphic representation of telomeres. Although they do not contain genes, telomeres are important for replication or duplication of the chromosomes during cell division.

Telomere length recently has been described as a marker of cellular aging. The originator of the theory is Olovnikov.[13] In 1973, he proposed that cells lose a small amount of DNA following each round of replication due to the inability of DNA polymerase to fully replicate

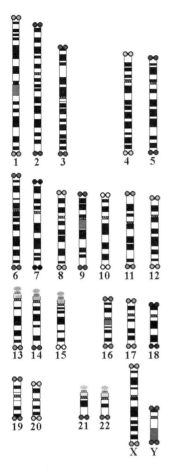

FIGURE 2–2. Graphic representation of telomeres, located at end of chromosomes.

chromosome ends (telomeres) and that eventually a critical deletion causes cell death. Telemetries are specialized DNA sequences located at the end of eukaryotic chromosomes. In humans, telomeres are composed of repeats of the sequence TTAGGG reiterated in tandem for up to 15 kilobases at birth. Telomeres are synthesized by telomerase, a ribonucleoprotein reverse transcriptase enzyme that maintains the lengths of chromosomes. Loss of telomeres can lead to DNA damage. This association of telomere shortening and senescence in vitro has been established. Cells that have been supplied with an exogenous source of telomerase maintain a youthful state and proliferate indefinitely.

CLINICAL IMPLICATIONS

It is a notion that in normal human organs with a capacity for cell replacement, the telomere clock may allow enough divisions for normal growth, repair, and maintenance. However, this may not be enough to enable additional cell replications needed during chronic disease. A potential remedy may be found by increasing

the life span of tissue cells, by telomerase. Another possibility may involve taking cells from an individual, extending the life span of the cells in vitro by telomerase, and then reintroducing the cells into the organ that requires help. At this point, these possibilities are still experimental, and genetic engineering may prove useful for some diseases in the future, and perhaps for extending life.

Endocrine Theory of Aging

Aging is followed by a fall in neuroendocrine functions, resulting in a decreased secretion of sex steroids and growth hormone. When we are young, hormone levels tend to be highest. This accounts for, among other things, menstruation in women and high libido in both sexes. As we age there is a decline in the function of the hypothalamic-pituitary adrenal axis. This results in age-related changes. For example, decline in growth hormone may result in loss of muscle mass. Drops in levels of testosterone and thyroxine can increase the fat-to-muscle ratio.

With aging, cortisol may be inadequately secreted upon stress challenges. This could be due to deficient functioning of central glucocorticoid receptors. In combination, these endocrine perturbations probably result in changes in psychological factors such as energy and well-being, altered body composition, and insulin resistance, as well as other risk factors for diseases characteristic of the aging man.

CLINICAL IMPLICATIONS

Hormones such as estrogens are frequently prescribed for women in the postmenopausal state. The Women's Health Initiative's recently published study had raised some doubt about the use of an estrogen/progesterone combination.[14] Increasingly, testosterone is prescribed for men as they age. Preventive and therapeutic interventions with hormones can indeed improve quality of life. Whether hormone replacement results in longevity is yet to be determined.

Free Radical Theory of Aging

Dr. R. Gerschman first introduced this theory of aging in 1954, and later Dr. Denham Harman of the University of Nebraska College of Medicine further developed this theory. *Free radical* is a term used to describe any molecule that differs from conventional molecules in that it possesses a free electron. This is a property that makes molecules react with other molecules in highly volatile and destructive ways. They can attack cell membranes, creating metabolic waste products, including substances known as lipofuscins. An excess of lipofuscins in the body is shown as darkening

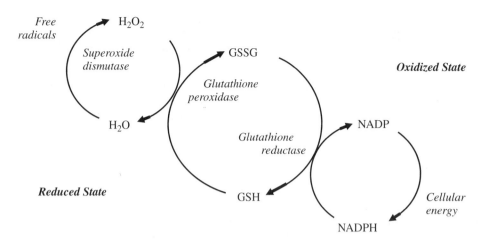

FIGURE 2–3. Besides lipid peroxidation, changes in status of antioxidant compounds such as glutathione and others could be measures of oxidative stress. GSSG, oxidized glutathione; GSH, reduced glutathione; NADP, nicotinamide adenine dinucleotide phosphate; NADPH, reduced NADP.

of the skin in certain areas, so-called aging spots. Lipo-fuscins, in turn, interfere with the cells' ability to repair and reproduce themselves. They disturb DNA and RNA synthesis of protein, lower our energy levels, prevent the body from building muscle mass, and destroy cellular enzymes, which are needed for vital chemical processes. Substances that prevent the harmful affects of free radicals are called antioxidants. Natural antioxidants include vitamin C, E, and A. As shown in Fig. 2–3, lipid peroxidation and changes in status of antioxidant compounds such as glutathione and others could be measures of oxidative stress.

CLINICAL IMPLICATIONS

Most commonly, measurements of oxidative stress are by-products of lipid peroxidation, but changes in status of antioxidant compounds such as glutathione, protein and DNA oxidation products, and antioxidant enzyme activities have also been used. These are all indirect measures of free radical activity. Electron spin resonance, a direct measure of free radicals, has been used predominantly in in vitro studies, but it recently has been used to detect free radicals in blood.

In addition, many patients take vitamins as antioxidants. Some small clinical trials suggest that patients taking vitamins like E and C are less likely to develop illnesses. Even if antioxidants could provide the benefits suggested by epidemiology studies, smoking cessation and other lifestyle factors would have a far greater effect on the rates of lung cancer and coronary heart disease. Overall, the benefits of taking high doses of vitamin E remain to be established. At present, there is no convincing evidence that taking supplements of vitamin C prevents any disease.

Low Calorie Theory of Aging

In the 1930s, researchers discovered that they could extend the life of rats by 33% if they limited them to a very low-calorie diet. Oxidative stress is a major factor in aging and cellular senescence. It has been theorized that dietary restriction without malnutrition protected rats and mice against oxidative stress, decreased oxidative damage, and decreased accumulation of oxidatively damaged proteins within the mitochondria. The animals lived longer, suffered fewer late-life diseases, and appeared more youthful, and their bodies' biological aging processes were slowed.[15] Since that time, scientists have produced similar life-extending results with many other creatures, ranging from fruit flies to fish.

Is caloric restriction per se responsible for the observed benefits, or is some other factor that is reduced when calories are restricted? Studies show that limiting fat, protein, or carbohydrate, without accompanying caloric reduction, does not seem to increase maximum life span. Nor does supplementation with extra antioxidants and multivitamins. Varying the types of fats, carbohydrates, and proteins ingested also had no effect. In fact, no other intervention except caloric restriction has yet been shown to slow aging.

At this point, it is unclear that human life spans can be increased with calorie restrictions. Humans live much longer than many other creatures, and thus such studies are difficult to do. Studies with primates provide some clues. Investigations on monkeys have been underway since 1987, and preliminary results suggest that caloric restriction increases both health and life span in primates. Biomarkers of aging, such as insulin levels, glucose levels, and blood pressure, have led researchers to conclude that monkeys eating less age more slowly. In most such studies, calories are

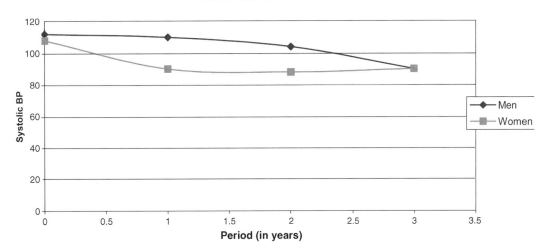

FIGURE 2–4. Human participants of Biosphere-2 demonstrating lowering of systolic BP with calorie restriction.

restricted to 30 to 50% of what the animal would normally eat. Care is taken to see that enough vitamins, minerals, protein, and fat are ingested for the proper functioning of tissues. The participants of the Biosphere-2 experiment in Arizona were forced to eat a low-calorie diet for 2 years because their food production was less than projected. They experienced the same anti-aging trends in biomarkers as were found in the monkey experiments. Fig. **2–4** shows blood pressure changes in the Biosphere-2 participants on this diet.

CLINICAL IMPLICATIONS

The clinician is often asked to advise on weight loss by means of a low-calorie diet. A low-calorie diet is one that is below 1000 calories per day. Often, the aim of a low-calorie diet is to lose weight so as to lower blood pressure and cholesterol levels and to ameliorate arthritis. Most people do not realize that a low-calorie diet, with the appropriate nutrients, may actually lead to a longer and better life. Patients may not only have better health outcomes, but they also feel better as well. Losing weight is one of the best anti-aging antidotes. Patients who actively lose weight often do better in the long run than those who rely just on medications or nutraceuticals.

Immunologic Theory of Aging

The thymus may be the master gland of the immune system. The size of this gland shrinks from 250 g at birth to around 3 g by age 60. The reduction in size and function of the thymus is associated with senescence. Some studies have shown that thymic factors are helpful in restoring the immune system of children born without them as well as rejuvenating the poorly functioning immune system of the elderly. Dr. R.L. Walford first proposed the immunologic theory of aging. Thymic hormones may also play a role in stimulating and controlling the production of neurotransmitters and brain and endocrine system hormones, which implies that they may be the pacemakers of aging itself besides being key regulators responsible for immunity.

CLINICAL IMPLICATIONS

Some health stores sell thymic extracts, suggesting that their intake may lead to life extension. Thymic extracts are essentially protein, and they get broken down in the digestive tract and absorbed as amino acids. They are not proven to be useful clinically in extending life or fighting infections. However, it is well recognized that poor immunity leads to illness, morbidity, and mortality.

Discussion of Case History

Dr. Terror, introduced at the start of this chapter, requested to be on testosterone replacement therapy. He also inquired about growth hormone replacement therapy, but changed his mind after finding out the cost of the treatment. He wondered about the advertised growth hormone secretagogues, but was advised that they rarely work. Quite a bit of time was spent discussing diet and exercise with him. As he was fairly thin, strength-training exercises were suggested for him. He decided to join a gym and exercised 45 minutes three to five times a week. He increased protein intake, fresh vegetables, and fruits (potential antioxidants). He was prescribed testosterone gel 5 mg/day. After 6 months, on a routine follow-up visit, he was very satisfied. Overall, he said, his mood is better, and he is sharper and more energetic. He said the program restored some of his lost years. Objectively, it was noticed that despite weighing the same at 150 pounds, his body fat was reduced to 19%, implying muscle growth. His skin was less dry, his libido better, and he thought that testosterone was the elixir of youth for him. He was advised that testosterone was only supplementary and that it was the combination of testosterone with exercise and the right nutrition that seemed to have made him feel five years younger.

Conclusion and Key Points

Men on average live about 7 or 8 years less than women. Genetics and lifestyle account for the difference. Paying attention to simple issues like weight, smoking cessation, nutrition, and exercise can alter longevity in men. The role of hormone replacement, at least in the short term, has been shown to improve the quality of life of not only women but also of men. The keys to longevity can be summarized in the following few points:

- Lifestyle habits from youth, if altered in midlife, can alter outcomes in later life.

- Modest food intake, antioxidants, hormones, and exercise can alter the outcome for some individuals.

- Positive mood, being married, and not smoking are predictors of longevity.

- There are many theories of aging. Care must be taken in interpreting those that are based on animal studies. The human body is more complex, and applying some of these theories to humans can be erroneous.

- The best antidotes for aging are weight loss through proper nutrition, exercise, smoking cessation, alcohol moderation, and a positive outlook.

REFERENCES

1. Harman D. Aging: phenomena and theories. Ann NY Acad Sci 1998;854:1–7

2. Olshansky SJ, Carnes BA, Desesquelles A. Demography. Prospects for human longevity. Science 2001;291:1491–1492

3. Waldron I. Causes of sex differential in longevity. J Am Geriatr Soc 1987;35:365–366

4. Moore SL, Wilson K. Parasites as a viability cost of sexual selection in natural populations of mammals. Science 2002; 297: 2015–2018

5. Perls TT, Fretts RC. The evolution of menopause and human life span. Ann Hum Biol 2001;28:237–245

6. Hitt R, Young-Xu Y, Silver M, Perls T. Centenarians: the older you get, the healthier you have been. Lancet 1999;354: 652

7. Hybels CF, Pieper CF, Blazer DG. Sex differences in the relationship between subthreshold depression and mortality in a community sample of older adults. Am J Geriatr Psychiatry 2002;10:283–291

8. Masoro EJ, Austad SH. In: *Handbook of the Biology of Aging.* San Diego: Academic Press; 2001

9. Olshansky SJ, Hayflick L, Carnes BA. Position statement on human aging. J Gerontol A Biol Sci Med Sci 2002;57: 292–297

10. Takata H, Suzuki M, Ishii T, Sekiguchi S, Iri H. Influence of major histocompatibility complex region genes on human longevity among Okinawan-Japanese centenarians and nonagenarians. Lancet 1987;10:824–826

11. Ivanova R, Henon N, Lepage V, et al. HLA-DR alleles display sex-dependent effects on survival and discriminate between individual and familial longevity. Hum Mol Genet 1998;7: 187–194

12. Zahn RK. DNA status in brain and heart as prominent co-determinants for life span? Mech Ageing Dev 1996;89:79–94

13. Olovnikov AM. Telomeres, telomerase, and aging: origin of theory. Exp Gerontol 1996;31:443–448

14. Rossuw JE, Andersen GL, Prentice RL, et al. Risks and benefits of estrogens and progestin in healthy postmenopausal women. JAMA 2002;288:321–323

15. Kolestky S, Puterman DI. Effect of low calorie diet on hyperlipidemia, hypertension and life span of genetically obese rats. Proc Soc Exp Biol Med 1976;151:368–371

3

Diagnostic Concerns and Evaluation of the Andropause in Aging Males

ROBERT S. TAN

Case History

A 50-year-old man who has a long history of depression, and also suffers from sexual dysfunction. He has a past medical history of hypertension, coronary artery disease, and hypercholesterolemia. He has also attended Alcoholic Anonymous meetings. He is currently on Prozac, Lipitor, atenolol, several vitamins, and various herbal preparations thought to increase sex drive. He reports that he has difficulty with erections. His wife complains that he has lost the desire for sex. On direct questioning, he admits that he has lost the ability for early-morning erections. He also feels easily tired, and unable to stay awake right after dinner. For the past 3 months he has also complained of periodic night sweats, and he would wake up in the morning drenched. Examination revealed that he weighed 225 pounds, his blood pressure (BP) was 135/100 mm Hg, and he had loss of hair in the armpits. His testicular size was compared with an orchidometer and his left testes had a reduced volume compared with the right. He scored 8/10 on the Androgen Decline in Aging Males (ADAM) questionnaire. His laboratory results revealed that the 8 A.M. total testosterone was 148 ng/dL (range 260–1000), bioavailable testosterone was 88 ng/dL (range 84–403), estradiol 26 pg/dL, luteinizing hormone (LH) 1.8 mIU/mL, prolactin 19.2 ng/dL, prostate-specific antigen (PSA) 1.5 ng/dL, leptin 19.2 ng/dl (range 1.2–9.5). The course of treatment of the patient is discussed at the end of the chapter.

Is There Indeed an Andropause and If So, Can There Be Symptoms?

It is not unusual for physicians to treat men in their 40s and older who complain of loss of libido, erectile dysfunction, fatigue, and depression. Psychological problems and medical illness are often confounders to the andropause. Knowledge of the patient's history and a careful examination combined with laboratory tests are keys to an accurate diagnosis of symptomatic hypogonadism, or the andropause syndrome.[1]

Androgens are a group of hormones that include testosterone, dehydroepiandrosterone (DHEA), androstenedione, 3 alpha-androstanediol glucuronide (AAG), and others. It is a misnomer to classify them as "male hormones" as they are present in both males and females, albeit in different amounts. There is undeniable evidence that aging results in a lowering of androgens. When total testosterone is measured, 20% of men above 55 years are hypogonadal.[2,3] When bioavailable testosterone is measured, however, 50% of men above 50 years are defined as hypogonadal.[4] Ninety-eight percent of circulating testosterone is bound to plasma proteins; the remaining 2% of free testosterone is responsible for biological activity. Approximately 40% of the bound testosterone is bound to sex hormone–binding globulin (SHBG). The rest is weakly bound to albumin and is readily available to tissue when needed. Bioavailable testosterone includes free testosterone and that loosely bound to albumin.[5]

As the decline in androgens is gradual, the alternative term of *androgen decline in aging males* (ADAM) has been used. *Partial androgen decline in aging males* (PADAM) has also been suggested because the androgen deficiency in older men is generally moderate and not a complete deficiency. There is often confusion that andropause is a symptomatic state. It must be emphasized that like menopause, there could be the presence or absence of symptoms. Transitory symptoms can include changes in mood and sexuality. The long-term effects of hypogonadism can result in osteoporosis, muscle atrophy, and cognitive changes. Symptomatic hypogonadism is sometimes referred to as the "andropause syndrome."[6]

TABLE 3–1. The Definition of Andropause

A hypogonadal state in older males resulting from gradual partial androgen deficiency, alternately termed androgen decline in aging males (ADAM) or partial androgen decline in aging males (PADAM)

Decreased sensitivity to androgens in target organs (thus absolute serum levels of testosterone can be misleading)

Symptoms include decreased energy and well-being, changes in sexual function: symptomatic andropause or the "andropause syndrome"

In the long term, androgen deprivation can affect bone, muscle, lipids, and cognition

Symptoms can develop with the andropause syndrome. Our previous study in 302 male subjects revealed that loss of libido and erectile dysfunction (46%), fatigue (41%), and memory loss (36%) were dominant symptoms, in that order.[1] The correlation of symptoms to levels of testosterone is very variable and is the subject of ongoing investigation by my team. Most laboratories give a normal range of 260 to 1000 ng/dL for total testosterone, 50 to 210 pg/dL for free testosterone, and 66 to 417 ng/dL for bioavailable testosterone.[7] The International System of Units (SI) conversion factor is ~35. The range often is not age adjusted and poses dilemmas for physicians. Patients can have low-normal levels and yet display symptoms, which are reversed after androgen supplementation. This suggests the possibility of "relative hypogonadism" in which levels are appropriate for each individual.[8] Another dilemma is that testosterone levels can vary in the course of a day. Frequent sampling of testosterone in a study of 20 normal men revealed ranges from 105 to 1316 ng/dL between subjects.[9] Table 3–1 summarizes definitional aspects of andropause.

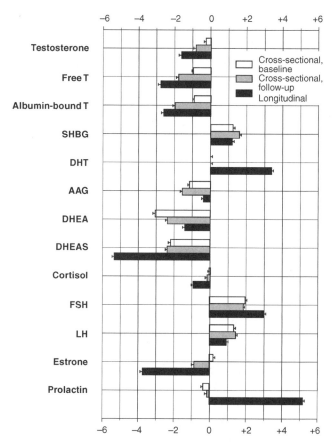

FIGURE 3–1. Age trends in the level of serum testosterone and other hormones in middle-aged men: longitudinal results from the Massachusetts male aging study. (Adapted from Feldman HA, Longcope C, Derby CA, et al. Age trends in the level of serum testosterone and other hormones in middle-aged men: longitudinal results from the Massachusetts male aging study. J Clin Endocrinol Metab 2002; 87:589–598, with permission.)

Andropause and Menopause: Similarities and Contrasts

Andropause refers to a lowered state of androgens. Total testosterone declines ~1.6% per year, free testosterone decreases 2% per year, and bioavailable testosterone decreases 2 to 3% per year between the age of 40 and 70 years. On the other hand, SHBG increases 1.6% per year (Fig. 3–1).[10]

Testosterone declines gradually in the andropause. In contrast, menopause usually has a narrower window of 5 to 10 years, with the ultimate shutdown of the ovaries. The production of sperm in males is primarily regulated by LH on Leydig cells in the testis. Testosterone has a supplementary role in regulating spermatogenesis. Fertility in males may not be affected in the andropause, although the numbers of sperm with normal motility and morphology can be altered.[11] In contrast, menopause brings about a cessation of reproductive capability.

It is important to realize that long-term deprivation of androgens and estrogens in both males and females leads to similar outcomes: osteoporosis, muscle loss, and cognitive changes. The precise mechanism whereby androgens affect bone and cognition is unclear, and it is postulated that the effect is both from testosterone itself and the conversion to estradiol.[12–14]

Hypogonadism or Low Testosterone Results from Aging

Hypogonadism or low testosterone level with aging is the result of both primary gonadal failure and hypothalamic-pituitary failure. Primary hypogonadism refers to a decrease in the numbers of Leydig cells, reduction of testosterone production, and decreased secretion of testosterone in response to the stimulation of human chorionic gonadotropin. The decline in testosterone levels with aging is associated with increase in follicle-stimulating hormone (FSH) and, to a lesser extent, LH. Low testosterone level with abnormal LH pulse suggests an age-associated impairment of hypothalamic gonadotropin-releasing hormone secretion.[15,16]

Physiology and Regulation of Androgens

In the normal male, the testes produce 95% of androgens, namely testosterone. About 5 to 10 mg of testosterone is produced on a daily basis by the testes. The adrenals produce the rest of the androgens mainly in the form of DHEA. The pituitary gland secretes LH and regulates the production of testosterone in Leydig cells from cholesterol. FSH mainly affects spermatogenesis. In turn testosterone is metabolized to dihydrotestosterone (DHT) by 5α-reductase and aromatized to estradiol by aromatase. DHT is linked to prostate hypertrophy and male alopecia (Fig. 3–2).[17]

Influence of Age and Obesity on Testosterone and Sex Hormone-Binding Globulin

In the uterus during the first trimester, fetal testes produce testosterone. Levels peak at about age 20 years, and there is a gradual decline thereafter. Total testosterone declines at the rate of 110 ng/dL, typically after 40 years of age.[18] Bioavailable testosterone declines much more dramatically especially with aging. Bioavailable testosterone is thought to be the active component of androgen acting directly at the cellular level. It consists of free testosterone and that loosely bound to albumin. SHBG binding increases with the aging process and hence decreases available free testosterone.[19,20] Overall, the impact of aging on testosterone is negative. Obesity in males is accompanied by a significant decrease in testosterone levels. In moderate obesity, free testosterone levels can be apparently "normal" because of the decrease of SHBG. In massively obese males, there is "real" hypogonadotrophic hypogonadism with deceased free testosterone levels.[21]

Clinical Assessment, Screening Questionnaires, and Confounding Factors for Diagnosis of Andropause

It is important for the physician to take a careful history and to do a physical examination when assessing a patient for hypogonadism. The physician should inquire about symptoms of loss of libido and distinguish it from erectile dysfunction. A loss of early-morning erection can be indicative of hypogonadism. Stress and chronic illness can depress testosterone levels, and so can medications including cimetidine, digoxin, and spironolactone. Diabetes, insulin resistance, and obesity have been associated with hypogonadism.[2,11,22] As such, one should look for symptoms and signs associated with diabetes. A history of chronic alcoholism must be ascertained as it can suppress the production of androgens. Rare conditions can be associated with hypogonadism including Prader-Willi syndrome and Klinefelter's syndrome, as well as Kallmann's syndrome and they have to be excluded in the clinical examination. The physical examination should include weight, body mass index, waist/hip ratio, and a measurement of body fat.[23] The

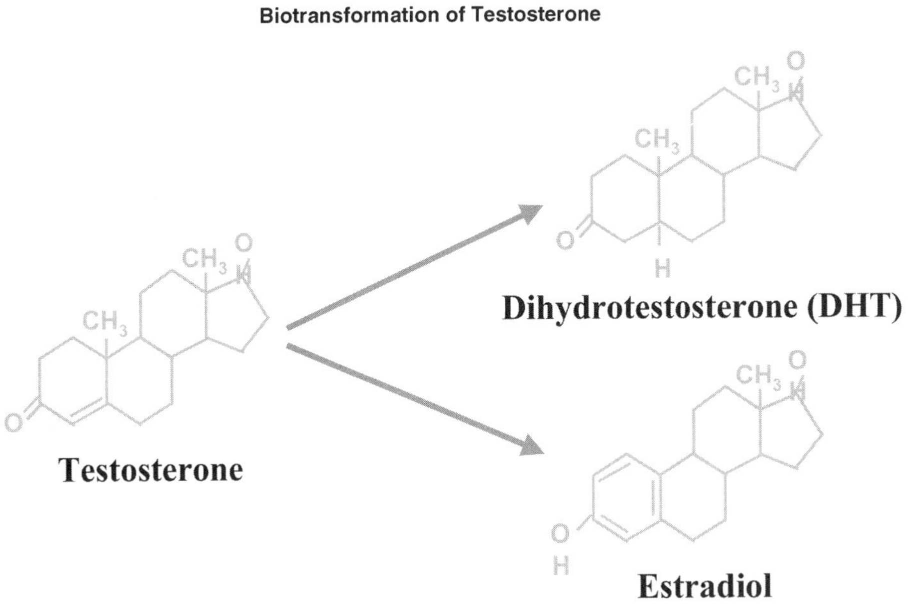

Biotransformation of Testosterone

Testosterone

Dihydrotestosterone (DHT)

Estradiol

FIGURE 3–2. Biotransformation of testosterone.

skin should be examined for evidence of hyperestrogenism such as spider telangiectasia. The face, axilla, and groin should be inspected for hair loss. Testicular size can be measured using an orchidometer. A prostate examination should also be done as part of the screening.

Screening tools like the ADAM questionnaire (Table 3–2) can be helpful.[24] Screening for depression can be done with the Geriatric Depression Scale.[25] The Folstein Mini–Mental State Examination can be used to screen for cognitive problems.[26] Unfortunately, memory loss in the andropause can be very subtle, and more sophisticated neuropsychological tools may sometimes be needed.[27] There are several confounders to the andropause syndrome, and these are summarized in Table 3–3.

Laboratory Assessments for Diagnosis of Andropause

Laboratory assessments should be used only in conjunction with a good history and examination before deciding that the patient has the andropause syndrome. Reliance on laboratory tests alone can often lead to misdiagnosis. Bioavailable testosterone is preferred in older patients above 65 years because of alterations in binding with SHBG. Most laboratories

TABLE 3–2. The Androgen Decline in Aging Males (ADAM) Questionnaire

Do you have a decrease in libido (sex drive)?
Do you have a lack of energy?
Do you have a decrease in strength and/or endurance?
Have you lost height?
Have you noticed a decreased enjoyment in life?
Are you sad and/or grumpy?
Are your erections less strong?
During sexual intercourse, has it been more difficult to maintain erection to completion of intercourse?
Are you falling asleep after dinner?
Has there been a recent deterioration in your work performance?

TABLE 3–3. Confounders to the Andropause Syndrome

Chronic illnesses (e.g., diabetes, chronic renal failure, cirrhosis)
Depression
Decrease in albumin from alcohol and poor nutrition
Acute stress (e.g., surgery, severe burn, acute myocardial infarction)
Medications (e.g., cimetidine, spironolactone, antidepressants)
Other endocrinopathies including hypothyroidism, Cushing's syndrome
Hypothalamic-pituitary tumor, hemochromatosis
Circadian rhythm of testosterone
Kallmann's syndrome, Klinefelter's syndrome

do not offer an actual free testosterone and calculate free testosterone based on a formula after measuring total testosterone. Radioimmunoassay remains the most common method to assess testosterone levels. Saliva testing for testosterone is a novel approach, allowing convenience for the patient. If done correctly, it has a good correlation to free testosterone levels.[28] Bioavailable testosterone levels < 60 ng/dL or total testosterone levels < 300 ng/dL with consistent clinical symptoms and signs are diagnostic of androgen deficiency.[3,4] The physician must also be aware that testosterone is secreted in spurts, and levels are highest after awakening. However, the daily fluctuation in serum testosterone levels is attenuated in older men.

If the initial testosterone level is low, the evaluation should determine whether there are reversible causes of low testosterone. LH and FSH should also be measured. Some men in andropause demonstrate a slight rise in LH. However, the rise is modest in comparison to the menopause. Prolactin measurements are useful to screen for rare causes of hypogonadism secondary to prolactinoma. As there may be overlapping conditions mimicking the andropause syndrome, it would be prudent to measure TSH and cortisol levels if indicated. Routine laboratory tests, for example, complete blood count, liver function, renal function, and PSA, are also indicated. In geriatric patients whose nutrition may pose a problem, a measurement of zinc levels may be useful. Zinc has antiaromatase action and can be used to treat hypogonadism.[29] A repeated serum testosterone level should always be obtained to confirm androgen deficiency before starting testosterone replacement. If the initial free or bioavailable testosterone is within normal range in a man with symptoms and signs of androgen deficiency, the patient's clinical status and testosterone levels should be monitored on follow-up visits.[30]

Discussion of Case History

The patient was diagnosed to be hypogonadal based on his symptoms and laboratory results. It was hypothesized that his sexual dysfunction, fatigue, and depressed moods were related to his hypogonadal state, resulting in the andropause syndrome.

A therapeutic trial of compounded testosterone 5 mg with 100 mg zinc in a topical gel vehicle was prepared for him. Zinc was added as it has some antiaromatase activity and may decrease his conversion of testosterone to estradiol. As obesity results in increased aromatase activity, he was put on an exercise and diet regimen. The exercise advised was a combination of strength training and aerobics four times a week. He was advised to stop drinking completely. After 3 months of follow-up, the patient returned reporting significant improvement in quality of life. His sexuality had returned, he felt less tired, and his mood had improved, so that it was decided that his Prozac should be discontinued. Prozac in itself can contribute to sexual dysfunction. His blood pressure was better controlled, and so the atenolol was stopped completely. Part of this improvement was because of his 20-pound weight loss, and perhaps the vasodilatory effects of testosterone. Testosterone has also been known to modulate leptin, thereby reducing weight by decreasing satiety. The total testosterone remeasured was 275 ng/dL, and PSA was 1.5 ng/dL. In retrospect, the patient wonders if his depressed moods were related to his hypogonadal state, but it is difficult to ascertain, and the patient will be followed up, and his mood assessed. The lesson learned in this case is that if androgen supplementation is combined with lifestyle modifications, the outcome is usually better in the long run.

Conclusion and Key Points

- The physician should have knowledge of the multiple factors that can affect testosterone levels.

- Some of the factors that are associated with the andropause syndrome results from lifestyle habits. As such, counseling may play a big role in these patients. Weight loss may potentially reverse hypogonadism in some obese men, although this is an area for further research. Exercise, in particular weight training, may even increase testosterone in older men.[31]

- There has been overreliance on laboratory assessment of the andropause syndrome, and as such many patients are underdiagnosed or even overdiagnosed at times. The physician must be aware that the testosterone level is not static and that it changes with time. Listening to the patient carefully is the key to a successful treatment plan.

REFERENCES

1. Tan RS, Philip P. Perception and risk factors for the andropause. Arch Androl 1999;43:97–103

2. Smith KW, Feldman HA, McKinlay JB. Construction and field validation of a self-administered screener for testosterone deficiency (hypogonadism) in aging men. Clin Endocrinol (Oxf) 2000;53:703–711

3. Tenover JS. Androgen administration to aging men. Endocrinol Metab Clin North Am 1994;23:877–892

4. Sih R, Morley JE, Kaiser FE, et al. Testosterone replacement in older hypogonadal men: a 12-month randomized controlled trial. J Clin Endocrinol Metab 1997;82:1661–1667

5. Basaria S, Dobs AS. Hypogonadism and androgen replacement therapy in elderly men. Am J Med 2001;110:563–572

6. Morales A, Heaton JP, Carson CC. Andropause: a misnomer for a true clinical entity. J Urol 2000;163:705–712

7. DeGroot LJ, Jameson JL. Endocrinology. 4th ed. Philadelphia: WB Saunders; 2001

8. Tan RS. Andropause: introducing the concept of "relative hypogonadism" in aging males. Int J Impot Res 2002;14:319

9. Spratt DI, O'Dea LS, Crowley WF, et al. Neuroendocrine-gonadal axis in men: frequent sampling of LH, FSH and testosterone. Am J Physiol 1988;254:658–666

10. Feldman HA, Longcope C, Derby CA, et al. Age trends in the level of serum testosterone and other hormones in middle-aged men: longitudinal results from the Massachusetts male aging study. J Clin Endocrinol Metab 2002;87:589–598

11. Gallardo E, Guanes PP, Simon C, et al. Effect of age on sperm fertility potential: oocyte donation as a model. Fertil Steril 1996;66:260–264

12. Siemenda CW, Loncope C, Zhou L, et al. Sex steroid and bone mass in older men: possible associations with serum estrogens and negative association with androgens. J Clin Invest 1997; 100:1755–1759

13. Pietschmann P, Kudlacek S, Grisar J, et al. Bone turnover markers and sex hormones in men with idiopathic osteoporosis. Eur J Clin Invest 2001;31:444–451

14. Barrett-Connor E, Goodman-Gruen D, Patay B. Endogenous sex hormones and cognitive function in older men. J Clin Endocrinol Metab 1999;84:3681–3685

15. Vermeulen A, Kaufman JM. Aging of the hypothalamo-pituitary-testicular axis in men. Horm Res 1995;43:25–28

16. Morley JE, Perry HM III. Androgen deficiency in aging men. Med Clin North Am 1999;83:1279–1289

17. Nieschlag E, Behre HM. Testosterone: Action, Deficiency, Substitution. 2nd ed. Berlin: Springer; 1998.

18. Morley JE, Kaiser FE, Perry HM, et al. Longitudinal changes in testosterone, luteinizing hormone, and follicular-stimulating hormones in healthy older men. Metabolism 1997;46: 410–413

19. Gray A, Feldman HA, McKinlay JB, et al. Age, disease, and changing sex hormones levels in middle-aged men: results of the Massachusetts Male Aging Study. J Clin Endocrinol Metab 1991;73:1016–1025

20. Winters SJ, Kelley DE, Goodpaster B. The analog free testosterone assay: are the results in men clinically useful? Clin Chem 1998;44:2178–2182

21. Vermeulen A. Decreased androgen levels and obesity in men. Ann Med 1996;28:13–15

22. Marin P, Holmang S, Jonsson L, et al. The effects of testosterone treatment on body composition and metabolism in middle-aged obese men. Int J Obes Relat Metab Disord 1992; 16:991–997

23. Ukkola O, Gagnon J, Rankinen T, et al. Age, body mass index, race and other determinants of steroid hormone variability: the HERITAGE Family Study. Eur J Endocrinol 2001; 145:1–9

24. Morley JE, Charlton E, Patrick P, et al. Validation of screening questionnaire for androgen deficiency in aging males. Metabolism 2000;49:1239–1242

25. Sheikh JI, Yasavage JA. Geriatric Depression Scale (GDS): recent evidence and development of a short version. In: Brink TL, Clinical Gerontology: A Guide to Assessment and Intervention. New York, NY: Haworth Press; 1986

26. Folstein MF, Folstein SE, McHugh PR. Mini-mental state: a practical method for grading the cognitive state of patients for clinician. J Psychiatr Res 1975;12:189–198

27. Tan RS, Pu SJ. The andropause and memory loss: is there a link between androgen decline and dementia in the aging male? Asian J Androl 2001;3:169–174

28. Klee GG, Heser DW. Techniques to measure testosterone in the elderly. Mayo Clin Proc 2000;75:S19–S25

29. Fuse H, Kazama T, Ohta S, et al. Relationship between zinc concentrations in seminal plasma and various sperm parameters. Int Urol Nephrol 1999;31:401–408

30. Matsumoto AM. Andropause: clinical implications of the decline in serum testosterone levels with aging in men. J Gerontol A Biol Sci Med Sci 2002;57A:M76–M99

31. Zmuda JM. Exercise increases serum testosterone and SHBG in older men. Metabolism 1996;45:935–939

4

Laboratory Assessments: Implications in Diagnostics for the Andropause

CHRISTINA HO AND ROBERT S. TAN

Case History

A 32-year-old man presented with depression, loss of libido, and extreme tiredness. He would fall asleep at work, but had difficulty sleeping at night. He was concerned that his testosterone levels were falling after reading about this possibility in a men's magazine. He said that he actively worked out and had a steady relationship. His exercise regimen included running in marathons and weight lifting as well. Physical examination revealed a well-built 6-foot-tall male weighing 180 pounds. His blood pressure was 120/80 mm Hg. He had no overt signs of hypogonadism, and was hairy in his chest. His gonads were of normal size. His laboratory results were as follows:

Total testosterone = 590 ng/dL (260–1000 ng/dL)
Free testosterone by dialysis equilibrium
    method = 32 pg/dL (34–194 pg/dL)
Bioavailable testosterone by ammonium precipitation method = 65 ng/dL (84–402 ng/dL)
Luteinizing hormone (LH),
    immunoassayed = 3.8 mIU/mL
    (1.5–9.3 mIU/mL)
Follicle-stimulating hormone (FSH), immuno-
    assayed = 2.5 mIU/mL (1.4–8.1 mIU/mL)
Prolactin = 4 ng/mL (2–18 ng/mL), thyroid-
    stimulating hormone (TSH) = 1.38 mIU/mL
    (0.4–5.5 mIU/mL)
Cortisol levels and magnetic resonance imaging
    (MRI) of the pituitary were normal, and sperm
    sample was normal; the patient was given a
    clomiphene challenge of 100 mg for 5 days, and
    his total testosterone went to 980 ng/dL.

Different Tests to Measure Testosterone

Many clinicians and laboratories are confused as to the correct test to use when determining androgen status in aging men. Although opinion leaders have agreement in some areas as to which test to use, there are also areas of disagreement. This is in part because symptoms of androgen deficiency are not proportional to androgen levels and also because of the pulsatile nature of secretion of hormones. In any event, it is important for the practitioner to determine which test to order based on the clinical assessment of the patient.

Most laboratories measure the three domains of testosterone: total testosterone, free testosterone, and bioavailable testosterone. Fig. 4–1 represents the different fractions of testosterone. Total testosterone refers to all the testosterone that is measurable including those bound and unbound portions. Testosterone is bound to proteins like albumin and sex hormone–binding globulin (SHBG). As mentioned, changes in protein concentrations can alter the true levels and give false impressions. Testosterone is loosely bound to SHBG, and as such comes off easily, making it "free." The actual free amount and that bound to SHBG are referred to as bioavailable testosterone. Laboratories can measure free testosterone using analog ligand radioimmunoassay methods, or they can sometimes calculate it based on a formula.

In older men, the binding of testosterone to SHBG is increased, making it less likely for it to be released and become free testosterone. As such, total testosterone in older men is much less reliable, and bioavailable testosterone is recommended instead. Bioavailable testosterone represents the "active form" of testosterone, and has a satisfactory correlation with androgenicity. It is also reduced more rapidly than total testosterone and as such approaches a real-time view of the androgen status of the patient. Unfortunately, most laboratories charge more for this test as it is more difficult to perform.

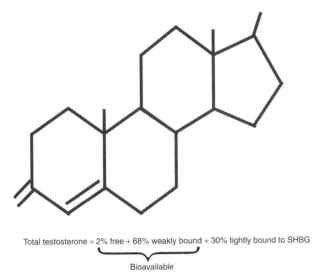

Total testosterone = 2% free + 68% weakly bound + 30% tightly bound to SHBG

Bioavailable

FIGURE 4–1. Fractions of testosterone.

Different Methods of Assays

Equilibrium dialysis is based on the separation of molecules of interest from solutions containing a mixture of cells and cellular products. Because of its physical simplicity, equilibrium dialysis is one of the key tests used in the study of protein binding. For equilibrium dialysis, neither fluorescent nor radioactive tags are needed. It is difficult to distinguish between bound and free ligand (small ions/molecules) in any mixture of ligand and macromolecule; therefore, equilibrium dialysis uses an indirect method to estimate the amount of ligand bound to a macromolecule. If, however, the free ligand can be dialyzed through a membrane until its concentration across the membrane is at equilibrium, the free ligand concentration can be measured easily.[1] To perform an equilibrium dialysis, two half-cells of equal volume separated by a semipermeable dialysis membrane are used. This membrane is typically made of polysulfones or cellulose materials. Important system parameters are the pore size (of the membrane) and spatial distribution of the pores across the silicon substrate, as well as adhesion to the silicon surface.[2] The dialysis is continued until equilibrium is reached and the concentration of free ligand is the same in both chambers. Because a portion of the ligand and macromolecule bind and are prevented from diffusing across the membrane into the assay chamber, the amount of free ligand in the assay is markedly reduced, and additional tests are required to provide information about specific binding sites. Although equilibrium dialysis is one of the more accurate methods, it is very time consuming.

Ultracentrifugation, on the other hand, is a technique that utilizes speed and sedimentation to separate macromolecules from solution. This process uses the approaches of sedimentation velocity and sedimentation equilibrium to characterize several macromolecular components. Properties such as homogeneity, macromolecular self-association, and macromolecule hydrodynamic properties (size, shape, and hydration) are ascertained through sedimentation velocity. Sedimentation equilibrium experiments, however, are conducted at significantly slower speeds "so that sedimentation is balanced by diffusion, yielding an equilibrium radial distribution of the macromolecule."[3] From this method, molecular weight, stoichiometry, and equilibrium association constants are obtained. The particle immersed in liquid within the centrifuge tube is acted on by three forces: buoyant force, the frictional force between the particle and the liquid, and particularly the centrifugal force, which is proportional to the rotational rate of the rotor and the distance of the tube from the center of the rotor.[4] Reaching rotational speeds as high as 80,000 revolutions per minute (rpm), inert substances such as sucrose or cesium chloride, in which the concentration and density of the solution increase from top to bottom of the centrifuge tube, are often recommended for use in this procedure.[5] When using the ultracentrifugation method, time and temperature of centrifugation must be carefully monitored. Control of time of centrifugation is important because a pellet at the bottom of the tube can form or insufficient separation can occur. Incorrect temperature settings can yield sample degradation or cause an increase in viscosity of solution and affect aggregation.[6] One must also take note that with this method it is hard to assess purity of organelle preparations.

Radioimmunoassay (RIA) has been proven to be one of the most accurate tests and can be used to separate and measure even the smallest of molecules. In a typical RIA, the substance to be measured—the unlabeled sample antigen (also known as the ligand or analyte)—competes with radiolabeled antigen for a limited number of antibody binding sites.[7] Radioisotopes such as tritium (^{3}H), ^{57}Co, and the radioactive isotope of iodine (^{125}I) are used to tag proteins and peptides; however, because ^{125}I (which emits gamma radiation as it decays) attaches easily to most antibodies or antigens, it is most commonly used as an isotopic tag. ^{125}I also has a shorter half-life (60 days) as compared with tritium's half-life of 12.5 years, and thus, relieves the expenses of disposal. For this test to work, both a radiolabeled analog of the ligand and pure form of the liquid must be used. Both the labeled and unlabeled antigens are equally probable of binding to the antibody, and reflect the amount of each present; therefore, the amount of radioactivity present is inversely proportional to the amount of sample antigen present. When the Ag:Ab complex precipitates out of solution, a radiolabeled isotope is added and allowed time to bond. It is then separated from unbound reagents and measured in a gamma counter.[8] To determine the concentration of the unknown sample, it is compared with a standard curve in which the proportion of labeled to unlabeled antibody is known. The immunoradiometric assay (IRMA), also known as the *sandwich method* is a variation of the RIA method.[7] This technique uses two antibodies that bind to the antigen at two different sites: a capture antibody bound to a solid phase (typically a glass or plastic bead), and an antibody labeled with [125]I. These antibodies bind to the antigen and form the *Ab-Ag-Ab(^{125}I) sandwich*. The unbound reagents are washed away, and radioactivity measured. Because IRMA radioactivity signal is directly proportional to the amount of analyte, it is often more accurate with lower concentrations of the analyte; the IRMA method, however, fails to perform well with smaller peptides, as their size permits only one binding site. Health hazards and the costs of complying with the handling and disposal regulations of the radioisotope have caused researchers to seek alternative labeling methods such as chromogenic, fluorescent, or luminescent tagging. These nonisotropic methods, however, have not been able to acquire the high sensitivity and specificity of the RIA.

Total Testosterone

Most laboratories use automated machines to determine total amounts of testosterone. To measure total testosterone, these instruments must first displace bound testosterone from SHBG and albumin. Usually low pH buffers, surfactants, salicylates, or a competing steroid that does not bind to antitestosterone antibody is used in the immunoassay. However, the testosterone antisera used in commercial preparations often cross-react with other steroids including dihydrotestosterone (DHT). Solvent extraction and chromatography have been used to remove these interfering compounds prior to testosterone measurement. Unfortunately, these cannot be incorporated into the methods used by automated analyzers. Fortunately, the plasma levels of DHT are only ~10% of testosterone levels and cross-reactivity is usually less than 5%. Thus, in clinical application, the impact of DHT is minimal in most instances.

"Free" Testosterone

It has been generally agreed that free testosterone rather than total testosterone gives a better measure of androgenicity.[9] However, different laboratories may report free testosterone using various methodologies. It is prudent that the clinician be aware of the different methods of obtaining a free testosterone level, and interprets it in the context of the patient. Testosterone circulates in plasma and binds to SHBG and albumin. Testosterone binding to transcortin and orosomucoid is negligible. Albumin-bound testosterone is released into the plasma easily as compared with those bound to SHBG. There are several measures of free and bioavailable testosterone:

Apparent Free Testosterone Concentration (AFTC) as Measured by Equilibrium Dialysis

Apparent free testosterone concentration (AFTC) or testosterone as determined by equilibrium dialysis at 37°C is arguably the method of choice for measuring free testosterone in vivo. The measurement of free testosterone in the serum is technically demanding, as the free testosterone concentration is very low (2%). Routinely available assays are not sensitive enough to quantify free testosterone directly. Usually, free testosterone is estimated by indirect methods. In these indirect methods, titrated testosterone is added to the sample and allowed to come to equilibrium with testosterone in the serum at a physiological temperature (37°C). The amount of added radiolabeled testosterone must be low enough to guarantee that addition will not significantly increase the total testosterone concentration. When equilibrium is achieved, the free testosterone is separated from the bound by filtration through a membrane. The filtration is accomplished by equilibrium dialysis or centrifugal ultrafiltration. The radioactivity of the protein-free ultrafiltrate is measured and used to calculate the percentage of free testosterone. The concentration of free testosterone can then be calculated by multiplying the percentage of free testosterone by the total testosterone concentration. Measurement of free testosterone by these methods is not available in most clinical laboratories due to the complicated nature of the testing and the requirement of a scintillation counter to measure the titrated testosterone concentration. Overall, the results of equilibrium dialysis and centrifugal ultrafiltration methods have been shown to be quite comparable. Although equilibrium dialysis is often considered to be the "gold standard," centrifugal ultrafiltration is somewhat simpler to perform and may theoretically be more accurate due to the fact that the equilibrated sample is not diluted with dialysis buffer.

Free Androgen Index (FAI)

The concentration of testosterone in the different free and bound forms is really a function of total testosterone concentration and the relative concentrations of SHBG and albumin. It can be predicted that increased SHBG will decrease the concentration of both free and bioavailable testosterone for a given total testosterone concentration. Many clinicians use a calculated free androgen index to estimate physiologically active testosterone. This index is typically calculated as the *ratio of total testosterone divided by SHBG and multiplied by 100* to yield numerical results comparable in free testosterone concentration. Otherwise, more complicated mathematical algorithms can be used to approximate the percentage of free testosterone from the SHBG concentration alone or in combination with albumin concentration. The precision of these algorithms is subject to the combined errors of the individual tests performed.

Direct Immunoassay of Free Testosterone with a Labeled Testosterone Analog (aFT)

Several commercial kits are available for the direct estimation of free testosterone in serum. These kits use a labeled testosterone analog that has a low binding affinity for both SHBG and albumin, but is bound by antitestosterone antibody. Because the analog is unbound in the plasma, it competes with free testosterone for binding sites on an antitestosterone antibody that is immobilized on the surface of the well or assay tube. The first kits developed used a radiolabeled testosterone analog to compete with free testosterone for binding sites on an antibody-coated polypropylene tube. More recently developed kits employ an enzyme-labeled analog that can be measured after competitive binding to antitestosterone antibodies coated to microtiter wells. These analog methods are technically less demanding than equilibrium dialysis or centrifugal ultrafiltration and require substantially smaller blood samples. The analog methods also offer the benefit of direct estimation of free testosterone concentration without the need to measure total testosterone. Many laboratories can readily perform the enzymatic methods because they are nonisotopic.

Winters and colleagues[10] have found the analog method to correlate better with total testosterone levels than with bioavailable testosterone determined by the ammonium sulfate precipitation method. They suggested that the analog free testosterone results might be misleading in men with low SHBG concentration. Fig. 4–2 shows the correlation of total testosterone with bioavailable and free testosterone.

Ooi and Donnelly[11] suggested that the problems observed by Winters et al might be resolved, in large part, simply by using a more appropriate population-based reference interval. Vermeulen and colleagues[12] found that the analog-free testosterone method

correlated well with free testosterone by equilibrium dialysis but did not correspond with a free testosterone calculated from total testosterone and SHBG.

Free testosterone (FT) can also be calculated from total testosterone and immunoassayed SHBG.

Bioavailable Testosterone (BT)

Bioavailable testosterone (BT) is the fraction of serum testosterone not precipitated by 50% ammonium sulfate concentration. As in the free testosterone methods described previously, titrated testosterone is added to serum that is then allowed to come to equilibrium at physiological temperature. Testosterone

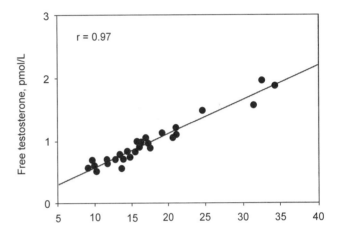

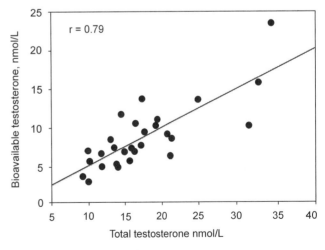

FIGURE 4–2. Winters et al study suggesting that the analog method of free testosterone rather than the bioavailable testosterone is better correlated to total testosterone. (Adapted from Winters SJ, Kelly DE, Goodpaster B. The analog free testosterone assay: are the results in men clinically useful? Clin Chem 1998;44:2178–2182, with permission.)

bound to SHBG is then selectively precipitated with 59% ammonium sulfate, leaving free and albumin bound testosterone in the solution. The percentage of titrated label not bound to SHBG is multiplied by the total testosterone to produce the bioavailable testosterone. Another method of measuring bioavailable testosterone is by direct radioimmunoassay in the supernatant after solvent extraction.

Care must be taken to ensure that the labeled tracer testosterone used for measuring the FT fraction is highly purified. In a study by Vermeulen et al,[12] AFTC was correlated against the other measures of testosterone. A coefficient of correlation of 1.0 would mean a perfect correlation. In that study with men, they found that the correlation of AFTC with FT (calculated free testosterone) = 0.987, aFT (immunoassayed free testosterone) = 0.937, FAI (free androgen index) = 0.848. In other words, calculated free testosterone approaches the accuracy of measuring testosterone by dialysis equilibrium. It has to be noted that conditions that alter SHBG may alter the results of not only total testosterone but also FT. In men, conditions like obesity, hypothyroidism, and acromegaly can lead to lowered levels of SHBG, and as such confound the results of FT. Incidentally, pregnancy also leads to altered levels of SHBG, and as a result leads to false levels of FT as well. Otherwise, calculated free testosterone (FT) may be a practical means for the clinician to measure free testosterone, as it is less time consuming and expensive than testosterone by equilibrium dialysis (AFTC). Bioavailable testosterone is a more expensive test, but will be more useful and accurate in older patients as SHBG binding increases with age, and BT measures only the free amounts and those loosely bound to albumin. In older patients, one often finds normal levels of total testosterone, but BT is often significantly depressed.

Salivary Testosterone

This form of testing is novel, especially for the patient, as it avoids a needle stick. It is also convenient for the patient, as it does not require coming to the office. The patient spits into a bottle and mails the sample to the laboratory. Fig. 4–3 illustrates a typical kit. This test may be useful for screening for hypogonadism but not for a diagnosis of the andropause syndrome. The history and physical examination should be weighted more than salivary tests or any other blood tests. Saliva testosterone does not give a real-time assessment of the androgen status, and patients on androgen therapy often have levels of testosterone in the thousands. It gives a picture of accumulated testosterone. However, when done properly, saliva testosterone has good correlation to free testosterone ($r = 0.90$), less for

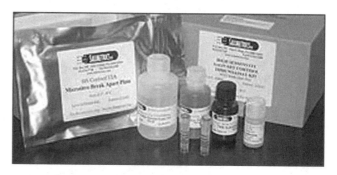

FIGURE 4–3. Saliva test kits. Patient typically spits into a test tube, which is mailed to a laboratory.

total testosterone ($r = 0.85$). The correlation for dehydroepiandrosterone (DHEA) is less at 0.70, and androstenedione 0.74. The patient has to rinse his mouth 5 minutes before testing and avoid food and tooth brushing for 30 minutes. He chews on a gum and spits into a bottle and mails the specimen to the laboratory.

Dynamic Testing for Testosterone

HCG Stimulation Test

Human chorionic gonadotrophin (HCG) is a glycoprotein with physiological actions similar to those of LH. After an intramuscular injection of HCG, the hormone binds to the LH receptors in the Leydig cells, and stimulates the production and secretion of testosterone. Typically, the test dose is 4000 IU for 4 days. This test is used to assess the viability of the axis and whether there is a gonadal disease. A positive response usually results in doubling of testosterone levels and improvement in symptoms. An alternate dosing is 5000 I.U. and measuring testosterone levels 3 days later. If there is no response, it is indicative of testicular failure, and if there is a response, it is indicative of pituitary-hypothalamic failure.[13]

Clomiphene Citrate Test

Clomiphene is a nonsteroid oral compound with estrogenic effects. It binds to estrogenic receptors in the body, and the hypothalamus responds by secreting LH. The test dose is 50 to 100 mg b.i.d. for 10 days, and both testosterone and LH levels are measured. By and large, healthy men have a 50% to 200% increase in LH and testosterone levels. If there is no response, it is indicative of testicular failure, and if there is a response, it is indicative of pituitary-hypothalamic failure.[13]

Variability in Testosterone Production and Measurements

Hormones are in constant fluctuation because of the pulsatile nature of their release. A single value is often misleading for the clinician, although single determinations may help differentiate between normal individuals and patients with severe hypogonadism. However, frequent samplings of testosterone may be necessary to guide the practitioner for an accurate diagnosis of the andropause syndrome. Stress can also lead to transient lowering of testosterone and to misdiagnosis. Testosterone is bound to various proteins including albumin and SHBG. As such, variations in concentrations of SHBG, for example, can alter levels of testosterone. Conditions that can lower SHBG concentrations include hypothyroidism, obesity, and acromegaly.

Spratt et al[14] studied the neuroendocrine gonadal axis in 20 men by frequent sampling of LH, FSH, and testosterone (Figs. **4–4** and **4–5**). They found that in normal men, serum total testosterone concentrations determined at 6-hour intervals ranged from 105 to 1316 ng/dL between subjects. When testosterone was measured at 20-minute intervals, marked intermittent declines in testosterone concentrations to levels well below the normal range were observed in 3 of the 10 subjects. The authors noted that testosterone secretion lagged behind LH secretion by ~40 minutes ($p < .02$). This suggests that LH's influence on Leydig cells is not immediate. LH in itself had great variability as well. Both mean LH concentrations and mean LH pulse amplitudes varied fourfold between individuals. LH interpulse intervals also varied from 30 to

480 minutes. The results suggested that there is a relative refractory period at the level of the hypothalamus or pituitary. In some subjects, there was a striking nighttime accentuation of LH, which is not seen with FSH.

In another study, Morley et al[15] compared the results of various testosterone assays in a cross-sectional study of 50 men age 28 to 90 years. The purpose of the study was to determine the relationship of the various testosterone assays to one another. In addition, they also determined the week-to-week variability in testosterone and bioavailable testosterone in 16 subjects. Hypogonadism may be diagnosed too frequently, and perhaps inappropriately especially in older men, due to an increase in their SHGB levels. There are often errors in the assays available to measure testosterone, thus leading to discrepancies in diagnosing hypogonadism. These tests include a total testosterone test, free testosterone test (by dialysis and ultracentrifugation), bioavailable testosterone, direct measurement of free testosterone by an analog immunoassay, and calculated free testosterone via a free androgen index (FAI) and a free testosterone index (FTI). Although measurement of free testosterone by equilibrium dialysis at 37°C (AFTC) is time-consuming, it is regarded as the best method for estimation of free testosterone, and is thus used as the "gold standard" in Morley's investigation. In the study, results of different testosterone tests were compared to determine the more accurate tests. First, blood samples from each subject were taken between 8 A.M. and 10 A.M. At the time of blood draw, none of the subjects was known to by hypogonadal, and none had disease or took medications that produce a decline in testosterone. None had an elevated LH level. No exclusion was made based on the serum SHBG level. The following tests were performed on each sample: the total serum testosterone, serum SHBG, and free testosterone by dialysis, bioavailable testosterone tests by radioimmunoassay (via a commercial available kit), and free testosterone by ultracentrifugation (via the method of Ekins). An analog radioimmunoassay was used to directly estimate the serum free testosterone. The FAI was calculated (total testosterone/SHBG), and the FTI was calculated using the method of Vermeulen et al.[12] A false-positive test meant that the subject was identified as hypogonadal by total testosterone, but eugonadal by BT or AFTC. A false negative represented the subject's classification as eugonadal by total testosterone, but hypogonadal by BT or AFTC. Using 300 ng/dL as a cutoff point in defining hypogonadism, the authors discovered that 42% of the men were misclassified using the total testosterone test. Similar results were observed when comparing total testosterone and AFTC. When comparing total testosterone with BT, total testosterone produced 16% false positives and

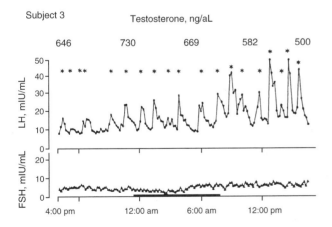

FIGURE 4–4. Variability of luteinizing hormone (LH) compared with follicle-stimulating hormone (FSH) in a 24-hour period in subject. (Adapted from Spratt DI, O'Dea LS, Schoenfeld D, et al. Neuroendocrine-gonadal axis in men: frequent sampling of LH, FSH and testosterone. Am J Physiol 1988;254: E658–666, with permission.)

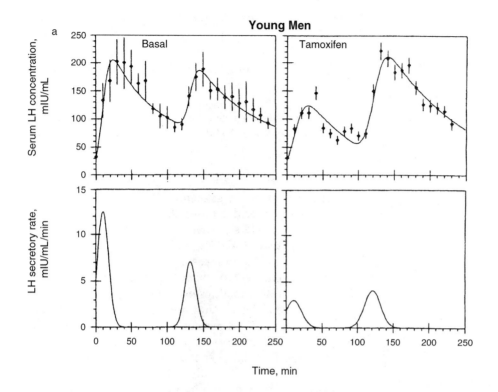

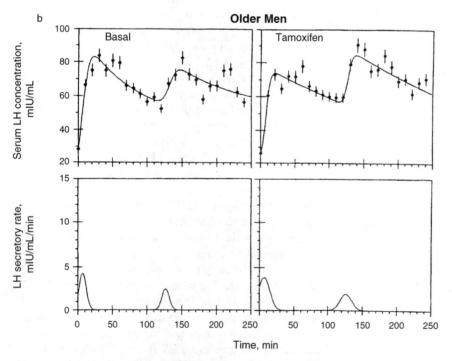

FIGURE 4–5. Augmentation of bioactive LH secretion and enhancement of bioactive LH and enhancement of plasma bioactive/immunoreactive ratios by antiestrogens (tamoxifen) were significantly reduced or absent in older men. (Adapted from Spratt DI, O'Dea LS, Schoenfeld D, et al. Neuroendocrine-gonadal axis in men: frequent sampling of LH, FSH and testosterone. Am J Physiol 1988;254:E658–666, with permission.)

26% false negatives; however, a week-to-week comparison showed that BT varied more than total testosterone. Over an 8-week period, BT identified 10 of 16 men as hypogonadal at one point and eugonadal at another point, whereas in the total testosterone test, 8 of 16 men were considered hypogonadal at one point and eugonadal at another point. When comparing free testosterone by dialysis (FTD) with BT, there were 6% false positives and 30% false negatives. The final assessment of the tests show that aFT (the analog immunoassay procedure) corresponded with free testosterone and BT measurements, but did not correlate with AFTC. Also, although FAI connects with AFTC, the FAI/AFTC ratio is significant only in low levels of SHBG; thus, FAI may not be very accurate in older men with high SHBG. Morley et al's results were also consistent with Vermeulen et al's in that there was a close correlation between FTA and FTD and BT, making FTI a reliable measurement of unbound testosterone. Based on the discrepant results of total testosterone, the study concluded that *either BT or some measure of free testosterone* (AFTC, aFT, or FT) be used to determine hypogonadism; however, taking note of BT's week-to-week variability, a second assay must be used 1 or 2 weeks apart in light of a normal reading on the first measurement.

Measuring Luteinizing Hormone and Follicle-Stimulating Hormone in Aging Men

LH and FSH are glycoprotein gonadotrophins composed of α and β subunits secreted by the same cell. The biological activity of HCG, which is a glycoprotein from the placenta, resembles that of LH. LH and FSH bind to receptors in the testes and regulate gonadal function by promoting sex steroid production and gametogenesis. LH stimulates testosterone production from the interstitial cells of the testes (Leydig cells). Maturation of spermatozoa requires both LH and FSH. The secretion of LH and FSH is episodic with secretory bursts every hour and are mediated by a concomitant episodic gonadotrophin-releasing hormone (GnRH) release. Arguably, FSH is less pulsatile than LH and is sometimes used by clinicians as a measure of gonadal failure with aging. Testosterone is not the sole inhibitor of gonadotrophin secretion in men, as selective destruction of the testes by chemotherapy results in azoospermia and a rise in FSH only. Inhibin, which is secreted by the Sertoli cells of the seminiferous tubules, is the major factor that inhibits FSH secretion by negative feedback.

By and large, LH rises with aging and there is a concomitant drop in bioavailable testosterone levels. However, it is to be noted that the rise in LH seems to be inconsistent in aging men. In younger men, gonadal failure leads to fairly consistent rises in LH,

partly because the hypothalamic-pituitary-testicular axis is intact. With aging, this system is impaired and results in inconsistent rises. The disparity of low bioavailable testosterone and yet normal or low LH is consistent with a hypothesis of a relative hypogonadotropism, often seen in the andropause syndrome.

Urban et al[16] reported that some healthy older men exhibited evidence of neuroendocrine dysfunction, reflected by irregular bursts of bioactive LH release followed by transiently low plasma bioactive/immunoreactive (B/I) ratios. It has been reported that mean basal plasma bioactive LH concentrations, B/I ratios, and spontaneous LH pulse properties (peak, frequency, amplitude, duration, and enhanced B/I ratios within LH peaks) were not altered in older men. On the other hand, augmentation of bioactive LH secretion and enhancement of bioactive LH and enhancement of plasma B/I ratios by pulsed injections of exogenous GnRH were *significantly reduced or absent* in older men. In that study, authors conclude that bioactive LH reserve is markedly attenuated in older men challenged with either exogenous GnRH or antiestrogens.

In another study, Veldhius et al[17] explored an andropause state using ketoconazole, which is an antiandrogen. This study found a discrepancy between young and older men. In young men, the chemically induced andropausal state resulted in an elevated LH peak frequency, whereas there was a *reduced incremental LH pulse* area in older men. This study also supported the earlier study in that older men had an impoverished augmentation of LH pulse mass, impaired orderliness of LH release, and diminished 24-hour rhythmic LH secretion. Mitchell et al[18] also reported that there were age-related changes in the pituitary-testicular axis in normal men. However, they pointed out in their study that *immunoreactive* LH remained unchanged despite finding that levels of total testosterone and bioactive LH fell with age. The important lesson from this study is that as men age, there is a hypothalamic-pituitary defect that in turn leads to lower bioactive LH levels, which in turn is responsible for diminished gonadal steroidogenesis. In yet another study by Veldhius et al,[19] they found that age was a negative determinant of LH secretory burst amplitude and a positive predictor of LH secretory burst frequency as well as basal LH secretory rates. They suggested the attenuation of LH secretory burst amplitude as an approximate basis for hypoandrogenism of healthy aging in older men.

Should the clinician measure LH levels to determine if the patient has indeed undergone andropause? The LH surge in women undergoing menopause is not seen in men. If the LH levels were very high in older men, a pituitary cause for hypogonadism must be excluded. Often, men with andropausal symptoms have

LH levels within the normal range (0.5–15 ng/dL), but have low bioavailable testosterone. As such, the use of LH is more useful in excluding diagnosis of other pathological states. When ordering LH, the clinician should distinguish between bioavailable and immunoreactive LH. Some laboratories are not able to provide both tests, and do only the immunoreactive LH. Overall, LH tends to be secreted in spurts as well, and it is difficult to rely on single values. FSH is the gonadotrophin that stimulates spermatogenesis in the Sertoli cells. It has been suggested that FSH is less pulsatile, and that it may be a better measure of gonadal failure as a result of aging in men. This is yet to be verified in clinical trials.

Measuring Estrogens in Men

In recent years, the role of estrogens in men has become more important because of the discovery of human models of estrogen deficiency such as estrogen resistance or aromatase deficiency.[20] In men, testosterone remains the major source of plasma estradiol. The main biologically active estrogen is estradiol. The testes secrete 20% of men's estradiol. On the other hand, plasma estrone (5% of which is converted to plasma estradiol) originates from tissue aromatization of mainly adrenal androstenedione. The plasma concentration of estradiol in older men is ~20 to 30 pg/mL and its production rate in blood is 25 to 40 µg/24 hour. It is interesting to note that both of these values are actually significantly higher than in postmenopausal women. Plasma levels of estradiol do not necessarily reflect tissue-level activity, as peripherally formed estradiol is partially metabolized in situ. Not all enters the general circulation, with a fraction remaining only locally active. Of the factors influencing plasma estradiol levels, plasma testosterone remains most important. However, the age-associated decrease in testosterone levels is scarcely reflected in plasma estradiol levels, as a result of increasing aromatase activity with age and the age-associated increase in fat mass. Free and bioavailable estradiol levels decrease modestly with age as does the ratio of free testosterone to free estradiol. Estradiol levels are highly significantly positively related to body fat mass and more specifically to subcutaneous abdominal fat, but not to visceral (omental) fat. Indeed, aromatase activity in omental fat is only one-tenth of the activity in gluteal fat. Estrogens in men play an important role in the regulation of the gonadotropin feedback, cognitive functions, bone maturation, regulation of bone resorption, and lipid metabolism. Estrogens also affect skin metabolism and are important determinants of sexual interest in men.

Low Testosterone in Athletes/ Receptor Issues

For years researchers have known that exercise training in women often affect women's sex hormone levels. Exercise training in women produces various degrees of menstrual alterations, ranging from obvious clinical forms such as amenorrhea, to less obvious subclinical forms, such as luteal phase defects, or anovulation.[21] Monitoring sex hormone levels in males is more difficult; however, recent studies have shown male endurance-trained athletes (particularly runners) may experience similar hormone problems. These changes include a reduction in total and free testosterone, alterations in LH release, alterations in pituitary responses to GnRH, and other pharmacological perturbations.[22]

Leydig cells, which are regulated by the pituitary hormones FSH and LH, produce androgens such as testosterone, dihydrotestosterone, androsterone, androstenedione, etc. Some studies have shown significantly lower levels of testosterone[22,23] and free testosterone[24,25] in endurance-trained men compared with those of age-matched sedentary controls[22]; however, other researchers have not been able to support these results.[26–29] Investigators have also tried monitoring LH and FSH levels. LH and FSH are released in an episodic manner from the pituitary gland in response to the pulsatile release of GnRH.[30] Although researchers have again varied in their results, it is believed that discrepancies in results may lie in the variability in age, lifestyle, and type of exercise of the subjects. It must be clarified, however, that although some studies observed lower levels of free testosterone, these concentrations were still within the low normal range.

Prolactin is a stress hormone produced in the pituitary gland. Elevated prolactin levels (often found in male athletes) are detrimental to testosterone production, spermatogenesis, and fertility.[31,32] Much of the available information is reflective of animal studies. On the basis of these animal studies, it has been suggested that physical exercise affects the dopaminergic, noradrenergic, and serotonergic systems.[33–37] Hackney et al[24] corroborated the effect of exercise on the dopaminergic system through experimentation with the dopamine antagonist, metoclopramide. Their results suggested that elevated dopamine levels in endurance-trained men are responsible for fluctuations in GnRH and prolactin release. Other scientists believe that the levels of stress-related hormones such as prolactin, and several endocrine components of the hypothalamic-pituitary-adrenal (HPA) axis, such as corticotropin-releasing hormone (CRH), adrenocorticotropic hormone (ACTH), and cortisol, may participate in complex feedback mechanisms that disrupt the activity and release of hypothalamic GnRH and pituitary LH and FSH.[38–40] Although some researchers have observed a decrease in testosterone, further long-term studies are needed to clarify and confirm results.

Discussion of Case History

The patient is seemingly hypogonadal based on his history and laboratory findings. However, his age and physical examination did not quite fit the picture of a true state of hypogonadism. He had no obvious trigger to push him into premature andropause, like chemotherapy, testicular tumor or surgery, mumps orchitis, or trauma. However, arguably, he has participated in marathon races, which can suppress testosterone levels. It was unusual that he should be in andropause. He was challenged with clomiphene to see if there was true Leydig cell failure, and there was not, as his testosterone levels went up quite quickly. He was not hypothyroid, and there was no evidence of a pituitary tumor. In the end, his symptoms were attributed to clinical depression. He had several samplings of his testosterone, and they normalized. He even had his 24-hour urinary free testosterone measured, which was in the normal range. He responded to a course of antidepressants, switching from Zoloft to Paxil and finally to Wellbutrin. This case highlights the variability of testosterone measurements and that before committing a patient to testosterone therapy, a clinician must consider other physiological and pathological causes for the symptoms.

Conclusion and Key Points

- Laboratory assessments should only supplement and not replace a clinical assessment of a patient suspected to have the andropause syndrome.

- There are different measures of free testosterone, and the dialysis equilibrium method (AFTC) is arguably most exact, and the calculated (FT) or free androgen index (FAI) approaches the accuracy of AFTC.

- Bioavailable testosterone is the free testosterone portion plus that bound to albumin, which is loosely bound, and is useful especially in older patients with SHBG issues.

- Measuring LH is useful in excluding diagnosis.

- Estrogens in men have a physiological role, and could affect the balance of testosterone.

- To offset the circadian rhythms of testosterone and the variability, it may be necessary to measure urinary free testosterone over a 24-hour period to determine hypogonadal status in some patients.

REFERENCES

1. Harvard/Amika International Studies. Introduction to equilibrium dialysis. 2001. http://www.harvardapparatus.com/pdffiles/B2K_N121.pdf

2. Turner J. Equilibrium dialysis systems for selective molecular filtration. Cornell Nanofabrication Facility. 2001. http://www.nnf.cornell.edu/2001cnfra/200156.pdf

3. Brenowitz M. What is analytical ultracentrifugation? Albert Einstein College of Medicine. June 19, 2002. http://www.bioc.aecom.yu.edu/labs/brenlab/XL-I/XL-I.html

4. Kanjee U. Ultracentrifugation to separate cellular components. JLM349S, Eukaryotic Molecular Biology at the University of Toronto. June 19, 2002. http://www.cquest.utoronto.ca/botany/bio349s/techniques/assessingcells/15ultracentr.html

5. Rickford D. *Centrifugation.* New York: John Wiley; 1994;1–100

6. Block R. *Methods of Protein Chemistry.* New York: Pergamon Press; 1991:119–173

7. Krumm R. Radioimmunoassay: a proven performer in the bio lab. The Scientist. May 16, 1994. http://www.the-scientist. com/yr1994/may/tool_940516.html

8. Smith R, Simpson B. Radioimmunoassay (RIA). Department of Veterinary Pathobiology College of Veterinary Medicine Texas A&M University. Aug. 29, 2002. http://vtpb-www.cvm.tamu.edu/vtpb/vet_micro/serology/ria/default.html

9. Valcour A. Testosterone: free T or not free T. Adv Lab 2001; 11:3–10

10. Winters SJ, Kelly DE, Goodpaster B. The analog free testosterone assay: are the results in men clinically useful? Clin Chem 1998;44:2178–2182

11. Ooi DS, Donnelly JG. More on the analog free testosterone assay. Clin Chem 1999;45:715

12. Vermeulen A, Verdonck L, Kaufman JM. A critical evaluation of simple methods for the estimation of free testosterone in serum. J Clin Endocrinol Metab 1999;84:3666–3672

13. Greenspan FS, Gardener DG. In: *Basic and Clinical Endocrinology.* 6th ed. New York: Lange Medical Books; 2001

14. Spratt DI, O'Dea LS, Schoenfeld D, et al. Neuroendocrine-gonadal axis in men: frequent sampling of LH, FSH and testosterone. Am J Physiol 1988;254:E658–E666

15. Morley JE, Patrick P, Perry HM. Evaluation of assays to measure free testosterone. Metabolism 2002;51:554–559

16. Urban RJ, Veldhius JD, Blizzard RM, et al. Attenuated release of biologically active luteinizing hormone in healthy aging men. J Clin Invest 1988;81:1020–1029

17. Veldius JD, Urban RJ, Lizarralde G, et al. Attenuation of luteinizing hormone secretory burst amplitude as a proximate basis for the hypoandrogenism of healthy aging men. J Clin Endocrinol Metab 1992;75:704–706

18. Mitchell R, Hollis S, Rothwell C, et al. Age related changes in the pituitary testicular axis in normal men: lower serum testosterone results from decreased bioactive LH drive. Clin Endocrinol (Oxf) 1995;42:501–507

19. Veldhuis JD, Zwart A, Mulligan T, et al. Muting of androgen negative feedback unveils impoverished gonadotrophin releasing hormone/luteinizing hormone secretory reactivity in healthy older men. J Clin Endocrinol Metab 2001;86:529–535

20. Vermeulen A, Kaufman JM, Goemaere S, et al. Estradiol in elderly men. Aging Male 2002;5:98–102

21. Brocks A, Pirke KM, Schweiger U, et al. Cyclic ovarian function in recreational athletes. J Appl Physiol 1990;68: 2083–2086

22. Arce JC, De Souza MJ, Pescatello P, Luciano AA. Subclinical alterations in hormone and semen profile in athletes. Fertil Steril 1993;59:398–404

23. Ayers WT, Komesu Y, Romani T, Ansbacher R. Anthropomorphic, hormonal, and psychological correlates of semen quality in endurance-trained male athletes. Fertil Steril 1985;43: 917–921

24. Hackney AC, Sinning WE, Bruot BC. Reproductive hormonal profiles of endurance: trained and untrained males. Med Sci Sports Exerc 1988;20:60–65

25. Wheeler GD, Wall SR, Belcastro AN, Cumming DC. Reduced serum testosterone and prolactin levels in male distance runners. JAMA 1984;252:514–516

26. Bagatell CJ, Bremner WJ. Sperm counts and reproductive hormones in male marathoners and lean controls. Fertil Steril 1990;53:688–692

27. Gutin B, Alejandro D, Duni T, Segal K, Phillips GB. Levels of serum hormones and risk factors for coronary hear disease in exercise-trained men. Am J Med 1985;79:79–84

28. MacConnie SE, Barkan A, Lampman RM, Schork MA, Beitins IZ. Decreased hypothalamic gonadotropin-releasing hormone secretion in male marathon runners. N Engl J Med 1986;315: 411–417

29. Mathur N. Toriola AL, Dada QA. Serum cortisol and testosterone levels in conditioned male distance runners and non-athletes after maximal exercise. J Sports Med Phys Fitness 1986;26:245–250

30. Crowley WF Jr, Filicore M, Spratt DI, Santoro N. The physiology of gonadotropin-releasing hormone (GnRH) secretion in men and women. Recent Prog Horm Res 1985;41:473–531

31. Ambrosi B, Gaggini M, Travaglini P, et al. Hypothalamic-pituitary-testicular function in men with PRL-secreting tumors. J Endocrinol Invest 1981;4:309–315

32. Perryman RL, Thorner MO. The effects of hyperprolactinemia on sexual and reproductive function in men. J Androl 1981; 2:233–242

33. Brown BS, Van Huss WD. Exercise and rat brain catecholamines. J Appl Physiol 1973;34:664–669

34. Brown BS, Payne T, Kim C, et al. Chronic response of rat brain norepinephrine and serotonin levels to endurance training. J Appl Physiol 1979;46:19–23

35. Chaoloff F. Physical exercise and brain monoamines: a review. Acta Physiol Scand 1989;137:1–13

36. Chaoloff F, Laude D, Serrurier B, et al. Brain serotonin response to exercise in the rat: the influence of training duration. Biog Amines 1987;4:99–126

37. Elam M, Svenson TH, Thoren P. Brain monoamine metabolism is altered in rats following spontaneous, long-distance running. Acta Physiol Scand 1987;130:313–316

38. Martini L. The 5 alpha-reduction of testosterone in the neuroendocrine structures: biochemical and physiological implications. Endocr Rev 1982;3:1–25

39. Martini L, Zoppi S. Mode of action of androgens in neuroendocrine structures. In: Paulson et al, eds. *Andrology: MaleFertility and Sterility*. London: Academic Press; 1986:149–159

40. Negro-Vilar A, Valenca MM. Male neuroendocrinology and endocrine evaluation of reproductive disorders. In Lamb J et al, Eds. *Physiology and Toxicology of Male Reproduction*. San Diego: Academic Press; 1988:103–131

5

Androgens and Sexuality in Aging Men

Kew-Kim Chew and Robert S. Tan

Case History

A 64-year-old stockbroker was referred by his family physician for erectile dysfunction (ED), which had been getting worse over a 2-year period. He also complained of decreased libido, which was apparently first noticed after his vasectomy 20 years ago but had gotten worse. On occasions when he was able to just manage sexual penetration, he needed time and effort to achieve ejaculation and would often lose his erections before he could reach climax. His erections on waking from sleep had become infrequent and trivial. He felt frustrated, depressed, irritable, and easily tired, and thought that these might have been caused by the considerable work-related stresses because of recent fluctuations in the stock market. His sleep had been disturbed, and he recalled waking up struggling for breath. All these caused appreciable strain on his marital relationship of 30 years. Twelve months ago he had been prescribed an antidepressant in addition to paracetamol-codeine (acetaminophen-codeine) tablets for chronic low back pain. Since having coronary angioplasty with stenting 4 years ago, he had not had further episodes of angina, and he was taking daily low-dose aspirin, ramipril, atenolol, and simvastatin for secondary prevention. He was overweight and was advised on diet and exercise for his non-insulin-dependent diabetes. He was not a cigarette smoker and consumed some alcohol only on social occasions.

Apart from abdominal adiposity, the general physical examination was unremarkable. Blood pressure was 150/100 mm Hg, and neurovascular assessment of his lower limbs did not reveal any significant abnormality. He was well androgenized, and testes were equal in size (20 mL) and normal in consistency. There was no clinical abnormality in the penis. Blood biochemistry showed the following results: glycated Hb 7.2% (6.0–7.0 for good glycemic control), total cholesterol 5.4 mmol/L (<5.5),

triglycerides 3.2 mmol/L (<1.8), high-density lipoprotein (HDL) cholesterol 1.0 mmol/L (>0.9), and low-density lipoprotein (LDL) cholesterol 2.9 mmol/L (<3.5). Liver functions, urea, creatinine, and electrolytes were within normal limits. Hormonal evaluation showed a plasma thyroid-stimulating hormone (TSH) level of 1.5 mU/L (0.3–4.3), prolactin 270 mU/L (<340), total testosterone 9.5 nmol/L (10–35), luteinizing hormone (LH) 3 U/L (2–9), and follicle-stimulating hormone (FSH) 15 U/L (1–6).

Decline in Blood Androgen Levels in Aging Men

There has been criticism concerning the use of the term *andropause* to denote the state of age-related decrease of plasma testosterone (T) level in some elderly men.[1] Although andropause may indeed be a misnomer for a condition associated with a decline in the plasma testosterone level, in contrast to the term *menopause* for women going through the physiological transition signaling an almost complete cessation of ovarian function, it does provide, especially for the layman, a catchy term that depicts the predicament men are in.

Androgens comprise mainly testosterone (T) and include dihydrotestosterone (DHT), dehydroepiandrosterone (DHEA), androstenedione, and androstenediol, although, strictly speaking, DHEA is not a true androgen as it does not bind androgen receptors.[2] Sixty percent of T is bound to sex hormone–binding globulins (SHBG), whereas the bioavailable T (BT) consists of T bound to albumin (38%) and free T (FT) (2%).[3] There is an age-related decline in plasma T levels,[4] and this decline in the plasma FT and BT levels is even more significant and consistent.[4,5]

A single decreased plasma T result is euphemistically called hypotestosteronemia and has no clinical value unless it can be confirmed by one or more properly conducted and repeated tests. Standardized data on age-specific normal reference ranges for different ethnic, cultural, socioeconomic, and geographical groups are yet to be established. Currently, interpretation of plasma T levels in elderly men is based on a normal reference range provided by available values in young men.[6,7] Consensus has yet to be reached regarding whether total T, FT, BT, or calculated FT should be the measurement of preference.[6,8] Availability and accuracy of the tests selected or used are also matters requiring consideration.

The Physiological Effects of Androgens on Sexual Function in Men

Genetic sex is established when fertilization occurs. After 6 weeks of gestation, genetically directed sex differentiation commences with the formation of the testes from the mesonephric or wolffian duct. Initially, the process does not appear to be androgen-dependent. However, subsequent differentiation into the epididymides, vasa deferentia, and seminal vesicles requires the presence of androgens. At about 8 weeks of gestation, the fetal Leydig cells begin to secrete T, and the levels peak at 11 to 18 weeks. Conversion of T to DHT by 5α-reductase occurs at 13 weeks, and DHT promotes the masculinization of the external genital primordia.[9]

Male sexuality is the expression of the interaction between androgens and behavior. In humans, hormonal influences on behavior are much less potent than in animals, and the effects of social and cultural factors can be considerable. At puberty when the testes begin to secrete androgens, sex drive, sexual interest, sexual performance, and copulatory ability increase. The hypothalamus, the hippocampus, the medial preoptic area, and the limbic system become activated as the production of androgens rises.[10]

Shiavi et al[11-13] reported in 1990–1993 their investigation of factors that contributed to health, well-being, and marital satisfaction in a study of 77 healthy men aged 45 to 74 years in stable sexual relationships. There was a significant negative correlation between age and sexual desire, arousal, and activity, as well as a significant age-related increase in the prevalence of erectile dysfunction (Table 5–1). Significant age-related decreases in frequency, duration, and degree of nocturnal penile tumescence were also reported. Prospective and retrospective data on the prevalence of sexual dysfunction were found to be unrelated to hormonal variables.

In another study involving 57 young controls, 50 healthy potent older controls, and 267 men with erectile dysfunction, Korenman et al[14] found no difference in the plasma T and BT levels in men with and without erectile dysfunction, when these levels were adjusted for age and body mass index (Fig. 5–1). Hypogonadal BT levels were found to be associated with LH levels within the normal range, indicating hypothalamic-pituitary dysfunction. There was also correlation between low basal LH levels and hyporesponsiveness to gonadotropin release testing. These findings suggest that secondary hypogonadism and erectile dysfunction are in fact two independent conditions prevalent in older men.

TABLE 5–1. The Relationship Between Age and Sexual Behavior

Psychosexual Variable	Correlation Between Age and Sexuality	
	Coefficient (r_s)	p^a
Desire		
Frequency of sexual thoughts (1 = never; 8 = daily)	−0.61	.0004
Maximum time comfortable without sex (1 = >1 year; 4 = <1 week)	−0.43	.0008
Frequency of desire for sex arousal (1 = never; 8 = daily)	−0.54	.003
Degree of coital erections (1 = none; 10 = rigid)	−0.63	.0001
Degree of sleep erections (1 = none; 10 = rigid)	−0.54	.0001
Degree of masturbatory erections (1 = none; 10 = rigid)	−0.55	.0001
Ease of becoming aroused (1 = do not; 5 = very easily)	−0.49	.0001
Frequency of waking erections (1 = none; 8 = daily)	−0.43	.0003
Sexual activity		
Frequency of coitus (1 = none; 8 = daily)	−0.45	.0001
Frequency of masturbation (1 = none; 8 = daily)	−0.44	.0003

Adapted from Schiavi RC, Schreiner-Engel P, Mandeli J, Schanzer H, Cohen E. Healthy aging and male sexual function. Am J Psychiatry 1990;147:766–771, with permission.

Impotence Clinic and Health Fair Subjects

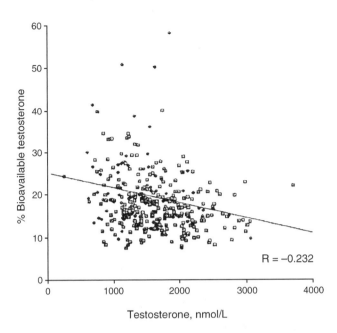

FIGURE 5–1. Distribution of bioavailable versus total testosterone in erectile dysfunction patients. (Adapted from Korenman SG, Morley JE, Mooradian AD, et al. Secondary hypogonadism in older man: its relation to impotence. J Clin Endocrinol Metab 1990;71:963–969, with permission.)

Using the International Index of Erectile Function (IIEF), Rhoden et al[15] showed that, although there was a clear association between erectile dysfunction and age, no correlation was observed between the occurrence of or the severity of erectile dysfunction and total T levels. There is thus no evidence that changes in plasma T levels contribute to erectile dysfunction in healthy aging men.

Although there are wide interpersonal variations in the age-related decline in plasma T and FT levels,[4] a decline of androgen-related sexual behavior in individual subjects was shown to be reproducible at certain critical plasma T levels in a study involving testosterone-treated hypogonadal patients and tamoxifen-treated eugonadal men.[16] In the same study, there was no evidence that androgen administration in excess of the individually determined critical levels would enhance sexual function.

The effect of T on sexual function is mainly centrally mediated through libido, which is regulated partly by T-dependent psychic factors. Testosterone influences the frequency of sexual thoughts and behavior.[11] However, only the very low end of normal range of testosterone is needed for normal libido and sexual performance.[11]

In the studies of Carani et al[17,18] using the Rigiscan to examine nocturnal penile tumescence and erectile response to visual erotic stimuli, the number of satisfactory nocturnal penile tumescence responses, in terms of both circumference increase and rigidity, was less in hypogonadal men than in normal controls. This was significantly increased by androgen replacement. However, such difference in erectile response and improvement with androgen replacement were not observed when the hypogonadal men and normal control were exposed to visual erotic stimuli. These findings suggest that nocturnal erections are T-dependent, but not the erections in response to visual erotic stimuli.

Testosterone has been reported to modulate the expression of nitric oxide synthase in the corpus cavernosum and the production of nitric oxide.[19,20] It acts on the cavernosal arterioles enhancing penile rigidity,[17] and influences, probably peripherally rather than directly through cognitive behavior, genital sensitivity and pleasurable enhancement of erectile activity.[21]

Knowledge regarding the effects of androgens on sexuality has come from research in other primates, observations of clinical outcome of castration in men, and studies of androgen replacement in hypogonadal men. T replacement and supplementation in hypogonadism increases the frequency of sexual fantasy, sexual arousal, desire and activity, nocturnal erections, ejaculation, and orgasm.[10] However, in a study examining the relationship between sexual function and pharmacologically manipulated T levels, treatment with T was shown to benefit only men with abnormally low plasma testosterone levels.[22] The findings not only indicated that relatively low T levels were needed for erectile function and sexual activity and feelings, but they also helped explain why some mildly hypogonadal men continue to have normal sexual function and why there is a lack of good correlation between sexual function and plasma T levels in the normal range.

In studies involving eugonadal men, administration of T produced an increase in sexual interest, but there was no change in sexual relationship or activity.[23–25] In a single-blind placebo-controlled study by Anderson et al[24] of 31 men treated with T for male contraception, trough plasma T levels were increased by 80% and peak plasma T levels by 400 to 500% with intramuscular T injections. There were no changes in frequency of sexual intercourse, masturbation, or penile erectile response. Only an increase in interest in sex was reported during treatment with T. The authors concluded that supraphysiological levels of T could promote some aspects of sexual arousability without stimulating sexual activity in eugonadal men in stable heterosexual relationships. This contrasts with the stimulation of sexual behavior in hypogonadal men treated with T.[10] Fig. **5–2** demonstrates the effect of testosterone on the Sexuality Experience Scale 2 (Psychosexual Stimulation) scores.

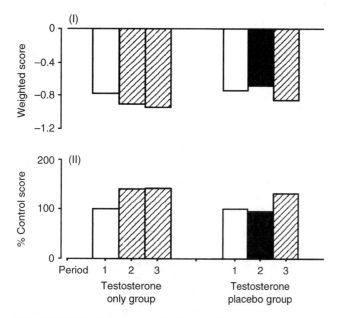

FIGURE 5–2. Effect of administration of testosterone on responses to the Sexuality Experience Scale 2 (Psychosexual Stimulation) scores (a and b). In treatment period 1, no testosterone is administered to either group; in treatment period 2, testosterone is administered to the testosterone-only group and a placebo to the testosterone placebo group; in treatment period 3, testosterone is administered to both groups. (Adapted from Anderson RA, Bancroft J, Wu FC. The effect of exogenous testosterone on sexuality and mood of normal men. J Clin Endocrinol Metab 1992; 75:1505–1507, with permission.)

The Role of Dehydroepiandrosterone (DHEA), Dihydrotestosterone (DHT), and Estrogens in Male Sexual Function

Dehydroepiandrosterone (DHEA) is generally regarded as an androgen, although, as alluded to earlier on,[2] it is not a true androgen as it does not bind androgen receptors. In spite of its widespread use in complementary medicine, studies on the role of DHEA in male sexuality are lacking. Anecdotal reports suggest that DHEA may increase libido. In a study of 280 healthy men and women aged 60 to 79 years,[26] a

small increase in plasma T and estradiol was noted with use of DHEA over 1 year. There was a significant increase in most libido parameters in the elderly women. No potentially harmful accumulation of DHEA sulfate (DHEAS) and active steroids was reported.

Overall, the physiological and pharmacological effects of T are the combined effects of T and those of its androgenic and estrogenic metabolites through respectively, 5α-reductase activity and aromatization.

The effects of dihydrotestosterone (DHT) on male sexual function may be inferred from the gamut of sexual disorders, which may occur in men receiving treatment with 5α-reductase inhibitor. These include decreased libido, erectile dysfunction, and ejaculatory disorders.[27]

Estradiol (E_2) is the main estrogen in men and is produced from the conversion of T under the influence of aromatase in the peripheral adipose tissues. Twenty percent of E_2 in men is secreted by the testes and indirectly derived from the adrenals.

Chronic exposure to estrogen and phytoestrogen was found to cause impairment of erectile function in animal studies.[28] Raised plasma E_2 levels had been reported in men with erectile dysfunction (ED) who were receiving treatment for epilepsy,[29] liver cirrhosis, or chronic renal failure.[30–32] However, in a study involving 75 consecutive adult male patients with ED, raised plasma E_2 level was observed in only one patient (1.3%). These findings seem to suggest that E_2 is probably of relevance in ED only in clinical conditions predisposed to hyperestrogenism.[33]

On the other hand, hypoestrogenism in men with aromatase deficiency appeared to be associated with difficulty in sexual functioning although not with problem in gender identity.[34] Treatment of hypoestrogenism with E_2 led to an increase of libido, erotic fantasy, and frequency of coitus and masturbation.[34] Table 5–2 summarizes the positive effect on depression in a patient with aromatase deficiency. The role of estrogen in male sexual function/dysfunction, therefore, still remains speculative and unclear.

TABLE 5–2. Case Report Showing Positive Effect of Estrogen on Depression in a Man with Aromatase Deficiency

Treatment	BDI	STAI	GRISS
Phase C: before estradiol treatment	17	52	5
Phase D: during estradiol treatment (estradiol 50 μg twice weekly)	8	31	4
Phase E: during estradiol treatment (estradiol 25 μg twice weekly)	9	29	1
Phase F: during estradiol treatment (estradiol 12.5 μg twice weekly)	5	27	1

BDI, Beck Depression Inventory; STAI, Spielberger Trait Anxiety Inventory; GRISS, Golombek-Rust Inventory of Sexual Satisfaction.
Adapted from Carani C, Rochira V, Faustini-Fustini M, Balestieri A, Granata AR. Role of estrogen in male sexual behaviour: insights from the natural model of aromatase deficiency. Clin Endocrinol (Oxf) 1999;51:517–524, with permission.

Discussion of Case History

The patient's plasma T level is decreased at 9.5 nmol/L (10–35). Other chapters of this book discuss the appropriate approach that should be taken to investigate this and similar results so as to exclude physiological fluctuations and to correctly establish a diagnosis of androgen deficiency. Suffice it to say that, although several factors are present in the patient's clinical profile that are known to contribute to androgen deficiency and hypogonadism, such as age, stress, depression, antidepressant medication, probable sleep apnea, overweight, and long-term use of narcotic analgesic, the presence of any or all of these factors in association with a single decreased plasma T level does not necessarily suggest androgen deficiency and hypogonadism.

Apart from decreased T, other age-related hormonal changes including declining growth hormone release and hypothyroidism may also contribute to sexual dysfunction. The diagnosis of androgen deficiency and hypogonadism, therefore, has yet to be made.

Given that androgen deficiency, medications, and psychological factors are among the main causes of decreased libido, there was evidently a significant overlay of psychological factors in the patient's decreased libido, as this was first noticed after his vasectomy, long before other comorbid conditions came into play. It would hence be simplistic and clinically naive to attribute the patient's sexual problems to just decreased T, even if androgen deficiency were confirmed.

As regards the patient's retarded ejaculation, psychological stresses related to his strained marital relationship, antidepressant medication, and possible diabetic autonomic neuropathy are probable contributory determinants.

The main reason for the patient's referral was erectile dysfunction. There is little doubt that vascular factors (coronary artery disease), diabetes mellitus, depression, medications (particularly β-adrenergic blocker atenolol), possible androgen deficiency, stress, and probable obstructive sleep apnea and secondary anxiety had contributed to his inability to obtain and maintain his erections for satisfactory sexual intercourse.

The patient's history of coronary artery disease was relevant and significant. There has been, in fact, increasing evidence that erectile dysfunction may well be synonymous with endothelial dysfunction, and that erectile dysfunction is likely to be a predictor or harbinger of coronary artery disease.

Forty percent of men with erectile dysfunction with no cardiac symptoms had coronary artery disease diagnosed on investigation with coronary angiography.[35] Coronary artery disease was diagnosed in 16% of men with erectile dysfunction compared with 5% of men without erectile dysfunction.[36] The severity of erectile dysfunction was found to correlate with the number of occluded coronary vessels in patients with coronary artery disease.[37]

The management of the patient's sexual dysfunction (decreased libido, erectile dysfunction, and retarded ejaculation) will involve appropriate laboratory evaluation and consideration of a host of therapeutic options ranging from counseling to pharmacological intervention. The patient should understand and appreciate the value of lifestyle modifications, weight reduction, good control of his diabetes, lipid profile, and cardiovascular risk factors. It will be helpful too for him to be firmly reassured that, other than its possible psychological impact, uneventful vasectomy has no adverse bearing on either sexual function or androgen profile.

This case highlights the complex nature and multifactorial basis of male sexual function and dysfunction and the need for a multidisciplinary approach in its diagnosis and management. Nonpharmacological measures are no less important than the new and sophisticated pharmacotherapeutic options, which have engendered a paradigm shift in its treatment.

Conclusion and Key Points

- There is well-documented and well-recognized age-related decline in androgen levels in men.

- Decreased plasma T level has to be appropriately confirmed and diagnosis of hypogonadism collaborated with clinical symptom-complex of androgen deficiency.

- The physiological and pharmacological effects of T are the combined results of T and its androgenic and estrogenic metabolites.

- The influence of androgens on male sexuality is mainly centrally mediated through libido.

- T may influence erectile function by modulating the expression of nitric oxide synthase and the production of nitric oxide in the penile erectile tissues.

- Relatively low plasma T levels are sufficient for normal erectile function.

- Other hormonal and nonhormonal factors may also affect sexual function.

- Treatment with T would only improve sexual dysfunction in men who are truly hypogonadal.

- It would be simplistic and naive to attribute sexual dysfunction to just decreased plasma T levels, especially if these are the result of opportunistic investigation.

REFERENCES

1. Morales A, Lunenfield B. Investigation, treatment and monitoring of late-onset hypogonadism in males. Aging Male 2002;5:74–86

2. Roger M. DHEA: to have or hav'nt. ISSIR Newsbulletin 2001; 7:14

3. Horton R. Testicular steroid transport, metabolism and effects. In: Becker KL, ed. *Principles and Practice of Endocrinology and Metabolism.* Philadelphia: JB Lippincott; 1995:1042–1047

4. Vermeulen A. Androgens in the aging male—clinical review 24. J Clin Endocrinol Metab 1991;73:221–224

5. Pearson UJD, Blackman MR, Metter EJ, Waclawiw Z, Carter HB, Harman SM. Effect of age and cigarette smoking on longitudinal changes in androgens and SHBG in health men. *Abstracts of the 77th Annual Meeting of the Endocrinological Society* 1995; Abstract P2:129.

6. Vermeulen A, Kaufman JM. Diagnosis of hypogonadism in the aging male. Aging Male 2002;5:170–176

7. Beilin J, Chew G, Feddema P, O'Leary P. Testosterone levels in aging male: a population based study utilizing measurement of total and calculated bioavailable testosterone. *54th Annual Meeting & Clinical Laboratory Expo of the American Association of Clinical Chemistry* 2002; Presentation No. D-21.

8. Vermeulen A, Verdonck L, Kaufman JM. A critical evaluation of simple methods for the estimation of free testosterone in serum. J Clin Endocrinol Metab 1999;84:3666–3672

9. Quigley CA. The androgen receptor: physiology and pathophysiology. In: Nieschlag E, Behre HM, eds. *Testosterone—Action, Deficiency, Substitution.* 2nd ed. New York: Springer; 1998:33–106

10. Christiansen K. Behavioural correlates of testosterone. In: Nieschlag E, Behre HM, eds. *Testosterone—Action, Deficiency, Substitution.* 2nd ed. New York: Springer; 1998:106–142.

11. Schiavi RC, Schreiner-Engel P, Mandeli J, Schanzer H, Cohen E. Healthy aging and male sexual function. Am J Psychiatry 1990;147:766–771

12. Schiavi RC, Schreiner-Engel P, White D. Mandeli J. The relationship between pituitary-gonadal function and sexual behaviour in healthy aging men. Psychosom Med 1991;53:363–374

13. Schiavi RC, White D, Mandeli J, Schreiner-Engel P. Hormones and nocturnal penile tumescence in healthy aging men. Arch Sex Behav 1993;22:207–215

14. Korenman SG, Morley JE, Mooradian AD, et al. Secondary hypogonadism in older man: its relation to impotence. J Clin Endocrinol Metab 1990;71:963–969

15. Rhoden EL, Teloken C, Mafessori R, Souto CA. Is there any relation between serum levels of total testosterone and the severity of erectile dysfunction? Int J Impot Res 2002;14:167–171

16. Gooren LJG. Androgen levels and sex functions in testosterone-treated hypogonadal men. Arch Sex Behav 1987; 16:463–473

17. Carani C, Scrteri A, Marrama P, Bancroft J. The effects of testosterone administration and visual erotic stimuli on nocturnbal penile tumescence in normal men. Horm Behav 1990;24: 435–441

18. Carani C, Granata AR, Bancroft J, Marrama P. The effects of testosterone replacement on nocturnal penile tumescence and rigidity and erectile response to visual erotic stimuli in hypogonadal men. Psychoneuroendocrinology 1995;20:743–753

19. Zvara P, Sioufi R, Schipper HM, Begin LR, Brock GB. Nitric oxide mediated erectile activity is a testosterone dependent event: a rat erection model. Int J Impot Res 1995;7:209–219

20. Mills TM, Reilly CM, Lewis RW. Androgens and penile erection: a review. J Androl 1996;17:633–638

21. Davidson JM, Kwan M, Greenleaf WJ. Hormonal replacement and sexuality. Clin Endocrinol Metab 1982;11:599–623

22. Buena F, Swerdloff RS, Steiner BS, et al. Sexual function does not change when serum testosterone levels are pharmacologically varied within the normal male range. Fertil Steril 1993;59:1118–1123

23. O'Carrol R, Bancroft J. Testosterone therapy for low sexual interest and erectile dysfunction in men: a controlled study. Br J Psychiatry 1984;145:146–151

24. Anderson RA, Bancroft J, Wu FC. The effect of exogenous testosterone on sexuality and mood of normal men. J Clin Endocrinol Metab 1992;75:1503–1507

25. Bagatell CJ, Heiman JR, Matsumoto AM, Rivier JE, Bremner WJ. Metabolic and behavioural effects of high-dose exogeneous testosterone in healthy men. J Clin Endocrinol Metab 1994;79:561–567

26. Baulieu EE, Thomas G, Legrain S, et al. Dehydroepiandrosterone (DHEA), DHEA sulfate, and aging: contribution of the DHEA Study to a socio biomedical issue. Proc Natl Acad Sci USA 2000;97:4279–4284

27. McClellan KJ, Markham A. Finasteride: a review of its use in male pattern hair loss. Drugs 1999;57:111–126

28. Srilatha B, Adaikan PG, Ng SC. Chronic estrogen exposure causes erectile dysfunction. Int J Impot Res 2002;14:S7

29. Murialdo G, Galimberti CA, Fonzi S, et al. Sex hormones and pituitary function in male epileptic patients with altered or normal sexuality. Epilepsia 1995;36:360–365

30. Van Steenbergen W. Alcohol, liver cirrhosis and disorders in sex hormone metabolism. Acta Clin Belg 1993;48:269–283

31. Wang YJ, Wu JC, Le SD, Tsai YT, Lo KJ. Gonadal dysfunction and changes in sex hormones in postnecrotic cirrhotic men: a matched study with alcoholic cirrhotic men. Heptogastroenterology 1991;38:531–534

32. Griffin JE, Wilson JD. Disorders of the testes. In: Fauci AS, et al, eds. *Harrison's Principles of Internal Medicine.* 14th ed. New York: McGraw-Hill; 1998:2087–2097

33. Chew KK. Estradiol and Testosterone in male sexual dysfunction. *Second Asian ISSAM Meeting on the Aging Male Abstract Book.* 2003;S2–03

34. Carani C, Rochira V, Faustini-Fustini M, Balestieri A, Granata AR. Role of estrogen in male sexual behaviour: insights from the natural model of aromatase deficiency. Clin Endocrinol (Oxf) 1999;51:517–524

35. Pritzker MR. The penile stress test: a window to the hearts of man? Circulation 1999;100:711

36. Anderson M, Nicholson B, Louie E, Mulhall JP. An analysis of vasculogenic erectile dysfunction as a potential predictor of occult cardiac disease. J Urol 1998;159(suppl 5):118

37. Greenstein A, Chen J, Miller H, Matzkin H, Villa Y, Braf Z. Does severity of ischaemic coronary disease correlate with erectile function? Int J Impot Res 1997;9:123–126

6

Depression in Aging Men: Are There Unique Gender-Based Therapies?

CLAUDIA A. ORENGO, GINNY FULLERTON, AND ROBERT S. TAN

Case History

A 62-year-old man with a long history of depression has had multiple hospitalizations for alcohol abuse and depression and has received a diagnosis of bipolar II disorder after a history of occasionally experiencing hypomanic symptoms. He is currently taking Wellbutrin and hydroxyzine. He claims that antidepressant therapy and abstinence from alcohol have somewhat helped in reducing his depressive symptoms, but he still generally lacks interest, energy, libido, and desire for pleasurable activities. He states that he is often unmotivated to work, and he fears for the future of his business because of this. He has been aware of his low testosterone level for a couple of years and has tried dehydroepiandrosterone (DHEA) and other various methods in an attempt to boost his sex drive and energy level. For several years, he has had frequent difficulty falling asleep despite trying several different sleeping aids. Examination revealed that he weighed 263.4 pounds, indicating moderate obesity; blood pressure (BP) was 161/88 mm Hg. His 5 A.M. total testosterone was 210 ng/dL, and his prostate-specific antigen (PSA) 0.38 ng/mL. He scored in the moderate range (17) on the 21-item Hamilton Rating Scale for Depression (HAM-D) and endorsed a poor to fair level of life satisfaction on the Quality of Life Enjoyment and Satisfaction Questionnaire (Q-LES-Q).

The patient was enrolled in a double-blind study during which he received placebo for the first 12 weeks and testosterone for the second 12 weeks. During what was later discovered to be the placebo condition, his mood and energy level became increasingly worse. The date was approaching a trauma-related anniversary, and he claimed that his depression often worsened at that time of year. He also reported more difficulty sleeping, claiming that none of the previous medications were working. He was very irritable and easily aggravated, and he was sure he was being given a placebo. After 12 weeks of the initial treatment, he was crossed over to receive testosterone.

When contacted shortly after beginning testosterone treatment, he had noticed a dramatic increase in his mood and energy level. He stated that within 3 hours he began to feel "normal" as if a "fog had been lifted." His sleep and sex drive had also improved, although not as much as his depression. After 6 weeks of testosterone treatment, he reported depressive symptoms in the mild range (11) on the HAM-D, and after 12 weeks of treatment, his depression had improved even more (7), and ratings on the Q-LES-Q indicated increased life satisfaction. Laboratory results at 12 weeks concluded that his total testosterone had risen to 1060 ng/dL.

Epidemiological Perspectives of Depression in Aging Men

Overall, depression affects at least 3 million to 4 million men in the United States.[1] It has been estimated that one out of every 10 men will be diagnosed with depression in his lifetime.[2] Studies suggests that ~8 to 16% of community-dwelling seniors have significant depression.[3,4] However, the frequency of depression in the institutional setting such as in long-term care can reach as high as 20%.[5,6] Based on community studies, it is generally believed that men are less likely to suffer from depression than women. However, the community-based data may be skewed, as men are less likely to report symptoms. It is of significance that the prevalence rate of depression in *institutionalized men* equals that of women, especially if severity of illness and functional impairment are controlled.[5,6] Men are more likely than women to view a diagnosis of depression as a sign of weakness and therefore are less likely than women to seek treatment.[2]

However, the impact of depression on men seems to be greater with more tragic consequences. For instance, the *rate of suicide* in men is four times higher than women. More women than men attempt suicide but often do not complete suicide.[7] Many studies have also established that older white men are particularly prone to a risk of suicide.[8] Older men are also more likely to complete suicide, and resort to violent means such as a shooting or hanging themselves. The marital status of a man is also a factor for suicide. For instance, widowed and divorced men have the highest rates of suicide.[9] Younger widowed men are also very likely to kill themselves. It has been hypothesized that loneliness, comorbid physical illness, alcohol abuse, and impulsive behaviors can be contributing factors in these suicide attempts.

The physical impact of depression in men is also different from that in women. Depression can affect a man's physical health differently from how it might affect a woman's. A recent study showed that both men and women who suffer from depression are at an increased risk of coronary heart disease; however, only men suffer a higher rate of death.[10] Men are more likely to deal with depression by turning to drugs and alcohol or by being workaholics.[1] Depression may have *atypical presentations* in men. Symptoms like irritability, anger, and discouragement can occur because of depression, and depression can overlap with clinical situations including hypogonadism (andropause), thyroid disorders, and alcohol and substance abuse.

Etiological Perspectives of Depression in Older Men

In general, the causes of late-life depression can be divided into biological, psychological, and social factors.[11] Biological factors include hereditary factors, neuroanatomical changes, neurotransmitter abnormalities, dysregulation of endocrine function, and dysregulation of circadian rhythms, especially sleep.

Biological Factors

In considering hereditary factors, there is currently *no evidence* that older men are at greater or lesser risk for depression secondary to *genetic factors* than older women.[12] Previously, it had been postulated that women may be at greater risk than men for depression secondary to genetic factors, and men may be at greater risk than women for alcoholism secondary to genetic factors. This hypothesis has been expanded to suggest that men and women may carry the same gene, which expresses itself as alcoholism in men and depression in women. The notion that women may be more affected by genetics has also been suggested by assumptions

that the gene for depression may be carried on the X chromosome and transmitted as an X-linked dominant gene.[13] Evidence at present, however, does not support an X-linked dominant transmission.

Thirty percent of persons older than age 60 can exhibit patchy deep white matter lesions of abnormal signal intensity with magnetic resonance imaging (MRI).[14] These lesions have been associated with hypertension, and are noted more frequently among men.[15] It has also been noted that these lesions are commonly found in cases of clinical depression. Some investigators have suggested that these lesions may define a unique type of depression in the elderly, called *vascular depression*.[15] In general, patients with vascular depression are more likely to experience apathy. Vascular depression, if diagnosed, is more prevalent in men, but the costs of MRI often limit diagnosing this entity, which may be more common among older men.

A second major biological contribution to depression is neurotransmitter abnormalities, including, but not exclusively, the serotonin neurotransmitter system. Investigators have found reduced cerebrospinal fluid levels of 5-hydroxyindoleacetic acid in patients with depression. Research does not suggest neurotransmitter depletion or dysfunction to be more prominent among older persons than younger persons. In addition, there are *no gender differences in neurotransmitter function*.[16] Elevated levels of cortisol, which is arguably a biological indicator for stress, is associated with depression. However, no studies have suggested a gender difference in hypothalamic-pituitary-adrenal (HPA) axis functioning, baseline cortisol levels, or response of the HPA axis to stimulation with dexamethasone.

Psychological and Social Factors

According to cognitive and behavioral theories, depression results from three specific processes: the negative triad, underlying beliefs or schemas, and cognitive errors. An interactive set of negative views toward the self, experience, and the future defines the negative triad. As such, psychological factors may impact depression in men as well. Under this belief, psychotherapists work to relieve depression through psychotherapy. *Social factors* are also major contributors to depression in late life. Examples of stressful life events that can lead to depression in older men include death of a spouse, sudden onset of a physical illness, and retirement. Chronic stress can include situations that challenge or threaten the older man's well-being. Examples of chronic stress also include poverty, ongoing interpersonal difficulties, or living in a dangerous neighborhood. Daily hassles are the usual yet stressful events and interactions experienced by older persons on a regular basis. Examples of daily hassles include managing household finances, home maintenance, and unpleasant interactions with neighbors. Each of these may contribute significantly to depression in older persons. Impaired social support networks have been implicated as contributing to depression.[17] Social support can contribute to depression, but may vary by gender. In one study, women but not older men were sheltered against the impact of stressful life events by adequate social support.[18] Men are generally thought to have *fewer supports* available to them. Although men are more likely to live in a household with someone else, they typically report having fewer friends whom they see on a regular basis. Having fewer available supports contributes to the possibility of an etiological risk factor for depression in older men.

Diagnosis of Depression in Men

Men are less likely than women to report a depressed mood or to conceive of their dysphoric state as being associated with a psychiatric illness.[19] Direct questioning may sometimes elicit the diagnosis. The older man can have *atypical presentations* for depression, and can report insomnia, loss of appetite, lethargy, fatigue, difficulty concentrating, and even anger. Medical conditions like hypothyroidism, hypogonadism, and anemia can also have similar presentations. Sometimes patients report having thoughts that life is not worth living, but they rarely report overt thoughts of suicide. If the older man admits that he has considered taking his life, the clinician must deem this person at *significant risk for suicide* and take action accordingly. As this is a psychiatric emergency, hospitalization may be required until the risk of suicide diminishes with treatment.

In the severely depressed older man, laboratory studies are important, not so much for the diagnosis of depression as for the identification of secondary medical abnormalities that may either accompany or derive from the depressed mood. Therefore, the clinician should order a complete blood count, urinalysis, blood chemistry, thyroid profiles, and a drug screen. Testosterone levels can be measured if clinical hypogonadism is suspected, as low testosterone may be associated with depression. A baseline electrocardiogram (ECG) should be obtained, for two reasons: medications such as tricyclics may alter cardiac rhythms, and the depressed older man is often more likely to suffer concomitant heart disease. If the older man suffers from cognitive dysfunction along with the depressed mood, a workup for dementia should be performed.

Standardized Acceptable Treatments for Depression in Men

As the etiology of depression is divided into biological, psychological, and social components, treatments should also be directed to these components. Physicians often focus on the biological aspects of treatment and neglect the other components because of time and reimbursement issues. Nevertheless, effective treatment must include each of these components. Biological treatments include pharmacological therapy and electroconvulsive therapy (ECT).

The old adage of "start low, go slow" applies to the treatment of depression in older men as well. In general, the treatment should commence with an antidepressant medication such as selective serotonin reuptake inhibitors (SSRIs) such as fluoxetine (Prozac), paroxetine (Paxil), and sertraline (Zoloft). In general, SSRIs, as compared with tricyclic antidepressants, are equally valuable but produce fewer side effects during treatment. Also, issues of overdosing are less consequential with SSRIs. Sometimes, tricyclic antidepressants can be prescribed in the first instance, as when, for example, the patient on SSRIs complains of persistent nausea and vomiting, resulting in weight loss. When starting an antidepressant, the axiom has been "start low, go slow." For example, in elderly men, starting doses can be nortriptyline 50 mg/day and desipramine 50 mg/day. A tetracyclic antidepressant, like trazodone or nefazodone, can be effective. Tetracyclic antidepressants can be especially useful for agitated depression and depression with insomnia symptoms. In contrast to the SSRIs, both the tricyclic antidepressants and tetracyclics require increasing doses over time.

To ensure compliance, the clinician must pay special attention to the side effects of SSRIs. Unfortunately, one of the more common long-term side effects of SSRIs is sexual dysfunction. Older men may experience great concern regarding the loss of sexual function but yet not discuss this with the doctor. It is important to forewarn patients of this potential side effect. SSRIs may also decrease the appetite of the older man. Weight loss, even 5 pounds, can be most disturbing and problematic in the treatment of frail older men. If these medications amplify the potential for weight loss, they should be discontinued and replaced by tricyclic antidepressants, which paradoxically may potentiate weight gain. Another known side effect of SSRI is agitation. All of the SSRIs can produce agitation, but fluoxetine is most likely to do so. Also, fluoxetine has a long half-life, and this side effect may not arise until the drug has been taken over several days. There is no evidence to suggest that the SSRIs increase the risk of suicide in older men. However, severe agitation is a risk for suicide. Careful monitoring by the clinician of these medications is therefore essential.

Space limitations prevent us from discussing all the new antidepressants that have been marketed in recent years. The important ones will be mentioned briefly.[20] Venlafaxine (Effexor) is a structurally novel compound first approved by the Food and Drug Administration (FDA) in 1993 for the treatment of major depression. It is a *bicyclic* antidepressant that produces strong inhibition of both norepinephrine and serotonin. Initially, venlafaxine was released in an immediate-release (IR) form that is taken two or three times daily. However, in 1997, an extended-release form (Effexor XR) was approved by the FDA, which allowed for a once-a-day administration. The recommended venlafaxine XR starting dosage is 37.5 to 75 mg per day. The dosage may be increased in increments of up to 75 mg every 4 to 7 days, to a maximum daily dosage of 225 mg. Venlafaxine is the first antidepressant that has proved effective in treating patients with generalized anxiety disorder (GAD). The recommended dosage is similar to that in depression. Some patients can experience jitteriness with the usual starting dose of 75 mg per day, so beginning treatment at a dose of 37.5 mg per day for the first week is advisable. Overall, the side-effect profile is *comparable* to that of the SSRIs and *lower* than that of the tricyclic antidepressants. The most common side effects include nausea, dizziness, insomnia, somnolence, and dry mouth. Anticholinergic side effects are significantly less severe than those encountered with other antidepressants, and thus this medication is more suitable for older patients. Sexual side effects are similar to those caused by SSRIs.

Mirtazapine (Remeron) is a *tetracyclic* antidepressant unrelated to tricyclic antidepressants and SSRIs. It is unique in its action among the currently available antidepressants. Mirtazapine is a presynaptic α_2-adrenergic receptor antagonist plus a potent antagonist of postsynaptic 5-hydroxytryptamine 5-HT$_2$ and 5-HT$_3$ receptors. The net outcome of these effects is stimulation of the release of norepinephrine and serotonin. Current evidence suggests that mirtazapine is effective in the treatment of all levels of depressive illness. In addition, analyses of placebo-controlled trials in moderate and severe depression have shown mirtazapine to be effective in subgroups of depressed patients, particularly those with anxiety, sleep disturbance, and agitation, as well as mentally retarded patients. Mirtazapine has an onset of efficacy of 2 to 4 weeks. However, sleep disturbances and anxiety symptoms may improve in the first week of treatment. In a review of multiple double-blind studies comparing mirtazapine with SSRIs, the proportion of responders with onset of persistent improvement in week 1 was twice as great with mirtazapine (13% versus 6%).

Mirtazapine's unique pharmacological profile is virtually devoid of anticholinergic, adrenergic, and serotonin-related side effects. The most frequently reported adverse events were fatigue, dizziness, transient sedation, and weight gain. Sexual dysfunction is *not* a side effect of this agent. Drug interactions with mirtazapine have not been studied systematically. The recommended starting dosage is 15 mg at bedtime, which may be titrated up to 45 mg daily, if needed.

Although by now a fairly well-established antidepressant, bupropion (Wellbutrin) warrants mention based on its increasing use in the *augmentation* of SSRIs. Uses include a role as a possible *antidote* for SSRI-induced sexual dysfunction, as a favored agent for bipolar depression, as a stimulant-like agent that can produce beneficial cognitive and behavioral effects in pediatric and adult attention deficit hyperactivity disorder, and, more recently, because of its newly marketed repackaging as an antismoking aid. Bupropion, under the name Zyban, received FDA approval for smoking cessation. A sustained release form is now commonly prescribed (150 mg b.i.d., with a maximal dose of 200 mg b.i.d.). Bupropion apparently causes the least amount of sexual dysfunction in the treatment of sexual dysfunction.[21]

If the older depressed man does not respond to antidepressant medications and exhibits severe symptoms, ECT may be indicated. As not all psychiatrists are trained in ECT or have access to these facilities, the patient may need to be referred elsewhere. Today, ECT can be performed in outpatient settings. In general, ECT is used more often in treating older men than younger men because of the increased resistance to medications in older men. Also, *psychotic depression* is more frequent in late life, and this variant of depression is often responsive only to ECT. Although the response rate to ECT is excellent, frequently exceeding 80%, the relapse rate in many studies has been greater than 50%. By continuing treatments for weeks and even months at infrequent intervals, the relapse rate decreases dramatically.

Finally, psychotherapy can play an integral role in the treatment of moderate-to-severe depression in the older man. In general, there are two types of psychotherapy: cognitive-behavioral therapy and interpersonal therapy. However, time and reimbursement issues have made psychotherapy impractical in many clinical settings, particularly so in primary care. There is not much specific research on gender-based psychotherapy methods, but it is well known that psychotherapy in combination with medications can be additive in its effect.

Low Endogenous Testosterone and Depression

Often hypogonadal males exhibit significant psychiatric symptoms, including dysphoria, fatigue, irritability, and appetite loss, and these symptoms are generally alleviated after testosterone replacement. Davidson et al[22] found no effect on mood states when giving men with hypogonadism, age 32 to 65 years, two doses of testosterone enanthate (TE) (100 and 400 mg per month) for 5 months. By contrast, other studies have shown that administering TE improved mood.[23–26] Eight subjects recorded their mood states daily in diary format and completed visual analog scales rating 10 separate mood states. Significant TE-related improvements were found for several scales, including cheerful-happy, tense-anxious, and relaxed.[23] Burris and colleagues,[24] using self-reported measures, found that six hypogonadal men (age range 25–40 years) scored higher on ratings of depression, anger, fatigue, and confusion than did controls. After 200 mg every 2 weeks of TE treatment, mood scores improved, suggesting testosterone treatment had some capacity to elevate mood and decrease anger. In a similar study, Wang et al[25] studied 58 men (age range 22–60 years) with hypogonadism who were treated with either 200 mg of TE every 20 days or sublingual testosterone cyclodextrin (T) at 2.5 or 5 mg three times a day for changes in mood. These researchers concluded, after T replacement therapy for 60 days, that T treatment led to increased energy, good feelings, and friendliness, and decreased anger, nervousness, and irritability. Similarly, mood and energy improved with transdermal T administration using scrotal patches.[26]

Androgen replacement also has been applied to men with low T levels associated with Klinefelter's syndrome. No reported mood or energy differences were reported for four men given 160 mg testosterone undecanoate (TU) or placebo in a double-blind crossover study.[27] In a larger trial, 30 men with Klinefelter's syndrome were given T.[28] Follow-up telephone interviews revealed improved mood, less irritability, more energy and drive, less tiredness, more endurance and strength, less need for sleep, and better concentration.

Some studies report no difference in T levels between depressed and nondepressed men.[29,30] However, others have reported differences. For example, Vogel et al[31] reported that in 27 men with unipolar depression, total and free T were significantly lower by 30% than in 13 age-matched controls. Rubin et al[32] studied the hypothalamic-pituitary-gonadal (HPG) axis in 16 depressed men and 16 healthy age-matched controls. They found a statistically significant negative correlation between age and total T levels in depressed patients, suggesting that in depressed men the decline in total serum T is correlated with advancing age and depression. Davies et al[33] found a significant negative correlation between salivary T levels and severity of depression in 11 men with a mean age of 52.4 years. In summary, some men

with major depression may have significantly lower free and bioavailable T than their age-matched controls. Also, the levels are significantly lower as subjects age, and probably lower with greater severity of depression.

The Rancho Bernardo Study, a cross-sectional population-based study, examined the association between endogenous sex hormones and depressed mood in community-dwelling older men.[34] Figure **6–1** shows the relationship between low testosterone and depression. The study found that free T decreases with age, and the prevalence of depressive symptoms increases with age. Analyses showed that there was a stepwise decrease in free T with increasing levels of depressed mood, independent of age, weight change, and physical activity.

Plausibility of Adjuvant Testosterone Therapy for Depression

Adjuvant testosterone administration has been used *experimentally* to treat patients with well-diagnosed depression. The antidepressant effect of a synthetic weak androgen, mesterolone was compared with amitriptyline in men with depression.[35] *Mesterolone* relieved depression as effectively as amitriptyline. Additionally, mood enhancement was also found with the weak androgen dehydroepiandrosterone.[36]

Rabkin et al[37] have conducted two studies of T administration to HIV-positive men. In the first, an open clinical trial of 44 hypogonadal HIV-positive men who had mood problems and completed an 8-week trial of intramuscular T, 28 (64%) were much improved compared with baseline. In the second study, 74 HIV-positive men were enrolled in the double-blind, placebo-controlled, 6-week trial with biweekly testosterone versus placebo, followed by 12-week open-label treatment.[38] Based on Clinical Global Impression (CGI) Scale ratings, response to testosterone was 74%, and to placebo was 19% ($p < .001$). Among 26 study completers with Axis I depressive disorders, 58% of those randomized to testosterone versus 14% of placebo-treated patients were judged to have much or very much improved mood. There was no difference in response rate between men with T levels below versus within normal range. Grinspoon et al[39] also showed that hypogonadal men with AIDS wasting had increased depression scores. In these patients, administration of testosterone (300 mg intramuscularly [IM] every 3 weeks) resulted in significant improvement of depression inventory scores (Fig. **6–2**).

Intramuscular testosterone was successfully used as the only treatment of a suicidal depression in a case of bilateral cryptorchidism[29] and in two cases of

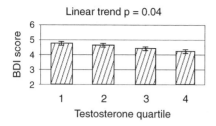

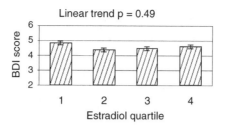

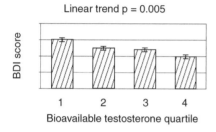

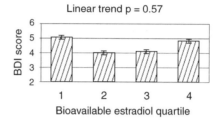

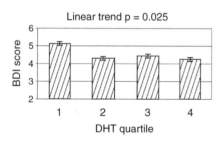

FIGURE 6–1. Data from the Rancho Bernardo study showing relationship of low testosterone to depression. BDI, Beck Depression Index. (Adapted from Barrett-Connor E, von Muhlen DG, Kritz-Silverstein D. Bioavailable testosterone and depressed mood in older men: the Rancho Bernardo study. J Clin Endocrinol Metab 1999;84:573–577, with permission.)

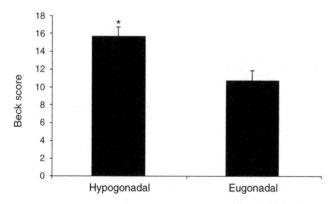

FIGURE 6–2. Human immunodeficiency virus (HIV) patients who are hypogonadal have higher BDI scores. (Adapted from Grinspoon S, Corcoran C, Stanley T, Baaj A, Basgoz N, Klibanski A. Effects of hypogonadism and testosterone administration on depression indices in HIV-infected men. J Clin Endocrinol Metab 2000;85:60–65, with permission.)

depression associated with Klinefelter's syndrome.[40] Seidman and Rabkin[41] administered testosterone openly to five men who had SSRI-refractory major depression and T level below 350 ng/dL. In the 8-week trial, all five patients had rapid dramatic recovery, with mean HAM-D decrease from 19.2 to 7.2 by week 2, and to 4.0 by week 8. This was an open trail with a small number of subjects. Seidman et al[42] also completed a randomized, placebo-controlled clinical trail examining the effects of testosterone enanthate 200 mg IM or placebo administration in 32 men with major depressive disorder, as defined in the *Diagnostic and Statistical Manual of Mental Disorders*, fourth edition (DSM-IV), and a low testosterone level. The authors found that the HAM-D score decreased in both the 13 subjects receiving testosterone and in the 17 subjects receiving placebo. There were no significant between group differences.

Perry et al[43] recently completed a 6-week randomized study with an initial single-blind 2-week placebo period evaluating the efficacy of testosterone therapy for 16 elderly eugonadal men with major depressive disorder (DSM-IV criteria) and HAM-D scores >18. Patients received testosterone cypionate in either a physiological dose of 100 mg/week or a supraphysiological dose of 200 mg/week. Results indicated an improvement in HAM-D scores in both groups, although the majority of the change was due to improvement in late-onset (≥ 45 years old) depression patients.

In another recent randomized, placebo-controlled trial, Pope et al[44] assigned 23 subjects to complete a 1-week single-blind placebo period, and then randomly assigned 22 subjects to receive 10 g testosterone gel or placebo for 8 weeks. Subjects included men aged 30 to 65 currently on antidepressant therapy with refractory depression and morning serum total testosterone levels of 350 ng/dL or less. Subjects receiving testosterone had significantly greater improvement in overall HAM-D scores and the vegetative and affective subscales of the HAM-D than did subjects in the placebo group (Fig. **6–3**).

In summary, T supplementation may improve mood in patients with depression and/or hypogonadism. Additional work that is focusing on T supplementation and depressive disorder type and depth will likely elucidate the apparent interaction.

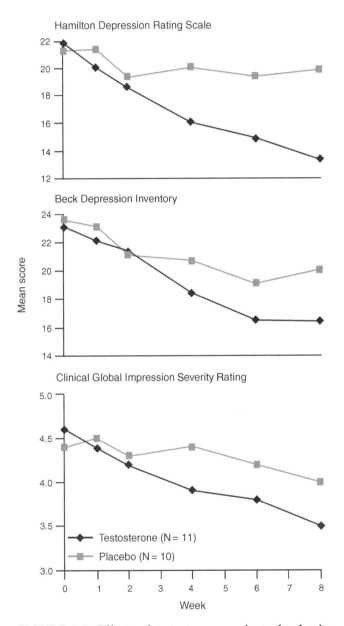

FIGURE 6–3. Effects of testosterone against placebo in treating depression. (Adapted from Pope HG, Cohane GH, Kanayama G, Siegel AJ, Hudson JI. Testosterone gel supplementation for men with refractory depression: a randomized, placebo-controlled trial. Am J Psychiatry 2003;160:105–111, with permission.)

The Influence of Stress as a Confounder in Hypogonadism and Depression

Can stress lead to temporary or permanent suppression of testosterone production? It is highly likely, but still speculative, although it has at least been observed in clinical settings. However, it can be stated that in general, the endocrine response to stress is complex.[45] Elevations in the serum concentrations of the classic stress hormones epinephrine and cortisol occur following many kinds of physiological challenge and are accompanied by elevations in corticotropin, growth hormone (GH), and glucagon levels. These changes are probably responsible for the *hyperglycemia* and *hypercatabolism* common to most critical illness. In addition, if volume depletion is present, vasopressin, renin, and aldosterone secretion are also likely to be stimulated. These hormones, if present in excess, may produce fluid retention and hyponatremia. We are aware that thyroid hormone metabolism is commonly affected by critical illness, which results in characteristic abnormalities of thyroid function testing known as the *euthyroid sick syndrome*.

The *reproductive axis* is also exquisitely sensitive to physiological stress; hypogonadotropic hypogonadism is a common finding in stress situations. In females, it is well established that stress can lead to menstrual abnormalities including amenorrhea. There is not much scientific literature on the observation of depression of androgen production with stress. However, it has been reported that male endurance runners can acquire a hypogonadotrophic hypogonadal state.[46] The mechanism whereby hypogonadism occurs is unclear and is postulated to be central. In space, where there is low gravity, temporary episodes of hypogonadism have been experienced in astronauts.[47] It can be argued that microgravity puts a stress on these astronauts and hence the hypogonadal state. In practice, stress is often encountered in patients. The challenge to the clinician is to determine whether seemingly abnormal hormone measurements in these patients reflect an appropriate homeostatic response to stress or whether they indicate an independent metabolic disorder that might actually cause or contribute to the patient's condition. The approach may be waiting and observing, as testosterone levels have been normalized in some patients without pharmaceutical interventions.

Discussion of Case History

This patient highlights the difficult area of the relationship between depression and hypogonadism. It can be argued in this case that he had clinical depression for many years, and that his hypogonadism added to his depressive feeling. It is unlikely that his depression was due to *only* his hyogonadal state. (In some instances, it could be the primary cause.) He was fortunate to have been crossed over and randomized to the drug during the clinical trial. It resulted in a dramatic increase in his mood and energy level. The added benefit was that his sexuality improved tremendously after androgen replacement. The placebo did not achieve the same effect for him. Alcohol abstinence also helped relieve his symptoms as well.

Conclusion and Key Points

Depression is a complex, multifaceted illness. It is sometimes a response to illness. Chronic illness in itself can suppress testosterone production. As a result, it is difficult to attribute depression to hypogonadism directly. Depression remains a huge public health issue, and has to be diagnosed before being treated. Many older men go undiagnosed, and this illness is confounded by multiple medical illnesses as well. There exists multiple screening tools for depression, and it is prudent for clinicians to spot depression in their interviews, as nontreatment can have tragic consequences.

- Community studies suggest a lower prevalence of depression in men than in women, but the data may be skewed as men in general report symptoms less frequently.

- The impact of depression is greater in men, as suicides tend to be more frequent. As such, screening for depression in men is important.

- There can be atypical presentations of depression in men, and it can include anger and irritability.

- There is no evidence of either genetics or neurotransmitter differences in depression. However, lack of social support in men can result in tragic outcomes in depressed men.

- As pharmacokinetics and pharmacodynamics are affected with aging, it is prudent to "start low and go slow" with antidepressants.

- Epidemiological studies support the finding of low testosterone and depression.

- Stress can result in lowering of hormones in men, including testosterone.

- Adjuvant testosterone therapy for depression in men has been studied, and results are encouraging but not conclusive. As such, standard antidepressives should always be used as first-line therapy.

- There may be a place for adjuvant testosterone therapy in the patient who is depressed and has concomitant hypogonadism.

REFERENCES

1. National Institute of Mental Health. Depression. NIMH publication No. 00–3561. Bethesda, MD: NIMH, 2000

2. Blehar MD, Oren DA. Gender differences in depression. Medscape Women's Health 1997;2:3. Revised from: Women's increased vulnerability to mood disorders: integrating psychobiology and epidemiology. Depression 1995;3:3–12

3. Blazer D, Hughes DC, George LK. The epidemiology of depression in an elderly community population. Gerontologist 1987;27:281–287

4. Weissman MM, Leaf PJ, Tischler GL, et al. Affective disorders in five United States communities. Psychol Med 1988;18:141–153. Erratum in Psychol Med 1988;18:following 792.

5. Koenig HG, Meador KG, Cohen HJ, et al. Depression in elderly hospitalized patients with medical illness. Arch Intern Med 1988;148:1929–1936

6. Parmelee PA, Katz IR, Lawton MP. Depression among institutionalized aged: assessment and prevalence estimation. J Gerontol 1989;44:M22–M29

7. Meechan P, Salsman L, Satin R. Suicide among older United States residents: epidemiologic characteristics and trends. Am J Public Health 1991;81:1198–1200

8. Baldwin R, Jolley DJ. The prognosis of depression in old age. Br J Psychiatry 1986;149:574–583

9. Smith J, Mercy JA, Conn J. Marital status and the risk of suicide. Am J Public Health 1988;78:78–80

10. Ferketich AK, Schwartzbaum JA, Frid DJ, Moeschberger ML. Depression as an antecedent to heart disease among women and men in the NHANES I study. National Health and Nutrition Examination Survey. Arch Intern Med 2000;160:1261–1268

11. Blazer D. Depression and the older man. Med Clin North Am 1999;83:1305–1316

12. Rice J, McGuffin P. Genetic etiology of schizophrenia and affective disorders. In: Michels R, ed. *Psychiatry*. Philadelphia: JB Lippincott; 1990

13. Winokur G, Clayton P. Family history studies, II: sex differences and alcoholism in primary affective illness. Br J Psychiatry 1967;113:973–979

14. Bradley W, Waluch V, Brandt D, et al. Patchy, periventricular white matter lesions in the elderly: a common observation during NMR imaging. Noninvasive Medical Imaging 1984; 1:35–41

15. Krishnan KR, Hays JC, Blazer DG. MRI-defined vascular depression. Am J Psychiatry 1997;154:497–501

16. Bissette G. Chemical messengers. In: Busse E, Blazer D, eds. *Textbook of Geriatric Psychiatry*. Washington, DC: American Psychiatric Press; 1996:73–94

17. Landerman R, George L, Campbell R, et al. Alternative models of the stress buffering hypothesis. Am J Community Psychol 1989;17:625–642

18. George LK, Blazer DG, Hughes DC, et al. Social support and the outcome of major depression. Br J Psychiatry 1989;154:478–485

19. Garrard J, Rolnick SJ, Nitz NM, et al. Clinical detection of depression among community-based elderly people with self-reported symptoms of depression. J Gerontol Series A Biol Sci Med Sci 1998;53:M92–101

20. Ables AZ. Antidepressants: update on new agents and indications. Am Fam Physician 2003;67:547–554

21. Masand PS. Sustained-release bupropion for selective serotonin reuptake inhibitor-induced sexual dysfunction: a randomized, double-blind, placebo-controlled, parallel-group study. Am J Psychiatry 2001;158:805–807

22. Davidson JM, Camargo CA, Smith ER. Effects of androgen on sexual behavior in hypogonadal men. J Clin Endocrinol Metab 1979;48:955–958

23. O'Carroll R, Shapiro C, Bancroft J. Androgens, behavior and nocturnal erection in hypogonadal men—a clinical research center study. Clin Endocrinol 1985;23:527–538

24. Burris AS, Banks SM, Carter CS, Davidson JM, Sherins R. A long-term, prospective study of the physiologic and behavior effects of hormones replacement in untreated hypogonadal impotent men. J Androl 1992;13:297–304

25. Wang C, Alexander G, Berman N, et al. Testosterone replacement therapy improves mood in hypogonadal men: a clinical research center study. J Clin Endocrinol Metab 1996;81: 3578–3582

26. Cunningham GR, Snyder PJ, Atkinson LE. Testosterone transdermal delivery system. In: Bhasin S, Gabelnick HL, Spieler JM, et al., eds. *Pharmacology, Biology, and Clinical Applications of Androgen*. New York: Wiley-Liss; 1996:437–447

27. Wu FCW, Bancroft J, Davidson DW, et al. The behavioural effects of testosterone undecanoate in adult men with Klinefelter's syndrome: a controlled study. Clin Endorcrinol 1982;16: 489–497

28. Nielsen FH, Hunt CD, Mullen LM. Effect of dietary boron on mineral, estrogen and testosterone metabolism in postmenopausal women. FASEB J 1987;1:394–397

29. Levitt AJ, Joffe RT. Total and free testosterone in depressed men. Acta Psychiatr Scand 1988;77:346–348

30. Sachar EJ, Halpern F, Rosenfeld RS, Galligher TF, Hellman L. Plasma and urinary testosterone levels in depressed men. Arch Gen Psychiatry 1973;28:15–18

31. Vogel W, Klaiber EL, Broverman DM. Roles of gonadal steroid hormones in psychiatric depression in men and women. Prog Neuro-Psychopharmacol 1978;2:487–503

32. Rubin RT, Poland RE, Lesser IM. Neuroendocrine aspects of primary endogenous depression VIII: pituitary-gonadal axis activity in male patients and matched control subjects. Psychoneuroendocrinology 1979;14:217–229

33. Davies RH, Harris B, Thomas DR, Cook N, Read G, Riad-Fahmy D. Salivary testosterone levels and major depressive illness in men. Br J Psychiatry 1992;161:629–632

34. Barrett-Connor E, von Muhlen DG, Kritz-Silverstein D. Bioavailable testosterone and depressed mood in older men: the Rancho Bernardo study. J Clin Endocrinol Metab 1999; 84:573–577

35. Vogel W, Klaiber EL, Braverman DM. A comparison of the antidepressant effect of a synthetic androgen (mesterolone) and amitriptyline in depressed men. J Clin Psychiatry 1985;46:6–8

36. Morales AJ, Nolan JJ, Nelson JC, et al. Effects of replacement dose of dehydroepiandrosterone in men and women of advancing age. J Clin Endocrinol Metab 1994;78:1360–1367

37. Rabkin JG, Wagner GJ, Rabkin R. Testosterone therapy for human immunodeficiency virus-positive men with and without hypogonadism. J Clin Psychopharmacol 1999; 19:19–27

38. Rabkin JG, Wagner GJ, Rabkin R. A double-blind, placebo-controlled trial of testosterone therapy for HIV-positive men with hypogonadal symptoms. Arch Gen Psychiatry 2000; 57:141–147

39. Grinspoon S, Corcoran C, Stanley T, Baaj A, Basgoz N, Klibanski A. Effects of hypogonadism and testosterone administration on depression indices in HIV-infected men. J Clin Endocrinol Metab 2000;85:60–65

40. Rinieris PM, Malliaras DE, Batrinos ML, Stefanis CN. Testosterone treatment of depression in two patients with Klinefelter's syndrome. Am J Psychiatry 1979; 136:986–988

41. Seidman SN, Rabkin JG. Testosterone replacement therapy for hypogonadal men with SSRI-refractory depression. J Affect Disord 1998;48:157–161

42. Seidman SN, Spatz E, Rizzo C, Roose SP. Testosterone replacement therapy for hypogonadal men with major depressive disorder: a randomized, placebo-controlled trial. J Clin Psychiatry 2001;62:406–412

43. Perry PJ, Yates WR, Williams RD, et al. Testosterone therapy in late-life major depression in males. J Clin Psychiatry 2002;63: 1096–1101

44. Pope HG, Cohane GH, Kanayama G, Siegel AJ, Hudson JI. Testosterone gel supplementation for men with refractory depression: a randomized, placebo-controlled trial. Am J Psychiatry 2003;160:105–111

45. Rolih CA, Ober KP. The neuroendocrine response to critical illness. Med Clin North Am 1995;79:211–224

46. Skarda ST, Burge MR. Prospective evaluation of risk factors for exercise–induced hypogonadism in male runners. West J Med 1998;169:9–12

47. Strollo F, Riondino G, Harris B, et al. The effect of microgravity on testicular androgen secretion. Aviat Space Environ Med 1998;69:133–136

7

Alzheimer's Disease in Older Men: Are There Gender-Specific Etiological Issues and Treatments?

Robert S. Tan and Ralph N. Martins

Case History

The classic thinking is that testosterone is an androgenic hormone responsible for normal growth and development of male sex and reproductive organs. It is produced from the testes and also converted from dehydroepiandrosterone (DHEA) in women. There is increasing evidence that androgens may modulate muscle growth, function, and possibly cognitive function as well. We applied this hypothesis to a patient we treated recently in an Alzheimer's center. This was a patient admitted to our Alzheimer's center a year ago for terminal care. The patient had a history of advanced Alzheimer's disease (AD) with associated agitation and delusion behavior. He also had asthma and diabetes mellitus type 2. On admission the patient seemed distant, and he did not answer any questions. On initial physical examination the patient weighed 138 pounds. The initial Mini–Mental State Examination (MMSE) score was 0/30 and the initial total testosterone 160 ng/dL [240–1000 ng/dL]. After informed consent from the family, he was started on an anabolic steroid primarily to increase functionality, but his cognitive parameters were observed.

Fig. 7–1 shows that after implementation of testosterone the MMSE score improved from 0/30 to 5/30, the clock drawing test (CDT) from 0/4 to 1/4, and the Tinetti scale, a fall assessment tool, from 5/2 to 20/28. This case study demonstrates that the use of testosterone in selected AD patients may possibly help with the improvement of cognition and functionality. AD is a terminal disease with little hope of improvement and any treatment modality that would improve the quality of life of a patient is to be investigated further. This patient tolerated the treatment well. It is difficult to conclude if the testosterone treatment alone or the good supportive care resulted in the improvement of the patient.

Do Men Have a Lower Probability of Alzheimer's Disease and How Can Men Protect Themselves Against It?

To the casual observer, it seems that AD is less common among men. A walk through a nursing home where AD is prevalent would more likely result in an encounter with more female residents than male residents. Men on average live 7 years less than women, and this may account for the seemingly higher prevalence of AD in women, as women tend to live longer and AD is a function of age.[1] However, a recent study conducted in Rochester argues that AD may be as common in men as in women if corrected for the age differences,[2] though gender-specific risk factors for AD cannot be excluded. For instance, smoking in men has been shown to increase the risk of AD.[3] Likewise, HIV-positive men who are carriers of the ε4 allele of apolipoprotein E (ApoE) have an increased risk of AD.[4]

The risk of AD in men may be reduced by several factors such as the use of antioxidants, moderate alcohol consumption, and exercise.[5,6] The Honolulu Heart Study examined 3385 men, and those who prophylactically consume vitamins C and E had an 88% reduction in risk for AD.[5] Another study from the Netherlands suggests that moderate alcohol intake may also decrease the risk for AD in men.[6] Several epidemiological studies have suggested that lifestyle is important in reducing risk of AD in men. For example, Honolulu Asia Aging Study found that men who walked and socialized regularly exhibited an 80% decreased risk of AD.[5]

Altered neuroendocrine regulation of cognition may also predispose men to AD.[7] Men, when they age beyond 50 years may suffer from androgen decline, sometimes termed andropause or androgen decline in

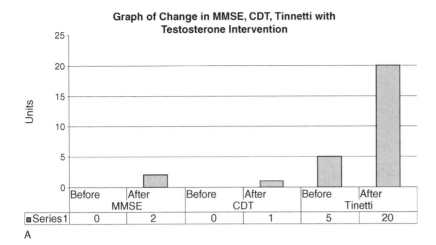

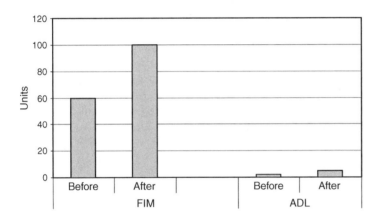

FIGURE 7–1. (A) Change in mini–mental state examination (MMSE), clock drawing test (CDT), and Tinneti (scale used to measure risk for falls) with testosterone intervention, and (B) change in activities of daily living (ADL) and functional instrument measure (FIM) before and after intervention with testosterone.

aging males (ADAM).[8] This decline in androgen occurs more gradually as compared with women in menopause. The decline is more marked with bioavailable testosterone than with total testosterone, partly because of the changing affinity of sex hormone–binding globulin (SHBG) with aging.[7] It has been observed that older men who have undergone androgen replacement often report improved memory function or report feeling "brighter." This suggests that androgen replacement may play an important role in improving memory in men entering into a transitional stage of lowered hormones. Low levels of androgen are not restricted to risk of AD per se but have also been reported for risk of vascular dementia.[9–11] Androgen replacement therapy was associated with improved cognition in vascular dementia patients.[12–14]

Alzheimer's disease is thought to occur less frequently in men than in women, and as such, it may seem logical to infer that androgens, which are more abundant in men, may be protective. This hypothesis has been tested in animal as well as human models, and indeed there may be a link. The following sections highlight some of the supportive evidence. AD is such a complex disease that it would be imprudent to attribute the association to be direct.

Another piece of the puzzle has been the observation that gonadotrophins such as the luteinizing hormone (LH) and the follicle-stimulating hormone (FSH) tend to be elevated in AD, and it is unclear if the hypogonadal state results in a rise in these gonadotrophins. Attempts have been made to use antigonadotrophins to treat AD, but this may sound irrational, as there might be further induction of the hypogonadal state, which in itself is suspected to be a risk factor for AD. As mentioned, AD remains a complex illness and has many facets yet to be defined. To date, the use of acetyl cholinesterase inhibitors for AD has not been very successful in terms of a cure. As such, it is prudent to prod other possibilities for therapeutics, as the disease is relentless and devastating.

Presently Recognized Drug Treatments for Alzheimer's Disease

As of this writing, there are five drugs that have been approved in the United States by the Food and Drug Administration (FDA) for treating the *cognitive* symptoms of Alzheimer's disease: memantine, galanthamine, rivastigmine, donepezil, and tacrine. Tacrine needs close liver function monitoring. These drugs belong to the class of *cholinesterase inhibitors*. Each acts in a different way to delay the breakdown of acetylcholine. It is believed that acetylcholine in the brain facilitates communication among nerve cells and is thus important for memory. It is not possible to measure patients' level of acetylcholine and as such treatment by and large is empirical and based on a clinical diagnosis. The reader is referred to other texts for a more detailed discussion of these treatments. Table **7–1** summarizes the drugs used to treat progression of Alzheimer's disease.

In general, these drugs are most effective when treatment is begun in early stages. They have all been shown to *modestly slow* the progression of cognitive symptoms and reduce problematic behaviors in some people. Unfortunately, at least 50% of patients do not respond to the acetylcholinesterase inhibitors. Although the effect of these medications is modest, studies show that when they do work, they can make a significant difference in a person's quality of life and day-to-day functioning (activities of daily living, ADL). The main differences between the drugs are the side effects they produce and the number of times they must be taken daily (Aricept is taken once daily, Exelon and Reminyl are taken twice a day).

One of the new treatments FDA approved for treatment of cognitive problems is memantine. Overstimulation of the *N*-methyl-D-aspartate (NMDA) receptor by glutamate is implicated in neurodegenerative disorders. In a double-blind, placebo-controlled trial, patients receiving memantine had a better outcome than those receiving placebo.[15] Memantine was not associated with a significant frequency of adverse events. *Antiglutamatergic* treatment reduced clinical deterioration in moderate-to-severe AD, a phase associated with distress for patients and burden on caregivers, for which other treatments are not available. Many drugs as well as nutraceuticals have been studies including vitamin E and gingko biloba, both of which look promising.

Laboratory Models of Androgens and Brain Function

Estrogens have been thought to be protective against AD in women, but treatment of AD patients with estrogens has by and large *not* been successful.[16] Animal studies have indicated that testosterone does play an important role in cognition. Many researchers have postulated that the action is through the androgen receptors (ARs). It has been discovered that AR and its associated messenger RNA is abundant in the rat brain. The cerebral cortex and hippocampus are areas that are rich in ARs.[17] In one study the blockade of ARs in mice expressing the major AD genetic risk factor, the ε4 of apolipoprotein E (ApoE 4) resulted in the development of prominent deficits in *spatial learning and memory*. Paradoxically, female neuron-specific enolase (NSE)—ApoE4 mice, which are prone to impairments in spatial learning and memory, exhibited memory improvement in response to androgen treatment (Fig. **7–2**).[18] In addition, cytosolic AR levels were decreased in NSE-ApoE4 mice and improved memory treatment in the NSE-ApoE4 female mice was associated with increased cytosolic AR levels.

Reproductive hormones influence the development of the central nervous system (CNS). For instance, androgens influence neuritic arborization and the receptive field of individual cells, whereas estrogen induces this communication by forming spines, synapses, and gap junctions. As such, estrogens and androgens act in *different but complementary* ways to modulate neural development and organization.[19] Castration of animals serves as a model to observe cognitive changes. For example, it has been shown that sexually intact male dogs were significantly less likely than neutered

TABLE 7–1. Different Drugs Used to Slow Progression of Alzheimer's Disease

Brand Name	Generic Name	Year of FDA Approval	Dosing
Cognex	Tacrine	1993	10–40 mg Q6H
Aricept	Donepezil	1996	5–10 mg once daily
Exelon	Rivastigmine	2000	1.5–6 mg twice daily
Reminyl	Galanthamine	2001	4–12 mg twice daily

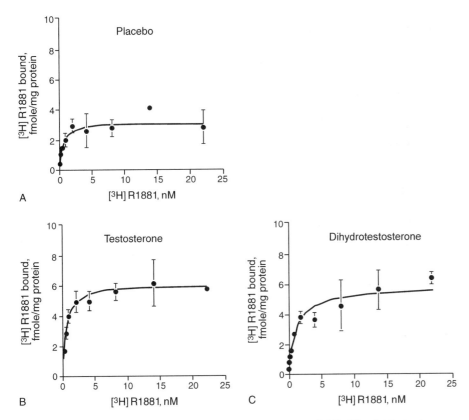

FIGURE 7–2. Effect of (A) placebo, (B) testosterone, and (C) dihydrotestosterone on androgen receptor (AR) saturation curves in female NSE-ApoE4 mice. Testosterone and dihydrotestosterone effects on androgen receptors, including those in the brain, are more significant than the placebo was. (Adapted from Raber J, Bongers G, Le Fevour A, Buttini M, Mucke L. Androgens protect against apolipoprotein E4-induced cognitive deficits. J. Neurosci 2002;22:5204–5209, with permission.)

dogs to progress from mild to severe impairment.[20] These observations are consistent with the notion that circulating testosterone in aging sexually intact male dogs slows the progression of cognitive impairment. The senescence-accelerated mouse (SAMP8) exhibits *age-related* learning and memory deficits in the ability to perform inferential tasks.[21] These memory deficits are associated with a corresponding decline in testosterone levels with age. Testosterone replacement improved age-related impairment of learning and memory.[22] It has been postulated that cognitive impairment of SAMP8 mice may be related to an interaction of aging and lowered testosterone levels.

Intracellular neurofibrillary tangle formation as a result of tau protein hyperphosphorylation is one of the major neuropathological hallmarks of AD. It has been demonstrated that heat shock–induced hyperphosphorylation of tau protein in the brain of orchiec-tomized male rats can in fact be reduced by testosterone.[23] Interestingly, this effect was not seen with estrogens, indicating it was a direct effect of testosterone or its metabolite dihydrotestosterone. It has also been demonstrated that treatment with testosterone

decreases the secretion of amyloid β (Aβ) peptide from cerebrocortical neurons of rats.[24] Evidence is accumulating to indicate that changes in Aβ precedes tau hyperphosphorylation.[25] In addition to reducing Aβ levels, testosterone and dihydrotestosterone can *protect* cultured hippocampal neurons of rats from cell death by directly blocking Aβ toxicity via an estrogen-independent mechanism.[26] Testosterone also exerts its neuroprotective effects through receptor-mediated mechanisms[27,28] by stimulating increased secretion of a metabolite of amyloid precursor protein (sAPP),[29] which has been demonstrated to exhibit neurotrophic properties.

Dehydroepiandrosterone sulfate (DHEAS) is synthesized in situ in the brain. Therefore, it has been termed a *neurosteroid*. Numerous animal studies have demonstrated the neuroprotective and memory-enhancing effect of DHEAS. DHEAS concentrations decline with age in various animals, including the rhesus monkey.[30] The data on DHEA and cognition is conflicting. One study demonstrated that cognitively impaired and unimpaired aged rhesus monkeys did not differ in their DHEAS levels.[31] DHEA replacement *enhanced memory in aged mice* in several experiments.[27-29] In experiments, DHEA has been shown to significantly reverse pharmacologically induced neurotoxicity with NMDA, dimethylsulfoxide kainic acid, H_2O_2, dizocilpine, ethanol and scopolamine.[4,32-34] Possibly, DHEA may be useful in treating memory impairment based on its *antioxidant and neuroprotective* effects in the hippocampus. It has been reported that DHEA increased the production of amyloid precursor protein (APP) and promoted the release of its metabolite sAPP from PC12 cells.[35,36] Although the increased production of APP per se may lead to acceleration of mild cognitive disorder presumably as a result of increased Aβ production, the increased APP produced in DHEAS-treated cells is processed through the non-amyloidogenic pathway to release APP, which is known to have neuroprotective properties that may be mediated in part by its ability to promote neurite outgrowth.[36-38] The effects of DHEA on cognition may be mediated through its nongenomic action on several neurotransmitter receptors, including cholinergic neurons.[39] Despite the promising data from animals, the effects of androgens on human cognition are less clear and will be discussed in the following sections.

Examining Human Studies of Androgens and Cognition

Older men often complain of memory problems. A Texas study of 302 older men found that 36% of patients identified themselves as experiencing andropause on a standardized questionnaire survey and reported memory loss as a symptom.[40] Another study in Perth, Western Australia, which employed a similar questionnaire given to 100 men treated for low testosterone, found initial symptoms included memory loss and lapses in concentration (85%), fatigue (96%), mood problems (83%), and loss of libido (77%) (Linda Byart, personal communication). Hormone replacement therapy in this cohort resulted in 86% of men mentioning an improvement in memory and attention span.

Investigators have examined the effect of androgen substitution or withdrawal upon various study populations. By and large, these studies are limited by population size, but are very encouraging. A double-blind study conducted by Janowsky et al[41] demonstrated that transdermal testosterone enhanced spatial cognition of healthy men aged 60 to 75 years when testosterone levels were raised for a 3-month period to those commonly found in young men. Endogenous production of estradiol was decreased in men receiving testosterone supplementation, and estradiol levels correlated inversely to performance on tests of spatial cognitive skills.[25] A 6-week randomized, double-blind, placebo-controlled study of healthy older men aged 50 to 80 years found that 100 mg of testosterone enanthate improved both spatial and verbal memory.[42] Another small randomized, double-blind study of healthy volunteers aged 61 to 75 years demonstrated improved working memory following intramuscular injections of testosterone enanthate 150 mg/week. Better performance on tasks involving frontal lobe–mediated working memory was related to a higher testosterone-to-estrogen ratio.[28] The subjects in these studies were healthy volunteers who did not have AD.

Studies involving testosterone substitution have historically been confounded by concurrent changes in estradiol levels and include patient samples that may not be generalized to the population of elderly male patients experiencing low-testosterone syndrome. A 12-month randomized, controlled trial of 32 hypogonadal men (bioavailable testosterone <60 ng/dL), aged 51 to 75 years, failed to demonstrate any difference in verbal or nonverbal memory with 200 mg testosterone cypionate biweekly; however, visuospatial ability was not specifically examined. The authors hypothesized that their inability to demonstrate the more generalized effects on cognition may have been due to the rise in estradiol levels in the treatment group.[43]

Sexual reassignments, because of the use of sex steroids, gave an opportunity for researchers to observe the effects of hormones on brain function. For instance, Slabbekoorn et al[44] demonstrated a profound effect of androgen treatment on spatial ability in female-to-male transsexuals (FMs) over a period of 18 months. After 3 months of cross-sex testosterone treatment, 25 FMs demonstrated a significant improvement on a three-dimensional rotated figures task of spatial ability. Untreated male-to-female transsexuals (MFs) had higher scores on visuospatial tasks than untreated FMs transsexuals.[44]

Recently, in a small 8-week randomized, controlled trial, testosterone was shown to exhibit a curvilinear dose-response effect for different verbal and spatial cognitive functions. Supraphysiological levels of testosterone (200 mg testosterone enanthate/week) reduced visuospatial ability and improved verbal fluency in 30 healthy young men aged 19 to 45 years.[45] This finding confirms a previous study that showed a curvilinear relationship between spatial performances and circulating testosterone concentrations with optimal effects on spatial performance, associated with intermediate plasma levels of testosterone.[38] This is consistent with the hypothesis first proposed by Janowsky et al[41] that high levels of testosterone may be aromatized to estradiol within the brain.

The development of new imaging technology has enabled investigators to explore neuropharmacology in a *functional* manner. A small study of six men with hypogonadotropic hypogonadism when given testosterone demonstrated enhanced cerebral glucose metabolism as assessed by using F18-deoxyglucose positron emission tomography (PET) with increased visuospatial capacity during a three-dimensional mental rotation task.[46] Brain localization on PET was consistent with those areas that were observed to be activated by such tasks in a previous study on eugonadal men.[23]

The hypothesis that gonadal replacement therapy might prevent or delay AD in postandropausal men has recently been supported by a study of the effect of gonadal hormone *withdrawal* on the levels of plasma (Aβ). This 4-kd peptide is the major protein component of the extracellular amyloid plaques, which is a neuropathological hallmark of AD[47] and is thought to play a key role in its pathogenesis.[48] Plasma Aβ levels increased in men with prostate cancer in response to flutamide (250 mg, three times daily) and leuprorelin acetate (22.5 mg, weekly for 12 weeks) treatment to lower plasma testosterone. This increase in plasma Aβ paralleled a rapid decrease in both plasma testosterone and 17β-estradiol, and remained elevated over a 6-month period while gonadal hormones remained at low detectable levels (Fig. 7–3).[49] However, regulation of Aβ production represents only one of several potential mechanisms that need evaluation. Some of these other potential candidate mechanisms are based on testosterone's ability to block neuronal apoptosis[50] as well as indirectly to prevent tau phosphorylation,[23] to increase the production of the neurotrophic agent nerve growth factor[51] and insulin growth factor-1,[52] and to reverse the age-related increases in glial fibrillary acidic protein (GFAP).[53] Increased GFAP reflects activation of astrocytes, which have been implicated in playing a more prominent role in the pathogenesis of AD.[54]

The adrenal steroid hormone DHEA is known to decrease with aging and is commonly used as an

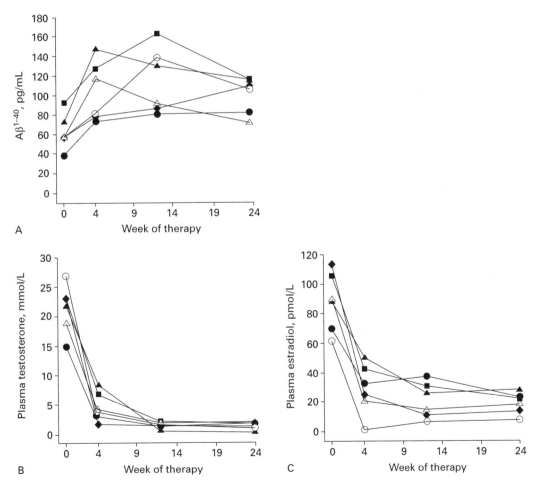

FIGURE 7–3. Plasma concentration of (A) β amyloid, (B) testosterone, and (C) estradiol before and after flutamide and leuprorelin acetate treatment. (Adapted from Gandy S, Almeida OP, Fonte J, et al. Chemical andropause and amyloid-B peptide. JAMA 2001;285:2195–2196, with permission.)

anti-aging nutritional supplement. Clinical studies of the effects of DHEA and its sulfate DHEAS on cognition have produced conflicting results. Fifty-two community-dwelling AD patients (mean age 76 years) were compared with a control group of age- and gender-matched healthy elderly men and women on a test of everyday memory. No differences were seen between the AD patients and controls in DHEAS levels. However, AD patients with higher levels of DHEAS scored better than those with lower levels.[55] A longitudinal study of 883 community-dwelling men (mean age at entry 53 years, range 22–91 years) was conducted for as long as 31 years (mean 11.5 years), with biennial assessments of multiple cognitive domains. Neither the rates of decline in mean DHEAS nor the mean DHEAS concentrations within individuals were related to cognitive status or cognitive decline.[56] It should be noted, though, that most of the previous studies are relatively small, and a definitive outcome must await properly controlled large-scale studies where dose effects are also evaluated, as recently reported by Racchi et al.[57]

A cross-sectional sample of 63 frail elderly patients aged 66 to 103 years (mean age 86 years) at a multilevel care facility actually found an opposite relationship to that postulated between DHEAS and cognitive impairment measured as performance on the Blessed Memory Information Concentration Test (BMIC). These findings suggest that the relationships between steroid hormones of the hypothalamic-pituitary axis may be different in the elderly nursing home population than in other groups, though this relatively small study needs verification with a larger cohort.

Antigonadotrophin Therapy for Alzheimer's Disease

The finding that gonadotrophins may be raised in AD has led to the possibility of using antigonadotropins such as leuprolide in the treatment of AD.[58] In individuals with AD, there is a twofold elevation in the serum concentrations of the gonadotropins, LH, and FSH compared with age-matched controls. Whether elevated gonadotropins actually cause AD or merely represent a feedback response to low testosterone levels remains to be determined. However, we have recently found that LH does promote increased Aβ secretion in cultured neurons (G. Verdile, C. Atwood, and R. Martins, unpublished results). LH has been localized in the cytoplasm of pyramidal neurons where it is found to be increased together with neurofibrillary tangles in AD brain compared with age-matched control brain. Although the functional consequences of increased neuronal LH are unknown, it is notable that LH is primarily localized to those neurons that are known to be vulnerable to AD-related neurodegeneration. Elevated serum and cortical neuron levels of LH, coupled with the decline in sex steroid production, could play important roles in the pathogenesis of AD. However, if a role for LH in AD is established, leuprolide treatment per se, while reducing LH, would also further deplete the already low testosterone levels, which are important for both reducing Aβ levels and simultaneously increasing the levels of the neurotrophic factor, sAPP. Thus testosterone replacement would be needed to complement the protective effect of leuprolide.

Discussion of Case History

Currently, there is no recommendation of the use of androgens for AD, and it is not FDA approved. The treatment of this patient with advanced AD was based on his hypogonadal state. The aim was to increase functionality with an anabolic hormone, and offer him a better quality of life. The observed improvement in cognition was a bonus, and he did become more communicative. He had improvements in his MMSE and CDT (which is a measure of visual-spatial skills). His functionality as measured by the functional instrument of measure (FIM) scale and the ADL scale also improved. Paradoxically, it has been reported in the literature that sexually disruptive male AD patients can benefit from estrogen treatment. This patient did not turn hypersexual, but that has been noted to occur in some AD patients. Carefully monitoring is needed.

Conclusion and Key Points

Many men become hypogonadal as a result of aging, although some men exhibit changes as early as their 40s. It has been established that androgen replacement may help restore energy and libido. An added major benefit may be the restoration of cognitive function. Many patients with advanced AD also suffer from the *wasting syndrome*, which can lead to falls. Testosterone being an anabolic steroid may help restore functionality in these patients.

- AD is seemingly less common in men because of the shorter life span of men, and AD is a function of aging.

- Men can reduce their risk of AD by consuming vitamin C and E.

- Moderate alcohol, exercise, and socializing may also decrease the risk for AD.

- In in vitro studies, androgens increase neuritic arborization and the receptive field of cells, whereas estrogens induce communication between cells.

- There is a link between amyloid and androgens, which suggests that AD is linked to androgens.

- Small clinical studies have suggested that cognition can be improved with androgens, in particular the visual spatial domain.

REFERENCES

1. Kranczer S. Continued U.S. longevity increases. Stat Bull Metrop Insur Co 1999;80:20–27

2. Edland SD, Rocca WA, Petersen RC, Cha RH, Kokmen E. Dementia and Alzheimer disease incidence rates do not vary by sex in Rochester, Minn. Arch Neurol 2002;59:1589–1593

3. Ott MM, Breteler, Harskamp F, Stijnen T, Hofman A. Incidence and risk of dementia. The Rotterdam Study. Am J Epidemiol 1998;147:574–580

4. Corder EH, Robertson K, Lannfelt L, et al. HIV-infected subjects with the E4 allele for APOE have excess dementia and peripheral neuropathy. Nat Med 1998;4:1182–1184

5. Masaki KH, Losonczy KG, Izmirlian G, et al. Association of vitamin E and C supplement use with cognitive function and dementia in elderly men. Neurology 2000;54:1265–1272

6. Ruitenberg A, van Swieten JC, Witteman JC, et al. Alcohol consumption and risk of dementia: the Rotterdam Study. Lancet 2002;359:281–286

7. Vermeulen A, Kaufman JM. Diagnosis of hypogonadism in the aging male. Aging Male 2002;5:170–176

8. Morley JE, Perry HM III. Androgen deficiency in aging men. Med Clin North Am 1999;83:1279–1289

9. English KM, Mandour O, Steeds RP, Diver MJ, Jones TH, Channer KS. Men with coronary artery disease have lower levels of androgens than men with normal coronary angiograms. Eur Heart J 2000;21:890–894

10. Pandey KN, Oliver PM, Maeda N, Smithies O. Hypertension associated with decreased testosterone levels in natriuretic peptide receptor-A-gene-knockout and gene-duplicated mutant mouse models. Endocrinology 1999;140:5112–5119

11. Rolf C, Nieschlag E. Potential adverse effects of long-term testosterone therapy. Baillieres Clin Endocrinol Metab 1998;12: 521–534

12. Azuman T, Nagai Y, Saito T, Fanauchi M, Matsubara TS. The effect of dehydroepiandrosterone sulfate administration to patients with multi-farct dementia. J Neurol Sci 1999; 162:69–73

13. Sih R, Morley J, Kaiser FE, Perry IHM, Patrick P, Ross C. Testosterone replacement in older hypogonadal men: a 12-month randomized controlled trial. J Clin Endocrinol Metab 1997;82:1661–1667

14. Webb CM, Adamson DL, De Zeigler D, Collins P. Effect of acute testosterone on myocardial ischemia in men with coronary artery disease. Am J Cardiol 1999;83:437–439

15. Reisberg B, Doody R, Stottler A, et al. Memantine in moderate to severe Alzheimer's disease. N Engl J Med 2003;348: 1333–1341

16. Grady D, Yaffe K, Kristof M, Liu F, Richards C, Barrett-Connor E. Effect of postmenopausal therapy on cognitive function: the Heart and Estrogen/Progestin Replacement Study. Am J Med 2002;113:543–548

17. Simerly RB, Chang C, Muramatsu M, Swanson LW. Distribution of androgen and estrogen receptor mRNA-containing cells in the rat brain: an in situ hybridization study. J Comp Neurol 1990;294:76–95

18. Raber J, Bongers G, Le Fevour A, Buttini M, Mucke L. Androgens protect against apolipoprotein E4-induced cognitive deficits. J Neurosci 2002;22:5204–5209

19. Lustig RH. Sex hormone modulation of neural development in vitro. Horm Behav 1994;28:383–395

20. Hart BL. Effect of gonadectomy on subsequent development of age-related cognitive impairment in dogs. J Am Vet Med Assoc 2001;219:51–56

21. Ohta A, Akiguchi I, Seriu NK, et al. Deterioration in learning and memory of inferential tasks for evaluation of transitivity and symmetry in aged SAMP8 mice. Hippocampus 2002;12: 803–810

22. Flood JF, Farr S, Kaiser FE, La Regina M, Morley JE. Age-related decrease of plasma testosterone in SAMP8 mice: Replacement improve age-related impairment of learning and memory. Physiol Behav 1995;57:669–673

23. Papasozomenos SC. The heat shock-induced hyperphosphorylation of tau is estrogen-independent and prevented by androgens: implication for Alzheimer disease. Proc Natl Acad Sci USA 1997;94:6612–6617

24. Gouras GK, Xu H, Gross RS, et al. Testosterone reduces neuronal secretion of Alzheimer's β-amyloid peptides. Proc Natl Acad Sci USA 2000;97:1202–1205

25. Lewis J, Dickson DW, Lin WL, et al. Enhanced neurofibrillary degeneration in transgenic mice expressing mutant tau and APP. Science 2001;293:1487–1491

26. Pike CJ. Testosterone attenuates β-amyloid toxicity in cultured hippocampal neurons. Brain Res 2001;919:160–165

27. Flood JF, Morley JE. Learning and memory in the SAMP8 mouse. Neurosci Biobehav Rev 1998;22:1–20

28. Frye CA, Lacey EH. The neurosteroids DHEA and DHEAS may influence cognitive performance by altering affective state. Physiol Behav 1999;66:85–92

29. Goodenough S, Engert S, Behl C. Testosterone stimulates rapid secretory amyloid precursor protein from rat hypothalamic cells via the activation of the mitogen-activated protein kinase pathway. Neurosci Lett 2000;296:49–52

30. Kemnitz JW, Roecker EB, Haffa AL, et al. Serum dehydroepiandrosterone sulfate concentrations across the life span of laboratory-housed rhesus monkeys. J Med Primatol 2000; 29:330–337

31. Herndon JG, Lacreuse A, Ladinsky E, Killiany RJ, Rosene DL, Moss MB. Age-related decline in DHEAS is not related to cognitive impairment in aged monkeys. Neuroreport 1999;10: 3507–3511

32. Kimonides VG, Khatibi NH, Svendsen CN, Sofroniew MV, Herbert J. Dehydroepiandrosterone (DHEA) and DHEA-sulfate (DHEAS) protect hippocampal neurons against excitatory amino acid–induced neurotoxicity. Proc Natl Acad Sci USA 1998;95:1852–1857

33. Reddy DS, Kulkarni SK. Possible role of nitric oxide in the nootropic and antiamnesic effects of neurosteroids on aging- and dizocilpine-induced learning impairment. Brain Res 1998; 799:215–229

34. Shen S, Cooley DM, Glickman LT, Glickman N, Waters DJ. Reduction in DNA damage in brain and peripheral blood lymphocytes of elderly dogs after treatment with dehydroepiandrosterone (DHEA). Mutat Res 2001;Sep 1:480–481

35. Danenberg HD, Haring R, Heldman E, et al. Dehydroepiandrosterone augments M1-muscarinic receptor-stimulated amyloid precursor protein secretion in desensitized PC12M1 cells. Ann N Y Acad Sci 1995;774:300–303

36. Danenberg HD, Haring R, Fisher A. Dehydroepiandrosterone (DHEA) increases production and release of Alzheimer's amyloid precursor protein. Life Sci 1996;59:1651–1657

37. Milward EA, Papadopoulos R, Fuller SJ, et al. The amyloid protein precursor of Alzheimer's disease is a mediator of the effects of nerve growth factor on neurite outgrowth. Neuron 1992;9:129–137

38. Sisodia SS, Koo EH, Beyreuther K, Unterbeck A, Price DL. Evidence that β-amyloid protein in Alzheimer's disease is not altered by normal processing. Science 1990;248:492–495

39. Wolf OT, Kirschbaum C. Actions of dehydroepiandrosterone and its sulfate in the central nervous system: effect on cognition and emotion in animals and humans. Brain Res Brain Res Rev 1999;30:264–288

40. Tan RS. Memory loss as a reported symptom of andropause. Arch Androl 2001;47:185–189

41. Janowsky JS, Oviatt SK, Orwoll ES. Testosterone influences spatial cognition in older men. Behav Neurosci 1994;108:325–332

42. Cherrier MM, Asthana S, Plymate S, et al. Testosterone supplementation improves spatial and verbal memory in healthy older men. Neurology 2001;57:80–88

43. Morrison MF, Redei E, TenHave T, et al. Dehydroepiandrosterone sulfate and psychiatric measures in a frail, elderly residential care population. Biol Psychiatry 2000;47:144–150

44. Slabbekoorn D, van Goozen SHM, Megens J. Activating effects of cross-sex hormones on cognitive functioning: a study of short-term and long-term hormone effects on transsexuals. Psychoneuroendocrinology 1999;24:423–447

45. O'Connor DB, Archer J, Hair WM, Wu FC. Activational effects of testosterone on cognitive function in men. Neuropsychologia 2001;39:1385–1394

46. Zitzmann M, Weckesser M, Schober O, Nieschlag E. Changes in cerebral glucose metabolism and visuospatial capacity in hypogonadal males under testosterone substitution therapy. Exp Clin Endocrinol Diabetes 2001;109:302–306

47. Masters CL, Simms G, Weinman NA, Multhaup G, McDonald BL, Beyreuther K. Amyloid plaque core protein in Alzheimer disease and Down syndrome. Proc Natl Acad Sci USA 1985;82:4245–4249

48. Martins RN, Robinson PJ, Chleboun JO, Beyreuther K, Masters CL. The molecular pathology of amyloid deposition in Alzheimer's disease. Mol Neurobiol 1991;5:389–398

49. Gandy S, Almeida OP, Fonte J, et al. Chemical andropause and amyloid-B peptide. JAMA 2001;285:2195–2196

50. Lue YH, Sinha Hikin AP, Swerdloff RS, et al. Single exposure to heat induces stage-specific germ cell apoptosis in rats: role on intratesticular testosterone on stage specificity. Endocrinology 1999;140:1709–1717

51. Tirassa P, Thiblin I, Agren G, Vigneti E, Aloe L, Stenfors C. High-dose anabolic steroids modulate concentration of nerve-growth factor and expression of its low affinity receptor (p75-NGFr) in male rat brain. J Neurosci Res 1997;47:198–207

52. Wyss-Coray T, Loike JD, Brionne TC, et al. Adult mouse astrocytes degrade amyloid-β in vitro and in situ. Nat Med 2003;9:453–457

53. Day JR, Frank AT, O'Callaghan JP, Jones BC, Anderson JE. The effect of age and testosterone on the expression of glial fibrillary acidic protein in the rat cerebellum. Exp Neurol 1998;151:343–346

54. Yoshizawa A, Clemmons DR. Testosterone and insulin-like growth factor (IGF) I interact in controlling IGF- binding protein production in androgen-responsive forsekin fibroblast. J Clin Endocrinol Metab 2000;85:1627–1633

55. Carlson LE, Sherwin BB, Cherkow HM. Relationships between dehydroepiandrosterone sulfate (DHEAS) and cortisol (CRT) plasma levels and everyday memory in Alzheimer's disease patients compared to healthy controls. Horm Behav 1999;35:254–263

56. Moffat SD, Zonderman AB, Harman SM. The relationship between longitudinal declines in dehydroepiandrosterone sulfate concentrations and cognitive performance in older men. Arch Intern Med 2000;160:2193–2198

57. Racchi M, Balduzzi C, Corsini E. Dehydroepiandrosterone (DHEA) and the aging brain: flipping a coin in the fountain of youth CNS. Drug Rev 2003;9:21–40

58. Bowen RL, Smith MA, Harris PL, et al. Elevated luteinizing hormone expression colocalizes with neurons vulnerable to Alzheimer's disease pathology. J Neurosci Res 2002;70:514–518

8

The Effects of Aging, Hypogonadism, and Nutrition on Endurance and Strength

THOMAS CURTIS NAMEY

"...come to the grave in full vigor, like sheaves gathered in season." Job 5:26

"Living systems are worn out by inactivity and developed by use." Albert Szent-Gyorgyi

"A hospital bed is a parked taxi with the meter running." Groucho Marx

The interaction between physical activity and aging is complex, but antithetical. In many instances the perception of aging is *due to inactivity* and self-perpetuating. The decline in testosterone and other trophic hormones with age contributes in a myriad of ways to the perceived acceleration of physical decline in men, not only by minimizing the benefits of activity, but also by contributing to an apathetic senescence, making the pursuit of physical activity more difficult. Nutritional issues complicating the elderly man further augment this maelstrom, and can push the older man into andropause and physical decline by minimizing testosterone production and enhancing estrogen formation (zinc deficiency) or increasing atherogenic substances, such as homocysteine (folate or B_{12} deficiencies). Which factors dominate this downward spiral are difficult to dissect, but the outcome is enhanced frailty, diminished quality of life, and accelerated senescence. What society wants is enhanced vitality with aging and natural augmentation of life expectancy, with eradication of diseases of lifestyle, even though we are not yet at the point of augmenting *life span*, the maximal obtainable age of a particular member of the human species. It remains to be seen if the optimization of hormonal status, enhanced physical activity, coupled with a scientific nutrition base, will make this happen.[1–4]

Working Case Report in Andropause, Activity, and Nutrition

E.B. was a 75-year-old retired university president who had had a successful prostatectomy 10 years ago for well-confined prostate cancer. His postsurgical prostate-specific antigens (PSAs) remained low and he had retained minimal sexual function. He had gained weight and was on a statin for lipid control. He did not exercise. He was seen by me for mild anemia (hemoglobin 13.2 g%) and lassitude, thought to be related, which previously had been evaluated at two university medical centers, but iron studies, B_6, B_{12}, folate, and bone marrow studies were normal.

The patient did not engage in regular aerobic activity. Why is this point important? Regular aerobic activity is essential in treating modest anemia. Athletes experience plasma volume expansion, with no decrement in red blood cell mass, secondary to a heightened renin-angiotensin-aldosterone activation. This is a principal cause of "runner's anemia," which is a normal adaptation to regular aerobic exercise. The patient did complain of loss of energy and some decline in sexual function over the past 3 years (one sexual episode/month versus two to three before that time). He attributed this to age. He was quite hirsute and had significant male pattern baldness. His body fat was 29% and mesenteric. His cholesterol/high-density lipoprotein (HDL) ratio was 4.2.

I initiated a hormonal evaluation. His total testosterone was 564 ng/dL and his free testosterone was 103.0 pg/mL (normal range is 50–210). The concept of *relative hypogonadism* is important. A person's free testosterone may have been very high years before he is seen for treatment. A man whose free testosterone falls from 200 to 100 pg/mL will experience more symptoms than will a man whose free testosterone falls from 70 to 50! The patient's follicle-stimulating hormone (FSH) was 62.9 (normal 1–8). His luteinizing hormone (LH) was 19.7 (normal 2–12). I found this result surprising. His baseline free testosterone may have been higher than 210 pg/mL! He is, by definition, in andropause. Both LH and FSH are extremely high, which tells you his hypothalamus detects a deficiency in androgens. He was placed on 5 g 1% of Androgel and started on a high-protein, low-carbohydrate (HPLC) diet. A regular aerobic and weight program was started.

With regard to his prostate cancer, no evidence exists that testosterone causes prostate cancer or that replacement makes it more likely. While controversial, one can argue that high testosterone levels prevent prostate cancer. The patient's initial PSA was 0.7. One year later it was 0.8. Mild increases in PSA do occur and reflect modest increases in prostate mass. Because this was a minimal increase, I was not concerned. His hemoglobin at this time was now 15.2, and his free testosterone was increased to 145 pg/mL. His LH was 4.6, but his FSH was 24.2. His body fat fell to 15.5%. His cholesterol/HDL ratio was now 3.0. I wondered if I should increase his topical testosterone. I decided not to do so, because he is doing quite well on an integrated hormonal replacement program, coupled with a diet and exercise. His body fat fell because he lost fat mass and increased muscle mass! His sexual activity had increased to twice monthly at 77 and his energy and vigor had substantially improved.

The Characteristics of Aging and the Axioms of Gerontology

The *characteristics of aging* seem harsh. They are universal, decremental, progressive, and intrinsic (Appendix 1). But the *axioms of gerontology* moderate this unpleasantness: (1) sustained exercise inhibits the decline of the physique with age; (2) sustained physical activity inhibits the decline of physical fitness with age; and (3) sustained physical activity inhibits the decline of mental functions with age. The net effect is moderation and slowing of the characteristics of aging, with routine exercise and physical activity, and with vitality and independence sustained![5–9]

Hormonal Issues Surrounding Aging and Activity

Most experts agree that the major hormones responsible for physical decline in men are falling levels of testosterone, human growth hormone (HGH), and a principal secondary hormone derived from the HGH-liver interaction, insulin-like growth factor-I (IGF-I) (also called somatomedin-C), and dehydroepiandrosterone (DHEA), which are coupled with rising levels of hydrocortisone (stress-induced) and insulin, secondary to increasing insulin resistance, and increasing levels of the testosterone-derived female hormones estradiol and estrone.[1,3] Many other hormones are involved to a lesser extent, so we will focus on the major players, while looking at strategies to optimize hormone levels, but not losing sight of nutrition, which directly affects hormone levels in men.[10]

Testosterone is responsible for the relatively larger muscle mass in males, their increased leanness, and aggression, as compared with females.[11–13] Conversely, lower levels of testosterone result in lower levels of muscle mass, increased body fat, and decreased strength, while increasing depressive symptoms and decreasing aggression to the point of apathy.[1,2,13,14] Supraphysiological amounts of testosterone administered to nonexercising males, by itself results in increased muscle mass, strength, and leanness compared with exercise alone. Combining both yields greater gains.[15] The aging man, however, because of rising levels of endogenous adrenal stress hormones (principally hydrocortisone), increasing obesity,[16,17] increasing insulin levels, increasing cyclic aromatase activity secondary to increasing mesenteric adiposity, and lower levels of certain nutrients, such as zinc, is victim to both a falling testosterone synthesis and rising levels of sex hormone–binding globulin (SHBG), the latter further decreasing the level of "free testosterone," in the milieu of falling total testosterone. The rise in SHBG is quite critical, because testosterone in highly bound (almost 99%) and very little is free and bioactive (Fig. 8–1). Conversely, estradiol and estrone are much more weakly bound, and therefore more "adroit" at shutting down the hypothalamic-pituitary-gonadal axis, resulting in a falling LH level and a further decrement in gonadal synthesis of testosterone. The high levels of estrogens further enhance hepatic synthesis of SHBG, augmenting the downward spiral of free testosterone.

Augmenting Testosterone Levels in the Older Hypogonadal Man

What strategies might be employed to circumvent these changes? Increasing the testosterone level, if low, is paramount. The goal is to bring levels in most patients to average or higher than normal levels of free testosterone. This can be accomplished by weight reduction, best promoted by the combination of increased levels of physical activity, including both strength and aerobic training, with an HPLC diet, which has the added benefit of lowering insulin levels and improving lipid levels, compared with traditional lower fat, low-calorie diets.[18–20] Energy efficiency on the HPLC diet is markedly decreased and it has been

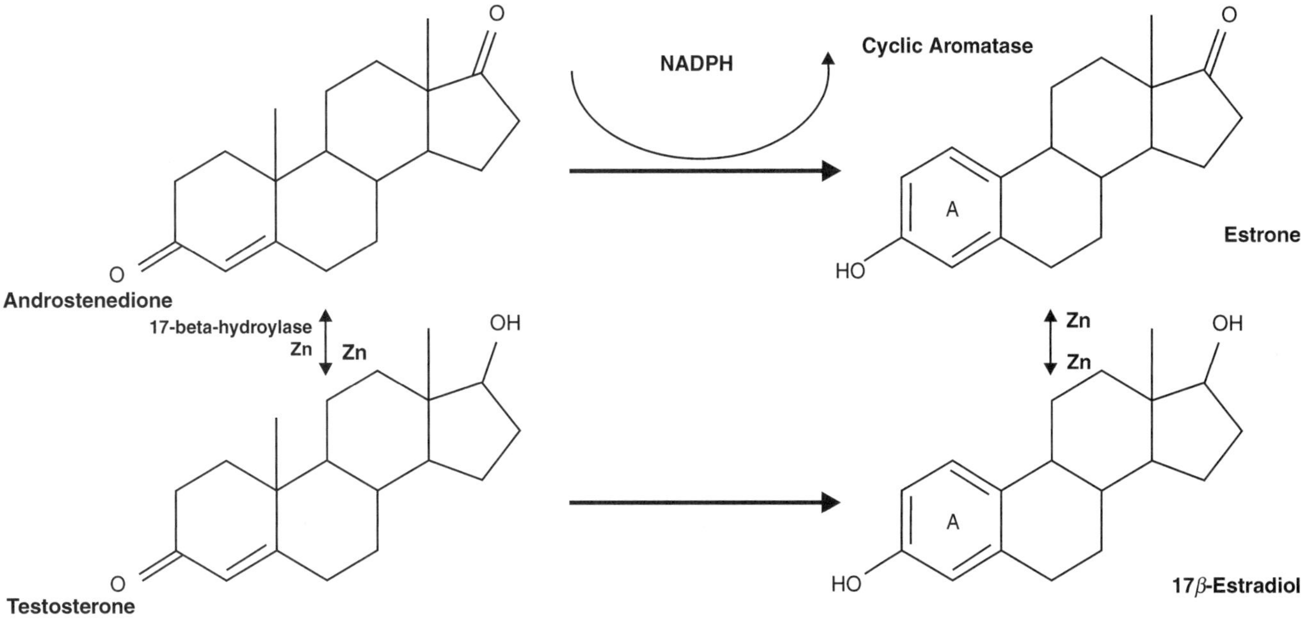

FIGURE 8–1. Testosterone metabolism (direct pathways). Low levels of zinc (Zn) impede synthesis of testosterone and estradiol from their precursors and increasing fat mass increases conversion (via aromatization) of androstenedione to estrone and testosterone to estradiol, lowering testosterone levels. Low levels of zinc and increased fat mass are commonly found in older men. NADPH, reduced nicotinamide adenine dinucleotide phosphate.

shown to reduce adipocyte fatty acid synthase (FAS) and FAS messenger RNA (mRNA) levels by over 50%, making energy storage in the fat cell much more difficult.[18] Decreasing fat mass directly decreases estradiol levels formed by cyclic aromatization of testosterone by the adipocyte, which in turn increases both LH and FSH levels, enhancing endogenous testosterone synthesis. Furthermore, decreased estrogen levels decrease SHBG levels, which raise free bioactive testosterone.

Administering testosterone to the hypogonadal man may be directly beneficial, but it potentiates a leaner body mass and enhances gains in strength through training. Furthermore, a testosterone replacement causes a rise in hematocrit[11,12] that enhances aerobic performance directly, by increasing oxygen delivery to peripheral muscles and vital organs, and reducing stress on the heart. Myocardial oxygen consumption is directly proportional to mean arterial pressure (MAP) times the heart rate (HR). Exercise lowers both the HR and the blood pressure for any given level of exertion. But increasing myocardial oxygen delivery by increasing hemoglobin levels augments this training effect.[21] One of the most surprising findings in recent studies is the antiatherogenic properties of normal testosterone levels in males, refuting earlier medical views that testosterone was responsible for differential rates of heart disease between males and females.[22–24] Testosterone's effect may even extend to

coronary artery vasodilatation, because its effects are immediately seen during stress testing in men with established coronary artery disease.[24] This body of research so contradicts the previous paradigm that a prospective, controlled, double-blind study on the replacement of testosterone in hypogonadal men will be needed to convince the unconvinced.

Exogenous Testosterone Administration in Men

The preferred routes of administration are transdermal (gels or patches), followed by intramuscular (injection). Transdermal administration results in more even levels of testosterone and avoids the peaks and valleys seen with injection and is associated with lower rates of benign prostatic hypertrophy and polycythemia.[1,3,11] There is *no* evidence that testosterone administration to hypogondal males potentiates the development of prostate cancer.[1,3,11]

Nutritional Issues Vital to Normal Hormonal Status

Zinc

It is prudent to check zinc levels in all male patients, because low zinc levels *cause* hypogonadism. Low zinc levels impair testosterone synthesis by the Leydig cells of the testes and increase aromatase activity. Indeed, low levels of zinc in utero will impair the testes irreparably, diminishing both testosterone synthesis and spermatogenesis.[25] A mild deficiency of zinc, affecting growth and development in children and adolescents, has been reported from developed countries as well. Zinc deficiency in humans, aside from hypogonadism in males, may manifest as severe, moderate, or mild. The manifestations of severe zinc deficiency include bullous pustular dermatitis, alopecia, diarrhea, emotional disorders, weight loss, intercurrent infections due to cell-mediated immune dysfunctions, neurosensory disorders, and problems with healing of ulcers. A moderate level of zinc deficiency has been reported in a variety of conditions. Clinical manifestations include growth retardation and male hypogonadism in adolescence, arrested penile development in puberty, rough skin, poor appetite, and mental lethargy. Even a mild level of zinc deficiency may present with decreased serum testosterone levels and/or oligospermia in males, and decreased lean body mass. The recognition of mild levels of zinc deficiency has been difficult.[4,25-27]

Dietary zinc restriction in normal young men was associated with a significant fall (>75%) in serum testosterone concentrations after 20 weeks of zinc restriction. Conversely, supplementation of marginally zinc-deficient normal elderly men for 6 months resulted in a doubling of serum testosterone.[28] Zinc is equally important in the postsynthesis disposition of testosterone. Zinc deficiency significantly decreases 5α-reductase conversion of testosterone to dihydrotestosterone (DHT). DHT is responsible for hair loss and prostatic hypertrophy. Zinc deficiency has been shown in animals to further enhance aromatase activity, increasing estradiol production, but also decreasing androgen receptors, while increasing estrogen receptor binding sites and SHBG. The negative impact on the aging male should be quite clear.[29,30] I recommend all older patients take between 50 and 100 mg of chelated zinc daily (zinc gluconate), not only to optimize testicular function, but to obtain the other additional benefits seen with immune function, enhanced healing, and sharpened senses of taste and smell.

Dehydroepiandrosterone

Dehydroepiandrosterone (DHEA) levels fall significantly with age, whereas adrenal cortisol production rises. The supplementation of 100 mg of DHEA to older men and women was studied and resulted in lower SHBG levels, higher free testosterone levels, increased IGF-I levels, and a youthful DHEA/cortisol ratio. No side effects were noted in either sex.[31] In another recent study dehydroepiandrosterone sulfate (DHEAS) serum levels were negatively correlated with patient age, sexual function score, total wellness score (an index of general well-being), and PSA.[32,33] DHEA is an independent variable apart from testosterone related to erectile dysfunction in diabetics.[33,34] DHEA is inversely correlated with bone mineral density in elderly men.[35] Because of the positive benefits of DHEA administration on free testosterone and a heightened IGF-I, and the correlations with male wellness and DHEA levels, I favor recommending 100 mg daily of DHEA for older men, irrespective of testosterone levels. Because DHEA taken orally affects hydrocortisone levels, taking it in the evening results in a modest lowering of cortisol levels compared with morning ingestion.[36]

As previously mentioned, decreased carbohydrate intake and increased protein intake, irrespective of fat (HPLC diet), lowers insulin secretion, which promotes higher testosterone levels, improved insulin sensitivity, and weight loss. I strongly recommend this diet for men, particularly for those with metabolic obesity syndrome. Whether or not adherence to this diet is possible, exercising men can optimize training benefits, particularly weight training, by ingesting 1.5 to 2.0 g of protein per kilogram of body weight daily. Chromium picolinate (200 to 400 μg/day) improves insulin sensitivity,[37] improves diabetic status, negates the adverse effects of glucocorticoids,[38] and even increases lean body mass.[39] I recommend that older men take 400 μg daily, irrespective of diet.

HGH levels can be increased with high-intensity training, although this effect is not seen with testosterone. Scattered reports suggest increases in testosterone levels might occur with exercise, but this is explained by the immediate hemoconcentration, secondary to transmigration of plasma water into exercising muscle, and secondary to increasing osmolality. HGH levels put out by the pituitary during exercise decrease with age, but still are significant. Higher intensity weight training, followed by high-intensity interval aerobic activity, results in substantial increases in HGH production, mirrored by higher IGF-I levels. IGF-I appears to be the principal mediator of the action of growth hormone, whereas IGF-II has more insulin-like activity. The principal anabolic actions of IGFs include stimulation of amino acid transport; stimulation of DNA, RNA, and protein synthesis; and induction of cell proliferation and growth. IGF-I is directly responsible for chondrogenesis, skeletal growth, and the growth of soft tissue. Linear growth is stimulated by affecting cartilaginous growth areas of long bones. Growth is also stimulated by increasing the number and size of skeletal muscle cells, influencing the size of organs, and increasing red cell mass through erythropoietin stimulation. These indirect effects are inhibited by glucocorticoids.

Low-level aerobic training has virtually no effect on HGH secretion.[40] Because DHEA levels augment hepatic synthesis of IGF-I, taking DHEA and engaging in high-intensity exercise are the best strategies for increasing the trophic benefits of HGH, short of taking the hormone itself, which has substantial benefits in older men with low levels, even enhancing insulin sensitivity.[41]

The Benefits of Regular Physical Activity on the Aging Process

The benefits of regular physical activity, consisting of both aerobic and anaerobic exercises, has gained scientific support in the medical literature.[42–45] The value of such programs extends beyond measures of disease prevalence and life expectancy. Exercise can markedly increase the quality of life for an individual by removing physical constraints and providing the necessary endurance and strength needed to participate in and enjoy numerous activities, even as one ages (Appendix 2). The aging process, as a naturally occurring phenomenon, does not place limitations on

TABLE 8–1. Physiological Changes That Are Common to Aging, Inactivity, and Weightlessness

Cardiovascular/respiratory
  Decrease in V_{O_2} max
  Decrease in cardiac output
  Decrease in stroke volume
  Decrease in maximum heart rate
  Increase in systolic blood pressure
  Decrease in body water
  Decrease in vital capacity and maximal ventilation

Blood
  Decrease in red blood cell mass and lymphocytes
  Decrease in fibrinolytic capacity
  Increase in total cholesterol and triglycerides

Body composition/musculoskeletal
  Decrease in lean body weight (muscle loss)
  Increase in fat weight
  Decrease in bone mineral mass
  Decrease in joint flexibility

Metabolic/regulatory
  Decrease in bowel function
  Disordered glucose tolerance
  Decrease in activity of sympathomedullary axis
  Decrease in serum androgen levels
  Decrease in immune system function

Nervous system
  EEG abnormalities
  Sleep pattern disorders
  Deterioration of sense organ function
  Deterioration of intellectual capacity (cognition, memory)

Adapted from Bortz WM. Disuse and aging. JAMA 1992;248:1203–1208, with permission.

the adaptations to training of cardiovascular endurance and muscular strength that were thought to exist. Rather, the decrease in these parameters may reflect more an inactive lifestyle (Appendix 3). Indeed the changes commonly associated with aging parallel the changes seen with inactivity or even with weightlessness in space (Table 8–1, Appendix 1).

Cardiovascular endurance has been established as a measure of fitness that allows one to make assumptions pertaining to the general level of physical activity an individual is able to endure. Higher levels of physical activity, as demonstrated by higher levels of physical fitness, appear to decrease all-cause mortality, primarily due to lower rates of cardiovascular disease and cancer.[46] Lower levels of physical fitness are associated with a higher risk of death from coronary heart disease and cardiovascular disease in clinically healthy men, independent of other major coronary risk factors.[43]

In addition to the positive effects of regular physical activity on morbidity and mortality, improved cardiovascular endurance and muscular strength have numerous practical implications when applied to lifestyle and physical activity. This concept is especially applicable to the aging population, which not only may lose the capability of participating in previously enjoyable activities, but also, more importantly, may lose the ability to maintain functional independence. The decline in strength and endurance with aging may also expose these individuals to significant dangers, such as falls, with their associated morbidity and mortality.[47,48]

After the age of 25 years, a decrease in the maximal capacity of the cardiovascular system to maintain an adequate oxygen supply to the exercising muscles occurs with advancing age, as evidenced by a progressive decline in maximal oxygen uptake (V_{O_2}max).[49,50] This regression occurs as a result of several factors found with aging: decreased maximum heart rate, decreased stroke volume, decreased cardiac contractility, increased body mass, and decreased level of physical activity (Table 8–1). Until recently, many people questioned the trainability of older individuals and the extent to which their physical fitness could be modified. The decrease in aerobic power was felt to be a naturally occurring phenomenon that explained the limitations placed upon the elderly and the limited range of activities in which they could comfortably and safely participate.

In a study using sedentary men between 49 and 65 years of age, Pollock et al[50] demonstrated a significant improvement in V_{O_2}max of 18%, from 31.0 to 36.8 mL/kg/min, following 20 weeks of aerobic training. These men trained for 30 minutes, 3 days a week, at an intensity of 80 to 90% of their maximum heart rate. A nonexercising control group showed no significant changes over the same time period. The investigators concluded not only that these men responded favorably to endurance exercise, but also that the improvement was similar in magnitude to that found in previous investigations of similar design with younger subjects.

The ability to endurance train individuals even older than the subjects used in the aforementioned study was previously not well understood, but felt to be negligible. Early investigations demonstrated nonsignificant to moderate changes in VO_2max in response to training in individuals over 60 years of age, depending upon the intensity of training and the length of the program.[51,52] Several studies have further examined these earlier findings and have demonstrated a significant improvement in VO_2max in subjects not only in the 60- to 69-year age range, but also even in older individuals.[53,54] In the study by Seals et al,[54] 11 subjects, with a mean age of 63 years, were evaluated before training, after 6 months of low-intensity training (~40% of heart rate reserve for 20 to 30 minutes, 3 days a week), and, finally, after an additional 6 months of higher intensity training (up to 85% of the heart rate reserve for 30 to 45 minutes, 3 days a week). VO_2max increased significantly, following low-intensity and high-intensity training (+30%). VO_2max was 25.2 mL/kg/min before training, 28.2 mL/kg/min after low-intensity training, and 32.9 mL/kg/min after high-intensity training. These findings demonstrated that individuals older than 60 could adapt to and benefit from prolonged endurance training with a large increase in aerobic power.

In a study of even older individuals, Hagberg et al[53] demonstrated improvement in cardiovascular endurance with training. Sixteen subjects, whose ages ranged from 70 to 79 years, participated in an exercising program consisting of three sessions per week for 26 weeks (40 minute per session at 50 to 70% VO_2max initially for the first 13 weeks and then at 75 to 85% VO_2max for the last 13 weeks of training).

These subjects increased their VO_2max by 16%, from 22.5 to 25.8 mL/kg/min, during the first 13 weeks of training and by a total of 22% after 26 weeks of training, with a final VO_2max of 27.1 mL/kg/min. Therefore, this study demonstrated that healthy men and women in their 70s can respond to prolonged endurance exercise training with adaptations similar to those in younger individuals.

Though improvements in cardiovascular endurance are clearly noted in elderly individuals following a conditioning program, the effects of a long-term routine of physical activity on the rate of decline in aerobic power requires further delineation. Several earlier studies demonstrated a decline in VO_2max despite habitual exercise, though fitness levels in the exercisers remained significantly higher compared with sedentary controls.[55,56] In a longitudinal study on 16 men initially between the ages of 32 to 56 years (mean 44.6), Kasch and Wallace[57] found essentially no change in VO_2max over a 10-year period of aerobic conditioning. These subjects trained 3 days a week, 12 months a year by running, swimming, or a combination of running and swimming. The duration of the activity session ranged from 20 to 105 minutes (mean 59 minutes) and intensity ranged from 60 to 95% (mean 86%) of maximal physical work capacity by heart rate.

In another study, Kasch et al[58] longitudinally followed two groups of men: 15 exercisers were followed for 23 years, from a mean age of 44.6 to 68.0, and 15 nonexercising control subjects were followed for 18 years, from a mean age of 51.6 to 69.7. Physical fitness, as measured by VO_2max, declined 13% in the exercisers, from 44.4 to 38.6 mL/kg/min and, in the nonexercisers, it declined 41%, from 34.2 to 20.3 mL/kg/min (Table 8–2). The combination of a higher level of fitness and a lower rate of decline would allow the active subjects to comfortably and safely participate in a wider range of physical pursuits when these results are compared with the MET (metabolic equivalent rate for any given person)

TABLE 8–2. Changes in Cardiovascular Endurance with Endurance Training in Older Individuals

	Age, yrs	Training, wks	Initial VO_2 max		Final VO_2max	
			L/min	mL/kg/min	L/min	mL/kg/min
Pollock et al, 1976	49–65	20	2.47 lumen	31.0	2.90 lumen	36.8
	Controls		2.77 lumen	31.5	2.82 lumen	32.2
Seals et al, 1984	63 ± 2	26	1.91 lumen	25.4		
	Low intensity				2.10 lumen	28.2
	High intensity				2.39 lumen	32.9
	Controls				1.61 lumen	26.3
Hagberg et al, 1989	70 ± 7	26	1.59 lumen	22.5	1.88 lumen	27.1
	Controls		1.51 lumen	22.2	1.48 lumen	22.0

TABLE 8–3. Changing Results with Age of Prolonged Endurance Training

	Age, yrs	Training, yrs	Initial V_{O_2} max		Final V_{O_2} max	
			L/min	mL/kg/min	L/min	mL/kg/min
Kasch and Wallace, 1976	32–56 (M-44.6)	10	3.376	43.7	3.303	44.4
Kasch et al, 1990	44.6 ± 6.9	23	3.366	44.4	2.793	38.6
	51.6 ± 6.2	18	2.743	34.2	1.685	20.3

requirements of selected activities (Fig. 8–1). The decline in V_{O_2}max of the nonexercisers was similar to that reported by Dehn and Bruce.[59] Habitual aerobic exercise retarded the usual loss of V_{O_2}max thought to characterize aging (Table 8–3).

A decline in muscle strength appears to be one of the more predictable characteristics of the aging process and creates difficulty performing the activities of daily living.[60] Decreased levels of trophic hormones, decreased neuromuscular coordination, decreased muscle mass, and low physical activity are several contributing factors to this decline (Table 8–2). This loss of strength and its associated loss of function was previously felt to be a natural consequence of aging and, with that assumption, the ability of a strength training program to improve strength in the elderly was thought to be minimal. As with endurance training, recent studies have further explored this previously held belief.

Using 12 healthy untrained volunteers (age range 60–72 years) participating in a 12-week strength-training program, Frontera et al[61] demonstrated significant increases in strength. The participants performed eight repetitions per set, three times a week at 80% of their one repetition maximum (1 RM) for extensors and flexors of both knee joints. At the end of the program, the subjects improved their strength by both isotonic (1 RM) and isokinetic measurements (Cybex II dynamometer measuring isokinetic peak torque). Upon further analysis, the rate of dynamic strength gain per training session of these elderly men was similar to the 4.4 to 5.6% increases in dynamic strength per session observed in young men (average age 28 years) after a similar training protocol.[62] A well-designed strength training program of appropriate intensity, similar in nature to standard rehabilitation techniques, caused a marked gain in strength in older men.

In a study using individuals in an older age group in an institutional setting, 10 frail, institutionalized volunteers aged 90 ± 1 years under took 8 weeks of high-intensity resistance training program (eight repetitions per set, three sets per session, three sessions per week, beginning at 50% 1 RM, then increasing to 80% RM after 1 week).[63] After 8 weeks of training, strength gains averaged 174% ± 31% (mean ± standard error of the mean, SEM) in the nine subjects who were able to complete the training program. Therefore, high-resistance weight training led to significant gains in muscle strength, size, and functional mobility among frail residents of nursing homes up to the age of 96 years (Table 8–4).[64]

Whether the improved muscle strength demonstrated in elderly individuals placed on a regular program of strength-training results in improved quality of life as indicated by a greater degree of independence and reduced rate of falls has not been studied, though one could postulate a positive correlation. Furthermore, the effects of a sustained program of resistive-training on the rate of decline in muscle strength is not yet known and would have significant clinical implications in terms of prescribing a well-rounded program of physical activity, combining aerobic as well as anaerobic exercise training.

Both cardiovascular endurance and muscle strength have been shown to increase in the elderly population

TABLE 8–4. Changes in Muscle Strength with Strength-Training in Older Individuals

	Age, yrs	Training, wks	Knee Muscle	Pretraining, kg	Midtraining, kg	Post-training, kg
Frontera et al, 1988	60–72	12	R ext	20 ± 2	31 ± 2	40 ± 2
			L ext	20 ± 1	32 ± 2	40 ± 2
Fiatarone et al, 1990	90 ± 1	8	R ext	8.0 ± 1	20.6 ± 2.4	
			L ext	7.6 ± 1	19.3 ± 2.2	

following a program of regular exercise. Regardless of the positive effects on improved rates of morbidity and mortality, the improvement in Vo_2max and strength allows individuals to actively participate in a broader range of physical activities as they age. The effect allows the elderly to maintain an active and independent role in society and decreases the burden of care on the younger generation. Most importantly, the positive effects of a well-designed exercise program remove the limitations placed on the elderly by a sedentary lifestyle and allows participation in some of the more pleasurable and vigorous activities of life. Exercise programs can be designed to minimize the effects of aging, both perceived and real, and also overcome some limitations imposed by preexisting disease, such as arthritis.

Conclusion

The synergy between optimal androgen status, optimal nutrition, and optimal physical activity in the aging male can be seen. An integrated approach has yet to be tried on a large scale. I believe better results for health can be expected in the aging male, when all three are addressed simultaneously rather than each individually. We are at societal edge where a brave few are testing the waters. Hopefully, a prospective study will confirm my belief.

Appendix 1: Physiological Changes During Aging

I.   Eyes
    A. Presbyopia: less able to focus on near objects
    B. Light reduction
       1. 33% as much light reaches retina
       2. Causes
          a. Yellowing of lens, sclera, vitreous humor.
          b. Reduction in size of pupils
          c. Entropion

II.   Ears
    A. Presbycusis
       1. Lose ability to hear higher frequencies
       2. Secondary to retrocochlear and inner ear degeneration
    B. Conductive hearing loss: stiffening of connective tissue

III. Cardiovascular
    A. Loss of elasticity in arteries as collagen cross-linkages stabilize with maturity
       1. Increased peripheral resistance
       2. Slight increase in blood pressure
          a. Age 20–24: females 116/70, males 122/76
          b. Age 60–64: females 142/85, males 140/85
    B. Heart
       1. Cardiac output
          a. Rate of decline
             (1) Decreases 40% between 3rd and 8th decades
             (2) A little less than 1%/year
          b. Components
             (1) Decreased responsiveness to β-adrenergic modulation: lower maximum heart rate
                (a) Young adult, 195 beats/min
                (b) Average 65 year old, max 170/min
             (2) Stroke volume decreases
       2. A-Vo_2 difference decreases
       3. Cardiac weight remains constant: left ventricular wall up to 25% thicker at age 80 than at age 30
       4. No significant increase in collagen or reticulin

IV.   Pulmonary—lungs
    A. Reduced expandability
    B. Abdominal musculature becomes responsible for much of the respiratory excursion of diaphragm
    C. Less efficient gas exchange at alveolar level
       1. Loss of alveolar tissue relative to air passages
       2. Stickier mucus
    D. Changes in lung volumes and capacities
       1. Residual volume: increases 100% between 3rd and 9th decades
       2. Vital capacity: decreases by 17 to 22 mL/yr
       3. Forced expired air volume in 1 second (FEV_1): falls progressively: males, 32 mL/yr; females, 25 mL/yr
       4. Maximal breathing capacity: reduced by 40% between 20 and 80 years

V.   Renal—kidney
    A. Weight: age 60, 250 g; age 70, 230 g; age 80, 190 g
    B. Renal function usually decreases
       1. Reduced renal blood flow: 2nd decade, 670 mL/min: 8th decade, 350 mL/min
       2. Decreased number of functioning nephrons: number of glomeruli per kidney
          a. Birth–age 40, 500,000–1 million
          b. By 7th decade ⅓ to ½ less

3. Glomerular filtration rate: decreases 46% from age 20 to age 90
4. Maximum specific gravity of urine: youth, 1.032; age 80, 1.024.
C. 33% of elderly persons have normal renal function

VI. Vascular compartment
   A. Markedly reduced total body water
      1. Volume of extracellular fluid remains about same
      2. Intracellular fluid volume reduced
      3. Red blood cell mass decreased, lower growth hormone, lower testosterone
   B. Decreased "tissue" percentage
   C. Increased body fat

VII. Central nervous system
   A. Difficult to define what constitutes average changes in brain function with increasing age
   B. Some loss of short-term memory and some decline in ability to learn new skills: given time, able to acquire new knowledge and skills
   C. Nerve conduction in peripheral nervous system slows
   D. Brain
      1. Weight decrease of 10 to 12% during normal long life
      2. Gray/white matter ratio
         a. Age 20, 1.28:1
         b. Age 50, 1.13:1
         c. Age 100, 1.33:1
      3. Decline in cerebral blood flow
         a. Age 17–18, 79.3 mL/min/100 g brain tissue
         b. Age 57–99, 47.7 mL/min/100 g brain tissue
      4. Neurofibrillary tangles and senile plaques
         a. Up to age 50, uncommon
         b. Age 90, found in 90%

VIII. Bone—skeleton
   A. Rate of bone loss/decade
      1. Men, 3%
      2. Women, 8%
   B. Average height loss: age 65–75, 1.5%; age 85–94, 3%
   C. Osteoporosis
      1. Generic term referring to a state of decreased mass per unit volume (density) of normally mineralized bone
      2. Risk factors
         a. Race and heredity

b. Nutrition
c. Endocrine factors
d. Weight-bearing activity
   (1) As much as 30 to 40% of initial total bone mass may be lost following 6 months of complete immobilization
   (2) Disuse osteopenia

IX. Musculoligamentous
   A. Cells making up voluntary muscle: decrease by 50% at age 80
   B. Weight of skeletal muscle
      1. Age 21–30, 45% of body weight
      2. Age 70+, 27% of body weight
   C. No change in muscle fibers or motor units
      1. Loss in muscle mass parallel to loss of strength
      2. Higher percentage of type I fibers
   D. Changes in collagen
      1. Increased stability to thermal denaturing and reduced solubility
      2. Collagen fibers possess all of its cross-linkages shortly after synthesis
         a. With maturation, reducible cross-linkages gradually stabilize
         b. Physical activity enhances rate of collagen turnover
   E. Tendons and ligaments
      1. Increased stiffness with decreased rate of collagen turnover
      2. Osseous resorption at sight of ligamentous attachment
      3. Decreased cellular component of ligaments and tendons
      4. Decreased collagen fiber bundle thickness
      5. Reduced capillarization
      6. Increased extensibility per unit length
      7. Decreased glycosaminoglycans
      8. Reduced water content
   Note: The final result is less elastic stiffer ligaments/tendons that have more cross-linked collagen.

Appendix 2: Exercise and Aging

I. Benefits reported in elderly who exercise
   A. Improved cardiovascular and respiratory function
   B. Reduced risk of coronary artery disease
   C. Decreased body fat
   D. Increased lean body and bone mass
   E. Better work capacity
   F. Greater flexibility

G. Reduced susceptibility to depression

H. Increased self-esteem and quality of life

II. Components of physical fitness
A. Cardiovascular endurance
B. Muscle strength
C. Muscle endurance
D. Flexibility
E. Body composition

III. Cardiovascular endurance and aging
A. Progressive decline in maximal oxygen uptake (Vo_2max) noted
1. Decreased maximum heart rate
2. Decreased stroke volume
3. Decreased cardiac contractility
4. Increased body mass
5. Decreased level of physical activity
B. Adaptability to training
1. Increase in Vo_2max noted in elderly individuals (Table 8–1)
2. Habitual aerobic exercise may retard the usual loss of Vo_2max characteristic of the aging process (Table 8–2)

IV. Muscle strength and endurance and aging
A. Predictable decline in strength
1. Decreased levels of trophic hormones
2. Decreased neuromuscular coordination
3. Decreased muscle mass
4. Low physical activity
B. Loss of strength with aging creates difficulties performing activities of daily living
C. Significant gains in strength noted in elderly following strength-training program (Table 8–4)

D. Association between improved strength and greater degree of independence/reduced rate of falls is not yet known: positive correlation with weakness and falling *has been* shown

Appendix 3: Average Biological Functional Changes Between the Ages of 30 and 70

I. Work capacity, −25 to 30%

II. Cardiac output , −30%
A. Maximum heart rate, –30 bpm
B. Blood pressure
1. Systolic, +10 to 40 mmHg
2. Diastolic, −5 to 10 mmHg

III. Respiration
A. Vital capacity, −40 to 50%
B. Residual volume, −30 to 50%

IV. Basal metabolic rate, −8 to 12%

V. Musculature
A. Muscle mass, −25 to 30%
B. Hand grip strength, −25 to 30%
C. Nerve conduction velocity, −10 to 15%

VI. Flexibility, −20 to 30%

VII. Bone
A. Women, −25 to 30%
B. Men, −15 to 20%

VIII. Renal function, −30 to 50%

REFERENCES

1. Shippen E, Fryer W, Evans S. *Testosterone Syndrome. The Critical Factor for Energy, Health, and Sexuality—Reversing the Male Menopause.* New York, NY: M. Evans and Co., Inc.; 1998

2. Barrett-Connor E, von Mhlen DG, Kritz-Silverstein D. Bioavailable testosterone and depressed mood on older men: the Rancho-Bernardo Study. J Clin Endocrinol Metab 1999;84:573–577

3. Lamberts SW, van den Beld AW, van der Lely A. The endocrinology of aging. Science 1997;278:419–424

4. Prasad AS, Mantzaros CS, Beck FW, Hess JW, Brewer CJ. Zinc status and serum testosterone levels of healthy males. Nutrition 1996;12:344–348

5. Fiatarone MA, O'Neill EF, Ryan ND, et al. Exercise training and nutritional supplementation for physical frailty in very elderly people. N Engl J Med 1994;330:1769–1775

6. Penninx BWJH, Messier SP, Rejeski WJ, et al. Physical exercise and the prevention of disability in activities of daily living in older persons with osteoarthritis. Arch Intern Med 2001;161:2309–2316

7. Singh MAF. Exercise comes of age: rationale and recommendations for a geriatric exercise prescription. J Gerontol A Biol Sci Med Sci 2002;57:M262–M282

8. Bortz WM 2nd. Disuse and aging. JAMA 1992;248:1203–1208

9. Aniansson A, Zetterberg C, Hedberg M, Henriksson KG. Impaired muscle function with aging. Clin Orthop 1984;191:193–201

10. Singh MAF. Combined exercise and dietary intervention to optimize body composition in aging. Ann NY Acad Sci 1998;854:378–393

11. Bagatell CJ, Bremner WJ. Androgens in men—uses and abuses. N Engl J Med 1996;334:707–714

12. Bardin CW. The anabolic action of testosterone. N Engl J Med 1996;335:52–53

13. Bhasin S, Storer TW, Berman N, et al. Testosterone replacement increases fat-free mass and muscle size in hypogonadal men. J Clin Endocrinol Metab 1997;82:407–413

14. Segal RJ, Reid RD, Courneya KS, et al. Resistance exercise in men receiving androgen deprivation therapy for prostate cancer. J Clin Oncol 2003;21:1653–1659

15. Bhasin S, Storer TW, Berman N, et al. The effects of supraphysiologic doses of testosterone on muscle size and strength in normal men. N Engl J Med 1996;335:1–7

16. Abate N, Haffner SM, Garg A, Peshock RM, Grundy SM. Sex steroid hormones, upper body obesity, and insulin resistance. J Clin Endocrinol Metab 2002;87(10):4522–4527

17. Tchernof A, Labrie F, Belanger A, et al. Relationships between endogenous steroid hormone,sex hormone-binding globulin and lipoprotein levels in men: contribution of visceral obesity, insulin levels and other metabolic variables. Atherosclerosis 1997;133(2):235–244

18. Morris KL, Namey TC, Zemel MB. Effects of dietary carbohydrate on the development of obesity in heterozygous Zucker rats. J Nutr Biochem 2003;14(1):32–39

19. Foster GD, Wyatt HR, Hill JO, et al. A randomized trial of a low-carbohydrate diet for obesity. N Engl J Med 2003; 348:2082–2090

20. Samaha FF, Iqbal N, Seshadri P, et al. A low-carbohydrate as compared with a low-fat diet in severe obesity. N Engl J Med 2003;348:2074–2081

21. Thompson WG, Namey TC. Cardiovascular complications of inactivity. Rheum Dis Clin North Am 1990;16(4):803–813

22. Alexandersen P, Haarbo J, Christiansen C. The relationship of natural androgens to coronary heart disease in males: a review. Atherosclerosis 1996;125(1):1–13

23. De Pergola G, Pannacciulli Niccone M, Tartagni M, Rizzon P, Giorgino R. Free testosterone plasma levels are negatively associated with the intima-media thickness of the common carotid artery in overweight and obese glucose-tolerant young adult men. Int J Obes Relat Metab Disord 2003;27(7): 803–807

24. English KM, Mandour O, Steeds RP, Diver MJ, Jones TH, Channer KS. Men with coronary artery disease have lower levels of androgens than men with normal coronary angiograms. Eur Heart J 2000;21(11):890–894

25. Rosano GM, Leonardo F, Pagnotta P, et al. Acute anti-ischemic effect of testosterone in men with coronary artery disease. Circulation 1999;99(13):1666–1670

26. Hamdi SA, Nassif OI, Ardawi MS. Effect of marginal or severe dietary zinc deficiency on testicular development and functions of the rat. Arch Androl 1997;38(3):243–253

27. Li Y. A study on the relationships among plasma testosterone, sex development and serum zinc in normal boys aged 10–15. Zhonghua Yu Fang Yi Xue Za Zh. 1992;3:148–150

28. Prasad AS. Zinc in growth and development and spectrum of human zinc deficiency. J Am Coll Nutr 1988;7(5):377–384

29. Prasad AS, Mantzoros CS, Beck FW, Hess JW, Brewer GJ. Zinc status and serum testosterone levels of healthy adults. Nutrition 1996;12(5):344–348

30. Om AS, Chung KW. Dietary zinc deficiency alters 5 alpha-reduction and aromatization of testosterone and androgen and estrogen receptors in rat liver. J Nutr 1996;126(4):842–848

31. Okubo T, Truong TK, Yu B, et al. Down-regulation of promoter 1.3 activity of the human aromatase gene in breast-zinc finger protein (SnaH). Cancer Res 2001;61(4):1338–1346

32. Morales AJ, Haubrich RH, Hwang JY, Asakura H. The effect of six months treatment with a 100 mg daily dose of dehydroepiandrosterone (DHEA) on circulating sex steroids, body composition and muscle strength in age-advanced men and women. Clin Endocrinol (Oxf) 1998;49(4):421–432

33. Ponholzer A, Plas E, Schatzl G, Jungwirth A, Madersbacher S. Association of DHEA-S and estradiol serum levels to symptoms of aging men. Aging Male 2002;5(4):233–238

34. Alexopoulou O, Jamart J, Maiter D, et al. Erectile dysfunction and lower androgenicity in type 1 diabetic patients. Diabetes Metab 2001;27:329–336

35. Clarke BL, Ebeling PR, Jones JD, et al. Predictors of bone mineral density in aging healthy men varies by skeletal site. Calcif Tissue Int 2002;70:137–145

36. Kroboth PD, Slalek FS, Pittenger AL, et al. DHEA and DHEA-S: a review. J Clin Pharmacol 1999;39:327–348

37. Anderson RA, Cheng N, Bryden NA, et al. Elevated intakes of supplemental chromium improve glucose and insulin variables in individuals with type 2 diabetes. Diabetes 1997;46: 1786–1791

38. Ravina A, Slezak L, Mirsky N, et al. Reversal of corticosteroid induced diabetes mellitus with supplemental chromium. Diabet Med 1999;16:164–167

39. Pasman WJ, Westerterp-Plantenga MS, Saris WH. The effectiveness of long-term supplementation of carbohydrate, chromium, fibre, and caffeine on weight maintenance. Int J Obes Relat Metab Disord 1997;21:1143–1151

40. Stuerenburg HJ, Jung R, Kunze K. Age effects on growth hormone and testosterone responses to endurance exercise in patients with neuromuscular diseases. Arch Gerontol Geriatr 1999;1:45–51

41. Svensson J, Fowelin J, Landin K, Bengtsson BA, Johansson JO. Effects of seven years of GH-replacement therapy on insulin sensitivity in GH-deficient adults. J Clin Endocrinol Metab 2002;87:2121–2129

42. Blair SN, Kohl HW, Paffenbarger RS, Clark DG, Cooper KH, Gibbons LW. Physical fitness and all-cause mortality: a prospective study of healthy men and women. JAMA 1989;262:2395–2401

43. Ekelund LG, Haskell WL, Johnson JL, Whaley FS, Criqui MH, Sheps DS. Physical fitness as a predictor of cardiovascular mortality in asymptomatic North American Men. N Engl J Med 1988;319:1379–1384

44. Paffenbarger RS, Hyde RT, Wing AL, Steinmetz CH. A natural history of athleticism and cardiovascular health. JAMA 1984;252:491–495

45. Paffenbarger RS, Hyde RT, Wing AL, Hsieh CC. Physical activity, all-cause mortality, and longevity of collge alumni. N Engl J Med 1986;314:605–613

46. Blair SN, Kohl HW, Paffenbarger RS, Clark DG, Cooper KH, Gibbons LW. Physical fitness and all-cause mortality: a prospective study of healthy men and women. JAMA 1989;262:2395–2401

47. Aniansson A, Zetterberg C, Hedberg M, Henriksson KG. Impaired muscle function with aging. Clin Orthop 1984;191: 193–201

48. Baker SP, Harvey AH. Fall injuries in the elderly. Clin Geriatr Med 1985;1:501–512

49. Heath GW, Hagberg JM, Ehsani AA, Holloszy JO. A physiological comparison of young and older endurance athletes. J Appl Physiol 1981;51:634–640

50. Pollock ML, Dawson GA, Miller HS, et al. Physiologic responses of men 49 to 65 years of age to endurance training. J Am Geriatr Soc 1976;24:97–104

51. DeVries H. Physicological effects of an exercise training regimen upon men aged 52 to 88. J Gerontol 1970;25:325–336

52. Souminen H, Heikkinen E, Parkatti T. Effect of eight weeks' physical training on muscle and connective tissue of the M. Vastus lateralis in 69-year-old men and women. J Gerontol 1977;32:33–37

53. Hagberg JM, Graves JE, Limacher M, et al. Cardiovascular responses of 70- to 79-yr-old men and women to exercise training. J Appl Physiol 1989;66:2589–2594

54. Seals DR, Hagberg JM, Hurley BF, Ehsani AA, Holloszy JO. Endurance training in older men and women, I: cardiovascualr responses to exercise. J Appl Physiol 1984;57:1024–1029

55. Pollock ML, Miller HS, Wilmore J. Physiological characteristics of champion American track athletes 40 to 75 years of age. J Gerontol 1974;29:645–649

56. Robinson S, Dill DB, Robinson RD, Tzankoff SP, Wagner JA. Physiological aging of champion runners. J Appl Physiol 1976;41:46–51

57. Kasch FW, Wallace JP. Physiological variables during 10 years of endurance exercise. Med Sci Sports 1976;8:5–8

58. Kasch FW, Boyer JL, Schmidt PK, et al. Ageing of the cardiovascular system during 33 years of aerobic exercise. Age Ageing (England) 1999;28:531–536.

59. Dehn MM, Bruce RA. Longitudinal variations in maximal oxygen intake with age and activity. J Appl Physiol 1972;3s3:805–807

60. Larsson L, Grimby G, Karlsson J. Muscle strength and speed of movement in relation to age and muscle morphology. J Appl Physiol 1979;46:451–456

61. Frontera WR, Meredith CN, O'Reilly KP, Knuttgen HG, Evans WJ. Strength conditioning in older men: skeletal muscle hypertrophy and improved function. J Appl Physiol 1988;64:1038–1044

62. Rutherford OM, Greig CA, Sargeant AJ, Jones DA. Strength training and power output: transference effects in the human quadriceps muscle. J Sports Sci 1986;4:101–107

63. Fiatarone MA, Marks EC, Ryan ND, Meredith CN, Lipsitz LA, Evans WJ. High-intensity strength training in nonagenarians. JAMA 1990;263:3029–3034

64. Namey TC. Exercise and arthritis. Adaptive bicycling. Rheum Dis Clin North Am 1990;16:871–886

9

Osteoporosis in Aging Men

MICHELLE IANNUZZI-SUCICH AND PAMELA TAXEL

Case History

An 80-year-old white man presented to the osteoporosis service 10 months after a fall with resultant L1 compression fracture. His past medical history was significant for two previous cerebrovascular accidents with secondary vascular dementia, seizure disorder, and balance disturbance. After his fracture, his primary care provider placed him on weekly alendronate therapy, which he was tolerating well. Additional significant history included Colles fracture at age 50, a height loss of approximately 3 inches since his mid-20s, poor erectile function, decreased energy level, and history of anticonvulsant therapy for 9 years.

Medications: Alendronate 70 mg weekly, Dilantin 200 mg daily, calcium 1500 mg daily, multivitamin tablet daily, atenolol 50 mg daily, Lescol 20 mg daily, quinine 325 mg daily

Social History: Married and living with wife; he has six children, all are alive and well; 100 pack per year smoking history, occasional beer on weekends

Family History: No known history of osteoporosis

Geriatric Review of Systems: Active; independent in activities of daily living; he ambulates with cane because of a fall 10 months ago, and is able to climb stairs and do yard work; he continues to drive.

Physical Exam: Height 64 inches; weight 116 lbs; body mass index (BMI) 20 kg/m^2

Cardiovascular System: Regular rate and rhythm; rate 70, BP 118/74

Neck: No thyromegaly

Back: Mild T-spine kyphosis; no pain on palpation of thoracic or lumbar spine

Genitalia: Testes 15 mL bilaterally, soft

Laboratory:

Hemoglobin 11.6 g/dL, hematocrit 34.3%, mean corpuscular volume 107

Erythrocyte sedimentation rate 33 (0–15 mm/hr), serum protein electrophoresis within normal limits, thyroid-stimulating hormone (TSH) 0.8 µU/mL, alkaline phosphatase 77

Creatinine 1.4 mg/dL, ionized calcium 1.33 mmol/L, (1.21–1.33 mmol/L), total calcium 9.8 mg/dL (8.3–10.2), phosphorus 3.1 mg/dL (3.0–4.8 mg/dL)

24-Hour Urine: Creatinine 0.86 g/d, calcium 0.072 g/d, phosphorus 0.143 g/d, creatinine clearance 43 cc/min

Parathyroid Hormone: Intact, 81 (11–54 pg/mL)

Total Testosterone (Total T): 220 ng/dL (200–1000 ng/dL)

25-Hydroxy Vitamin D: 32 ng/mL (15–65 ng/mL)

Bone Mineral Density: See Table **9–1**

Osteoporosis is an often unrecognized but important concern in the care of elderly men. The impact of osteoporosis on the health of the male population must be considered in the context of falls, which significantly increase the risk for fracture. Although osteoporotic fractures are more common in women, 25 to 30% of all hip fractures in the United States and Northern Europe occur in men. In 1995, total medical costs associated with all osteoporotic fractures were estimated at $14 billion, and fractures in men accounted for $2.7 billion, or ~20% of that sum.[1] As a result of the increasing longevity of older persons, it is estimated that the number of hip fractures worldwide

TABLE 9–1. Bone Mineral Density

Region	BMD, g/cm²	T-score	Young Adult, %	Adult Mean, %
L2–4	0.941	−2.49	76	89
L femur	0.644	−3.55	60	81

will increase from the present number, 1.7 million, to 6.3 million by 2050.[2] Thus, prevention as well as diagnosis and treatment of osteoporosis are essential components in the comprehensive care of aging men.

Pathogenesis

Bone remodeling is a process that continues throughout life, long after maximum height and bone mineral density (BMD) are achieved. Bone remodeling represents a "coupling" of resorption of old bone with the formation or laying down of new bone. On the endosteal bone surfaces osteoclasts resorb bone, leaving a cavity. Osteoblasts then lay down osteoid in the cavity, and this new bone is subsequently mineralized. If the amount of bone replaced by the osteoblasts equals the amount originally removed by the osteoclasts, then there is no net bone loss. However, if more bone is resorbed than is formed, as is most common with aging, then irreversible bone loss occurs. Bone loss begins as early as the third decade in both men and women and is thought to result primarily from an early decrease in bone formation rather than from an increase in resorption. The factors responsible for this

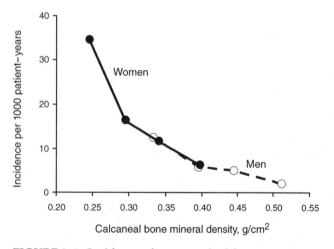

FIGURE 9–1. Incidence of new vertebral fractures per 1000 population per year as related to calcaneal bone mineral density. (Adapted from Melton LJ, Orwoll ES, Wasnich RD. Does bone density predict fractures comparably in men and women? Osteoporosis Int 2001;12:707–709, with permission.)

decrease in bone formation are not well understood. Bone loss later in life occurs primarily because of greater resorption relative to formation.[3]

Determinants of Bone Mineral Density and Fracture Risk in Men

The relatively lower rate of osteoporosis in men compared with women is likely attributable to multiple factors, including larger bone size and thicker cortices in men, shorter male life expectancy, and lack of precipitous decline in sex hormone level analogous to the female menopause.[4] At any given age, women are likely to have a lower BMD than age-matched men. For example, the mean femoral neck BMD for women age 60 to 69 is comparable to that of men over age 80.[5] Figure 9–1 shows the differential fracture rates between men and women.

Several predictors of overall bone density in men have been identified. Orwoll et al[6] studied 355 community-dwelling men (mean age 71.5 ± 7.4 years) and identified weight as one predictor of higher bone mass in multivariate analysis. Further, age, previous fracture, rheumatoid arthritis, gastrectomy, and hypertension were associated with lower bone mass.[6] Bone mass and fracture risk have been strongly correlated in women, and studies suggest a similar relationship in men.[4,7]

Diagnosis of Osteoporosis in Men

In 1993 the Osteoporosis Consensus Development Conference in Hong Kong defined osteoporosis as "a metabolic bone disease characterized by low bone mass and microarchitectural deterioration of bone tissue leading to enhanced bone fragility and a consequent increase in fracture risk." By correlating BMD with fracture risk, the World Health Organization (WHO) determined BMD values indicative of osteoporosis specifically in reference to postmenopausal Caucasian women. As BMD values have not yet been specifically validated by correlation with fracture risk in men, the WHO criteria for diagnosing osteoporosis in women are applied to men as well.

Application of the WHO criteria for women to the male population introduces inaccuracies. Faulkner and Orwoll[8] used multiple BMD measurement techniques that resulted in varied estimates of osteoporosis prevalence that may not accurately identify those men at risk for fracture. Further, these authors found that the established female diagnostic criteria of a T-score of ≤ -2.5 underestimates osteoporosis prevalence in men when comparing prevalence with fracture risk estimates.[8] Although the WHO criteria are imperfect, there are no other currently accepted criteria in men and, consequently, these criteria are routinely applied to men. Further study is needed to establish diagnostic criteria for osteoporosis specific to the male population, and a longitudinal study is now ongoing.

Screening for Osteoporosis in the Male Patient

Currently, there are no national guidelines for screening of osteoporosis in men. Burgess and Nanes[9] recommend screening with hip/spine dual-energy x-ray absorptiometry (DEXA) for asymptomatic men beginning at age 70 unless risk factors are present. They recommend screening men at age 55 if risk factors are identified, including previous fracture, current or past smoking history, family history of osteoporotic fracture, body mass index less than $18 \, kg/m^2$, or prior glucocorticoid use. Men of any age on glucocorticoids should be screened and receive calcium plus vitamin D supplementation.

In our own geriatrics clinic, a careful history is taken in all men to assess osteoporosis risk factors. Ideally, all men age 70 and over would receive a DEXA scan to determine their risk for future fracture. However, the issue of payment is a major obstacle to general screening for the population of older men. If osteoporotic risk factors exist or a history of a nontraumatic fracture is present, then Medicare will cover the cost of a DEXA in those men. If DEXA is performed, results guide the next step in management. If the initial DEXA is normal $(T > -1.0)$, then patient education is provided addressing adequate calcium plus vitamin D intake and exercise recommendations. DEXA may be repeated in 3 to 5 years. If the DEXA reveals osteopenia $(T = -1.0$ to $-2.5)$, calcium plus vitamin D intake and exercise recommendations are discussed, and consideration is given to pharmacological management, particularly if risk factors are present. When osteoporosis is present on DEXA $(T < -2.5)$, we evaluate for secondary causes, and guide treatment from these results. If an underlying cause is identified, then that must be treated first, and subsequently pharmacological therapy for osteoporosis should be considered. Patients with history of a fragility fracture resulting from minimal or no trauma are defined as osteoporotic regardless of BMD, and evaluation for secondary causes and pharmacological management is advised.

Differentiating Osteoporosis and Osteomalacia

The distinction between osteoporosis and osteomalacia is important, although both disorders may coexist. Osteomalacia is a histomorphometric diagnosis that is defined as a defect in bone mineralization and may appear as bone loss on bone density measurement.[10] Bone resorption may increase, and pseudofractures can develop. Vitamin D deficiency in the elderly may be attributable to poor nutrition, lack of exposure to sunlight, malabsorptive disorders, and decreased ability of the skin to convert 7-dehydrocholesterol to cholecalciferol (vitamin D3) by ultraviolet radiation.[10,11] Laboratory evaluation in osteomalacia reveals low 25-hydroxy vitamin D, low or normal serum calcium, low serum phosphorus, elevated alkaline phosphatase, and increased parathyroid hormone (PTH) levels.[10] Bone biopsy confirms the diagnosis, but is generally not required. If weight loss and/or symptoms of malabsorption are present, evaluation for celiac sprue or other malabsorption syndromes is indicated. Mild vitamin D deficiency can be associated with osteoporosis, and vitamin D is prescribed with the goal of increasing calcium absorption from the gut, as well as stimulating bone formation and mineralization.[10] Treatment consists of pharmacological doses of vitamin D administered orally, beginning with 50,000 IU weekly for 4 weeks followed by daily supplementation with 400 to 800 IU. Alternatively, intramuscular injections may be used, specifically in cases where absorption by mouth is poor or inadequate; however, the intramuscular formulation is no longer available.

Secondary Causes of Osteoporosis

A review by Taxel and Kenny[12] reported an incidence of secondary causes in osteoporotic men ranging from 50 to 80%, with steroid therapy and hypogonadism accounting for 13 to 17% and 14 to 16%, respectively. Secondary causes are numerous and are listed in Table **9–2**. Some of the more common etiologies are discussed in the following sections. Endocrine disorders are common in the aging population and should routinely be considered in the evaluation of secondary osteoporosis.

Primary Hyperparathyroidism

Primary hyperparathyroidism is defined biochemically as elevated serum calcium (ionized or total) with an inappropriately elevated PTH level. A common mistake is failure to diagnose primary hyperparathyroidism when an elevated calcium level is measured along with a PTH in the normal range. The "normal" PTH lvelel is inappropriate for the level of calcium (should actually be suppressed or at the low end of the normal range), and primary hyperparathyroidism is likely present.

Hyperparathyroidism dramatically impacts bone mineral density, resulting in selective cortical bone loss. The relationship between bone loss and PTH is incompletely understood: chronic, continuous excess PTH results in cortical bone loss; however, intermittent, exogenous PTH administration increases trabecular bone mass with minimal impact on cortical bone.[13] Two retrospective cohort studies suggest that overall risk of fracture is increased in patients with primary hyperparathyroidism.[14]

Management options in mild hyperparathyroidism include observation for asymptomatic patients, surgery, and medical management. Observation of patients with asymptomatic disease is a reasonable approach, and should be accompanied by biannual to annual surveillance for symptoms and laboratory criteria for surgery. The National Institute of Health Consensus Conference held in April 2002 issued guidelines for surgical intervention in primary hyperparathyroidism.[15] These recommendations advise surgery if serum calcium is greater than 1 mg/dL above the upper limits of normal; 24-hour urinary calcium is greater than 400 mg; creatinine clearance is reduced by more than 30% compared with age-matched subjects; BMD at the lumbar spine, hip, or distal radius is more than 2.5 standard deviations below peak bone mass; the patient is under 50 years of age; or when medical surveillance is either not desirable or not possible for an individual patient.[15] Studies reveal a 10 to 15% increase in bone mass following surgical management of primary hyperparathyroidism.[13] In studies of postmenopausal women with primary hyperparathyroidism, medical management via hormone replacement therapy (HRT) with estrogen resulted in decreased bone loss.[13] However, due to adverse events associated with HRT in the Women's Health Initiative trial, this is no longer a recommended treatment. Bisphosphonates offer another option for medical management of hyperparathyroidism. In one trial of 27 women and five men, subjects treated with alendronate for 2 years showed significant gains in bone density at the lumbar spine relative to controls.[16]

Finally, the recently cloned extracellular calcium-sensing receptor, which plays a central role in calcium homeostasis, represents an important target for therapeutic agents aimed at treating disorders of calcium metabolism. Compounds known as calcimimetics act as agonists of calcium at this receptor and can lower PTH and calcium concentrations. These agents hold promise as potential medical treatments for diseases of parathyroid hyperfunction, such as primary and secondary hyperparathyroidism.[17]

Thyroid Dysfunction

Long-standing, untreated hyperthyroidism leads to decreased bone density and confers an increased risk of osteoporotic fracture.[13] In patients with hypothyroidism, BMD generally remains normal and long-term replacement of thyroxine causes only minimal loss of BMD in the spine.[13]

Gonadal and Pituitary Dysfunction

It is well established that severe testosterone deficiency (hypogonadism) leads to significant bone loss. However, the gradual decline in testosterone (T) with age in men has not been definitively established as a cause of osteoporosis.[18] Studies of men over 65 years of age receiving testosterone replacement have demonstrated inconsistent results. In one randomized controlled study of testosterone replacement in 44 hypogonadal men age 65 and older, BMD was maintained in testosterone-treated subjects over 12 months (Fig. 9–2).[19]

TABLE 9–2. Secondary Causes of Osteoporosis

Endocrine Disorders
Primary/secondary hypogonadism
Endogenous/exogenous glucocorticoid excess
Hyperparathyroidism
Hyperthyroidism
Hyperprolactinemia
Diabetes mellitus type I

Malignancy and Disorders of the Bone Marrow
Systemic mastocytosis
Multiple myeloma
Lymphoproliferative diseases
Myeloproliferative diseases

Gastrointestinal Disorders
Primary biliary cirrhosis
Inflammatory bowel diseases
Celiac sprue
Postgastrectomy

Connective Tissue Disorders
Rheumatoid arthritis
Osteogenous imperfecta
Ankylosing spondylitis

Other Disorders
Chronic obstructive pulmonary disease
Homocystinuria
Hemochromatosis
Anorexia nervosa

Medications
Glucocorticoids
Heparin
Cyclosporin A
Warfarin
Thyroid hormone/in excess
Antiepileptics
Chemotherapy causing chemical castration
LHRH agonists
Lithium
Methotrexate

Adapted from Harper KD, Weber TJ. Secondary osteoporosis: diagnostic consideration. Endocrinol Metab Clin North Am 1998;27:325–348, with permission.

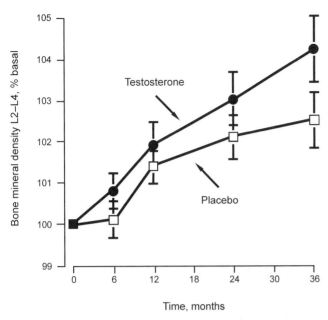

FIGURE 9–2 The effects of testosterone versus placebo in restoring bone mineral density over time.

Other studies have shown testosterone replacement to improve BMD in selected men with total testosterone levels below 200 ng/dL.[20] A randomized, double-blinded, placebo-controlled study of 96 men over 65 years of age showed testosterone treatment to have minimal impact on BMD in men with pretreatment serum testosterone concentrations ≥400 ng/mL, but BMD did improve in those subjects with pretreatment serum testosterone concentrations ≤200 ng/dL.[21] These results suggest that a serum testosterone level of 200 ng/dL may be a critical value below which osteoporosis risk increases, and consequently testosterone therapy may be beneficial under such circumstances.[22] Studies to evaluate the impact of testosterone replacement on fractures risk have not been reported.[20]

Recent "experiments of nature" have demonstrated the importance of 17-β-estradiol (E_2) in the acquisition of peak bone mass in men. Individuals with genetic defects in the estrogen receptor or in aromatase,[23–25] the enzyme that converts androgens to estrogens,[26] have demonstrated osteoporosis and increased bone turnover. Because all estrogens in men are derived from androgens through the aromatase pathway, defects in this enzyme lead to extremely low E_2 levels. A patient with an aromatase defect gained significant bone mass after the institution of E_2 therapy.[27] Several observational studies have demonstrated that serum E_2 levels are better predictors of BMD in older men than serum T levels.[28] In older men in whom both gonadotrophin secretion and aromatase conversion were suppressed, Falahati et al[29] demonstrated in a short-term study that

E_2 is the principal sex steroid regulating bone resorption. Thus, current evidence suggests an important role for estrogen in the pathogenesis of osteoporosis in men, and future research will help further clarify this role.

Growth hormone is now understood to play a role in maintenance of skeletal mass. Two studies evaluated growth hormone supplementation in growth-hormone-deficient adults and revealed improved BMD at the total body, hip, and spine. Research to determine if growth hormone administered to elderly patients can improve BMD or reduce fractures has not yet been completed.[13]

Gastrointestinal Disorders

Several gastrointestinal disorders have been associated with low bone mass. Osteopenia and fractures are more common in patients with primary biliary cirrhosis.[10] Patients with a history of gastrectomy have lower BMD and a higher rate of both femoral and vertebral fractures.[18] Calcium malabsorption leads to secondary hyperparathyroidism and, subsequently, to osteoporosis. Patients with mild malabsorption may have markedly low BMD. In one study, antigliadin antibodies were detected in 12% of patients with idiopathic osteoporosis and in 3% of healthy age-matched controls.[18] The incidence of biopsy confirmed sprue was 10 times higher in osteoporotic patients relative to controls.[18] Crohn's disease with involvement of the proximal small bowel leads to malabsorption of calcium and subsequent development of osteoporosis.

Malignancies

In malignant monoclonal gammopathy, plasma cells produce osteoclast-stimulating cytokines resulting in the uncoupling of resorption and formation with the ultimate development of osteoporosis.[18]

Rheumatic Diseases

Periarticular and generalized osteoporosis are associated with rheumatoid arthritis (RA), independent of the effect of medications used to manage this entity. Spine and hip BMD are decreased and have been correlated with cumulative corticosteroid doses. Patients with RA who have not received corticosteroid therapy also exhibit decreased bone mineral content at the distal radius that correlates with the duration of disease.[18]

Medications

Exogenous glucocorticoid excess is the most common cause of secondary osteoporosis and is a function of dose and duration of therapy. Glucocorticoid-induced osteoporosis is due to a combination of increased bone resorption and decreased bone formation. Secondary hyperparathyroidism and hypogonadism may play a role in resorption. Decreased bone formation is probably a direct effect of inhibition of osteoblast function.[18] The doses of glucocorticoids that produce osteoporosis are not well defined. More than 7.5 mg/day of prednisone or its equivalent is likely to cause bone loss, but lower doses in patients who have additional risk factors may lead to bone loss. The lowest dose and the lowest frequency of administration that are effective for the patient should be used.

The anticonvulsants diphenylhydantoin and phenobarbital can be important factors in bone loss due to their role in vitamin D metabolism. Although exact mechanisms are not thoroughly elucidated, these agents augment metabolism of vitamin D. Pharmacological doses of vitamin D supplementation may be required to maintain acceptable levels.

Long-term heparin therapy leads to osteoporosis, though some studies suggest low molecular weight heparin may cause less bone loss than unfractionated heparin.[18] The immunosuppressant cyclosporin A and glucocorticoids are administered after organ transplantation, and rapid decreases in bone density may occur, though the degree of bone loss attributable to cyclosporin A alone is difficult to determine.[18] Chronic methotrexate therapy leads to bone pain, swelling, and fractures.[18]

Luteinizing hormone–releasing hormone (LHRH) agonist therapy for prostate cancer was shown in two prospective studies to cause bone loss and increases risk of osteoporotic fractures.[30,31] This is primarily due to extreme lowering of both testosterone and estrogen by these agents.

Smoking and Alcohol

Epidemiological twin studies suggest a correlation between smoking and peak adult bone mass.[18] In a cohort study of men, smoking decreased bone mass; however, no research has determined the direct impact of smoking on fractures.[18] Conversely, chronic alcoholism has been shown to increase fracture rate.[18] Mild to moderate intake of alcohol has not been associated with bone loss.

Laboratory Evaluation

A directed algorithm for laboratory workup of secondary causes of osteoporosis is presented in Fig. 9–3.[10] Due to the high incidence of secondary causes of osteoporosis in men, all men diagnosed with osteoporosis should be evaluated for secondary causes, and this scheme provides a systematic approach to this evaluation. Initial laboratory studies should include measurement of calcium and creatinine in blood and urine, thyroid function screening with TSH, protein electrophoresis, total testosterone, phosphorus, and a complete blood count. If initial screening reveals abnormalities, further measurements of calciotropic hormones (PTH and 25-hydroxy vitamin D) and serum and urine phosphorus should be performed. If total T is < 200 ng/dL, then follicle-stimulating hormone (FSH) and luteinizing hormone (LH) are obtained to rule out pituitary or hypothalamic disease. If these are normal or low [inappropriate to the level of testosterone (T)], then magnetic resonance imaging (MRI) is indicated and consultation with an endocrinologist may be warranted. If T is between 200 and 400 ng/dL, then we repeat total T, free T, and sex hormone–binding globulin (SHBG) to determine gonadal status. Due to the normal diurnal variation in T levels, more than one laboratory assessment of T level is often required. Careful history and physical examination should direct additional testing such as tests of gastrointestinal and renal function to diagnosis specific secondary causes. The clinician should maintain a high clinical suspicion for secondary causes as these are frequently missed.[10]

Nonpharmacological Treatment of Osteoporosis in the Men

Patient education is an important component of osteoporosis management. Education should include information about nonpharmacological measures. Regular

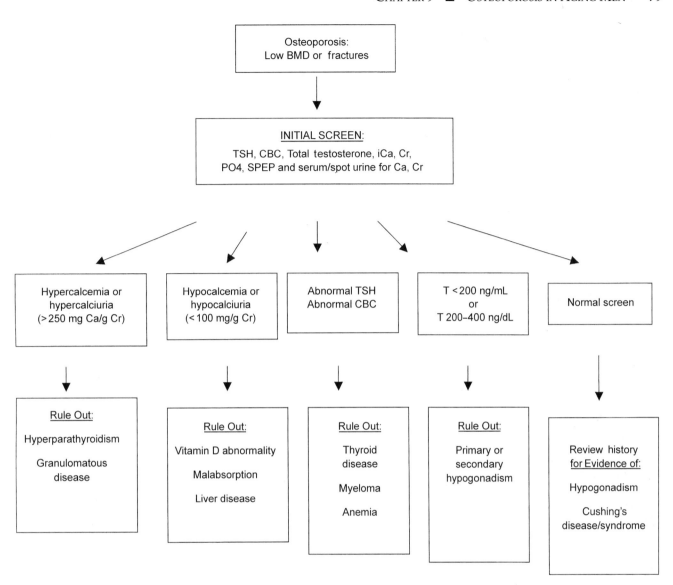

FIGURE 9–3. Algorithm for laboratory work-up for secondary causes of osteoporosis. Ca, calcium; CBC, complete blood count; Cr, creatinine; iCa, ionized calcium; PO4, phosphorus; SPEP, serum protein electrophoresis; TSH, thyroid-stimulating hormone. (Adapted from Taxel P, Kenny A. Differential diagnosis and secondary causes of osteoporosis. Clin Cornerstone 2000;2:11–21, with permission.)

weight-bearing exercise as well as smoking cessation and falls prevention should be discussed.

Falls

Falls in the geriatric population have serious public health implications, both for the individual and for society. In one study of 1103 community-dwelling men and women aged 72 and older, 49% of the cohort experienced at least one fall during a 1-year follow-up period.[32] Serious injury, including fracture, joint dislocation, other serious joint injury, serious laceration, and intracranial injury, occurred in 23% of those falls.[32] Additional consequences of falls include minor

injuries, inability to get up after a fall, and fear of future falls with resultant decreased activity.[33] From an economic perspective, in 1999 injuries among Medicare fee-for-service beneficiaries cost the system over $8 billion.[34] Fractures comprised 67% of injuries and cost over $5 billion.[34] Multiple studies have identified risk factors for falls in the elderly.[35] Modifiable risk factors for falls include mobility and transfer impairment, balance disturbance, multiple medications, sensory and perceptive deficits, postural hypotension and dizziness, foot and footwear problems, and environmental hazards. Identification of these risk factors should prompt subsequent intervention to reduce the probability of falls. According to the Cochrane Data-

FIGURE 9–4. Hip protectors (B and C) inserted beneath underwear (A) may prevent hip fractures in patients at high risk.

base of Systematic Reviews, hip protectors reduce the risk of hip fracture when used in a selected, high-risk population.[36] Cost effectiveness was not determined.[36] Patient compliance was an issue secondary to discomfort and practicality.[36] Nonetheless, hip protectors are a low-risk intervention and should be discussed with patients who are at high risk for osteoporotic hip fracture (Fig. 9–4). Specific environmental hazards contributing to the risk of falls include throw rugs in the home, inadequate lighting, and extension cords traversing floors in walking areas. Hence, patient education is an important adjunct to pharmacological management of osteoporosis.

Pharmacological Treatment of Osteoporosis in the Men

Pharmacological management should include calcium supplementation of 1.2 g/day and vitamin D supplementation of 400 to 800 IU/day.[9] Calcium can be obtained from food sources and in combination with calcium supplements that are available over the counter. When calcium supplements are used, we warn the older patient of side effects such as constipation, gas, or bloating. These effects may be remedied by switching brands or increasing fiber in the diet. Calcium carbonate requires stomach acid for absorption; therefore, it is best taken with meals. It should be avoided in patients with achlorhydria. Calcium citrate can be taken any time during the day or evening and does not require an acidic stomach pH. There are minimal data thus far to document that one calcium salt is superior to another.

Vitamin D is found in several food sources including milk, in combination with calcium in supplements or in multivitamins. We recommend 400 to 800 IU per day, and we check vitamin D levels in most of our older patients.

Bisphosphonates

Alendronate is Food and Drug Administration (FDA) approved for use specifically in men with osteoporosis. A double-blind placebo-controlled trial of 241 men (mean age 63) found alendronate significantly increases BMD for the spine, hip, and total body, as well as prevents vertebral fractures.[37] The effect was significant in men with and without hypogonadism. Alendronate and risedronate are FDA approved for both prevention and treatment of glucocorticoid-induced osteoporosis. Alendronate and risedronate can be dosed at 70 mg weekly and 35 mg weekly, respectively. The most commonly reported side effects of bisphosphonates are gastrointestinal and include esophagitis, abdominal pain, dyspepsia, nausea, vomiting, and diarrhea. Muscle pain has also been reported. Bisphosphonates should be avoided in patients with a history of significant gastroesophageal reflux, gastritis, or peptic ulcer disease. Patients should be instructed to take bisphosphonates upon rising in the morning with an 8-ounce glass of water; they should remain upright (sitting or standing) for a half-hour after taking these medications and should refrain from eating or drinking anything, as well as taking other medications, during that time.

Teriparatide (Human Parathyroid Hormone 1–34)

Teriparatide (human PTH) was recently FDA approved for treatment of osteoporosis in men. PTH is indicated to increase BMD in men with primary or hypogonadal osteoporosis who have high fracture risk, including those with history of osteoporotic fracture, those with multiple risk factors for fracture, and those who are failed or did not tolerate alternative therapies. In a randomized, placebo-controlled trial of 437 men with spine or hip BMD greater than 2 standard deviations below the young adult male mean, PTH was evaluated at a dose of 20 and 40 μg/day.[38] In this study, PTH significantly increased the bone mineral density of men for the spine, femoral neck, and whole body; radial bone mineral density did not change. Response to PTH was similar regardless of gonadal status, age, baseline BMD, BMI, smoking, or alcohol consumption.[38] PTH has been shown to reduce fracture rates in women, and recent data in men demonstrate a significant decrease in vertebral fracture.[39] Reported side effects include local injection site reactions, headaches, nausea, arthralgias, and hypercalcemia. Studies using PTH were terminated early due to findings of osteosarcoma in rats. Long-term safety and efficacy data on PTH are limited, and therefore the recommended duration of use is 2 years or less with subsequent antiresorptive therapy for maintenance of BMD.

Case Summary and Conclusion

The patient cited previously presented with an L1 vertebral compression fracture secondary to a fall from standing height, and therefore meets the criteria for severe osteoporosis. He takes adequate calcium and vitamin D in his diet (normal serum vitamin D level) and actively exercises, which is the base of any osteoporosis program. Because the total testosterone falls between 200 and 400 ng/dL, our next step would be to perform repeat laboratory evaluation to include a total and free T (or bioavailable) and SHBG. If these are low (total and free T), then the follicle-stimulating hormone (FSH) and LH are checked to rule out primary or secondary hypogonadism. If total and free T are low with normal SHBG, but elevated FSH and LH, the diagnosis is consistent with primary hypogonadism, a known risk factor for secondary osteoporosis. If T is low (either total or free) with a low or normal FSH and LH (inappropriate), then a further endocrine evaluation including MRI may be warranted. T replacement therapy would be important here not only for bone health and erectile function (which may or may not be improved on T replacement) but also for muscle mass, energy, and sense of well-being. Because there is no present evidence that T replacement decreases fractures, we would also recommend therapy with an oral bisphosphonate, if the patient was able to tolerate it. Long-standing anticonvulsant therapy with Dilantin may also be a contributor to this patient's bone loss, and a change of seizure medication that doesn't interfere with vitamin D metabolism could be considered. The patient also has a high-normal ionized calcium level with a normal total calcium and an elevated PTH, which may represent primary hyperparathyroidism. I would repeat these measures to verify and also include an albumin to adjust total calcium level. He has a low BMD, but otherwise does not meet the criteria for surgical intervention. We would likely opt for treatment with antiresorptive therapy to improve bone mass in patients with primary hyperparathyroidism. Finally, a repeat bone density test should be performed at 12 to 18 months to determine if therapy is stabilizing bone loss or increasing bone mass.

References

1. Ringe JD, Orwoll E, Daifotis A, Lombardi A. Treatment of male osteoporosis: recent advances with alendronate. Osteoporos Int 2002;13:195–199

2. Seeman E. Unresolved issues in osteoporosis in men. Rev Endocr Metab Disord 2001;2:45–64

3. Seeman E. Pathogenesis of bone fragility in women and men. Lancet 2002;359:1841–1850

4. Kenny A, Taxel P. Osteoporosis in older men. Clin Cornerstone 2000;2:45–51

5. Melton LJ, Orwoll ES, Wasnich RD. Does bone density predict fractures comparably in men and women? Osteoporos Int 2001;12:707–709

6. Orwoll ES, Bevan L, Phipps KR. Determinants of bone mineral density in older men. Osteoporos Int 2000;11:815–821

7. Melton LJ, Atkinson EJ, O'Connor MK, O'Fallon WM, Riggs BL. Bone density and fracture risk in men. J Bone Miner Res 1998;13:1915–1923

8. Faulkner K, Orwoll E. Implications in the use of T-scores for the diagnosis of osteoporosis in men. J Clin Densitom 2002;5:87–93

9. Burgess E, Nanes MS. Osteoporosis in men: pathophysiology, evaluation, and therapy. Curr Opin Rheumatol 2002;14:421–428

10. Taxel P, Kenny A. Differential diagnosis and secondary causes of osteoporosis. Clin Cornerstone 2000;2:11–21

11. Cobbs EL, Duthie EH, Murphy JB, eds. Geriatrics Review Syllabus: A Core Curriculum in Geriatric Medicine. 5th ed. Madlen, MA: Blakwell Publishing for the American Geriatrics Society; 2002:182

12. Taxel P, Kenny A. Differential diagnosis and secondary causes of osteoporosis. Clin Cornerstone 2000;2:11–21

13. Rosen CJ. Endocrine disorders and osteoporosis. Curr Opin Rheumatol 1997;9:355–361

14. Khosla S, Melton LJ. Fracture risk in primary hyperparathyroidism. J Bone Miner Res 2002;17:N103–N107

15. Bilezikian JP, Potts JT, Fuleihan GE, et al. Summary statement from a workshop on asymptomatic primary hyperparathyroidism: a perspective for the 21st century. J Clin Endocrinol Metab 2002;87:5353–5361

16. Parker CR, Blackwell PJ, Fairbairn KJ, Hosking DJ. Alendronate in the treatment of primary hyperparathyroid-related osteoporosis: a 2-year study. J Clin Endocrinol Metab 2002;87:4482–4489

17. Cohen A, Silverberg SJ. Calcimimetics: therapeutic potential in hyperparathyroidism. Curr Opin Pharmacol 2002;2:734–739.

18. Harper KD, Weber TJ. Secondary osteoporosis: diagnostic consideration. Endocrinol Metab Clin North Am 1998;27:325–348.

19. Kenny AM, Prestwood KM, Gruman CA, Marcello KM, Raisz LG. Effects of transdermal testosterone on bone and muscle in older men with low bioavailable testosterone levels. J Gerontol A Biol Sci Med Sci 2001;56:M266–M272

20. Iannuzzi-Sucich M, Kenny AM. Osteoporosis in older men. In: Martini L, ed. Encyclopedia of Endocrine Diseases, Vol. 3. San Diego: Academic Press; 2004:432–435

21. Snyder PJ, Peachey H, Hannoush P, et al. Effect of testosterone treatment on bone mineral density in men over 65 years of age. J Clin Endocrinol Metab 1999;84:1966–1972

22. Yialamas MA, Hayes FJ. Androgens and the aging male. Endocrinol Rounds 2003;2.

23. Smith EP, Boyd J, Frank GR, et al. Estrogen resistance caused by a mutation in the estrogen-receptor gene in a man. N Engl J Med 1994;331:1056–1061

24. Morishma A, Grumbach MM, Simpson ER, Fisher C, Qin K. Aromatase deficiency in male and female siblings caused by a novel mutation and the physiological role of estrogens. J Clin Endocrinol Metab 1995;80:3689–3698

25. Carani C, Qin K, Simoni M, et al. Effect of testosterone and estradiol in a man with aromatase deficiency. N Engl J Med 1997;337:91–95

26. Simpson E, Rubin G, Clyne C, et al. Local estrogen biosynthesis in males and females. Endocr Relat Cancer 1999;6: 131–137

27. Bilezekian J, Morishima A, Bell J, Grumbach MM. Estrogen markedly increases bone mass in an estrogen deficient young man with aromatase deficiency N Engl J Med 1998;339:599–603

28. Riggs LB, Khosla S, Melton J. A unitary model for involutional osteoporosis: estrogen deficiency causes both type I and type II osteoporosis in postmenopausal women and contributes to bone loss in aging men. J Bone Miner Res 1998;13:763–773

29. Falahati-Nini A, Riggs BL, Atkinson EJ, O'Fallon WM, Khosla S. Relative contributions of testosterone and estrogen in regulating bone resorption and formation in normal elderly men. J Clin Invest 2000;106:1553–1560

30. Daniell HW, Dunn SR, Ferguson DW, et al. Progressive osteoporosis during androgen deprivation therapy for prostate cancer. J Urol 2000;163:181–186

31. Diamond T, Campbell J, Bryant C, et al. The effect of combined androgen blockade on bone turnover and bone mineral densities in men treated for prostate carcinoma. Cancer 1998;83:1561–1566

32. Tinetti ME, Doucette J, Claus E, Marottoli R. Risk factors for serious injury during falls by older persons in the community. J Am Geriatr Soc 1995;43:1214–1221

33. King MB, Tinetti ME. Falls in community-dwelling older persons. J Am Geriatr Soc 1995;43:1146–1154

34. Bishop CE, Gilden D, Blom J, et al. Medicare spending for injured elders: are there opportunities for savings? Health Aff (Millwood) 2002;21:215–223

35. Tinetti ME, Speechley M, Ginter SF. Risk factors for falls among elderly persons living in the community. N Engl J Med 1988;319):1701–1707

36. Parker MJ, Gillespie LD, Gillespie WJ. Hip protectors for preventing hip fractures in the elderly. The Cochrane Database of Systematic Reviews 2003;3:CD0001255

37. Orwoll E, Ettinger M, Weiss S, et al. Alentronate for the treatment of osteoporosis in men. N Engl J Med 2000;343:604–610

38. Orwoll ES, Scheele WH, Paul S, et al. The Effect of triparatide [human parathyroid hormone (1–34)] therapy on bone density in men with osteoporosis. J Bone Miner Res 2003;18:9–17

39. Kaufman JM, Orwoll E, Goemaere S, et al. Teriparatide effects on vertebral fractures and bone mineral density in men with osteoporosis: treatment and discontinuation of therapy. Osteoporosis Int 2004. Online publication

10

Effect of Androgens on Muscle, Bone, and Hair in Men

Melinda Sheffield-Moore, Douglas Paddon-Jones, Melanie G. Cree, Darren W. Lackan, and Randall J. Urban

Considerable scientific effort has been directed toward the study of the aging man, with particular attention to the role of androgen replacement for the maintenance of muscle, bone, and hair. This chapter examines how the normal progression of aging affects these processes, and provides a clinical perspective for the use of androgen therapy to treat age-associated muscle, bone, and hair loss.

Case History

A 66-year-old man presents for his yearly examination. He has a medical history of hypertension and hypothyroidism, both diagnosed 5 years ago. He states that he is in his usual state of health but does notice worsening fatigue. On further questioning, he complains of not being able to play tennis as often because of exhaustion by mid-afternoon. He admits to depressive symptoms, decreased sexual desire, and inability to maintain erections during the last 9 months. His hypertension has been well controlled with a thiazide diuretic and his hypothyroidism has been managed by a stable dose of levothyroxine. On review of systems, the patient also complains of a 10-pound weight gain over 9 months, and he has noticed decreased beard growth and decreased muscle strength. He is happy that his frontal hair loss has slowed. Pertinent exam findings include a well-nourished Caucasian male that appears his stated age. He has frontal and vertex scalp hair thinning with occipital sparing, no thyroid tissue palpable, and appears noticeably more frail as compared with last year's exam. On testicular exam, no masses are palpable but his testicles are decreased in size and softer than previous exams. Laboratory data are listed in Table 10–1.

Initially, the patient's physical findings indicate hypothyroidism; however, laboratory data reveal his total and free testosterone levels are below normal.

TABLE 10–1. Laboratory Data

Test	Value	Normal Range
Total testosterone (mg/L)	220	280–880
Free testosterone (ng/L)	44	50–210
TSH (IU/mL)	2.52	0.49–4.70
LH (IU/L)	16.9	3.6–17.1
FSH (IU/L)	18.5	2.25–20
Prolactin (ng/mL)	6.5	1.6–18.8
Estradiol (pmol/L)	42	37–184
PTH (pg/mL)	35	18–73
Ca (mg/dL)	8.5	8.6–10.6
Fasting glucose (mg/dL)	87	<110

The patient is concerned that his recent loss of muscle mass and strength may impair his ability to perform his activities of daily living, and he is right to be concerned.

Androgens and Muscle

Muscle Loss and Aging

Aging is associated with a progressive deterioration of many physical processes. Changes include the loss of skeletal muscle mass and a decrease in functional capacity.[1] In the aging man, the mechanisms responsible for these physical changes, collectively referred to as sarcopenia, are multifactorial but are likely facilitated by the adoption of a more sedentary lifestyle, a less than optimal diet, and a gradual decline in gonadal function.[2,3]

The relationship between hypogonadism or andropause and the sarcopenic process poses a major health concern for aging males.[4–6] Although multifaceted, sarcopenia is ultimately the result of a chronic imbalance between muscle protein synthesis and breakdown.[7] The resultant loss of contractile protein mass and quality contributes to muscle weakness and is a key contributing factor in fall-related injury and disability. Consequently, the ability to counteract progressive losses in skeletal muscle mass and strength in older men has considerable health and economic implications. By attenuating or even preventing these losses, older individuals may have an improved quality of life, prolonged independent living, and a significantly reduced dependence on structured health care. There are also metabolic consequences of sarcopenia to be considered. These include, but are not limited to, thermoregulatory disruption,[8] glucose intolerance,[9] and altered lipid metabolism.[10,11] Thus, as the population of older men increases, the need to develop therapies to counteract losses in skeletal muscle mass and strength with aging takes on added significance.

A large number of strategies designed to ameliorate the effects of sarcopenia have been studied. Appropriately, many of these have focused on voluntary lifestyle changes. Older individuals have been encouraged to become more active, exercise regularly, modify their diets, and take nutritional supplements. When applied and monitored carefully, these lifestyle interventions often show promising results.[12–15] However, there is evidence that the effectiveness of strength training in older individuals is positively correlated with serum testosterone concentration,[16] and in many instances exercise and dietary interventions fail due to poor adherence and insufficient supervision.[17] Further, compromised joint integrity or balance, or frailty may restrict the intensity and type of physical activity that elderly individuals can perform.[1]

It is estimated that 50% of men over the age of 50 are hypogonadal.[18,19] Although a downregulation in gonadal function does not necessarily result in a symptomatic state, many individuals are symptomatic and may benefit from androgen replacement therapy. Positive outcomes may include a decrease in muscle protein breakdown,[20] an increase in muscle protein synthesis,[10,21] enhanced skeletal muscle mass,[20,22] and increased muscle strength.[23,24]

Unlike menopause in women, andropause does not occur in all older men and its presentation and pathophysiology may vary considerably.[5] Decreased potency and libido, increased fatigability, and decreased muscle strength are clinical features associated with andropause. However, a significant reduction in serum testosterone concentration represents the core physiological event and is a key diagnostic element in those affected by andropause.

Although age is undoubtedly a factor, advanced age per se is not the sole prerequisite for andropause. For example, when a gonadotropin-releasing hormone analog is administered to eugonadal young men, serum testosterone concentrations fall and lean body mass and muscle strength are also lost.[25] Further,

many chronically institutionalized older men have lower serum testosterone concentrations[26] compared with their healthy age-matched counterparts. Thus, although there is no definitive age of onset of andropause, a significant endogenous reduction in testosterone can occur as early as the fifth decade with the lowest androgen levels typically seen in men 70 years and older.[3,27,28] Although there is an abundance of evidence indicating a link between lower serum testosterone concentrations and reduced skeletal muscle function in older men, sarcopenia also occurs in elderly women. Clearly, testosterone alone is not the only factor implicated in muscle loss.

Androgens and Muscle Protein Synthesis

When combined with exercise, the ability of exogenous androgens to promote skeletal muscle hypertrophy in eugonadal men is clearly evident. Although much of the early evidence could be considered anecdotal, the ability of androgens to increase lean muscle mass and strength in an athletic context has been apparent for more than four decades. Although the successful use of androgens in athletic populations is predicated on a rigorous concomitant diet and exercise regimen, it has been demonstrated that both testosterone and one of its synthetic analogs (oxandrolone) are capable of reducing muscle protein breakdown and inducing myotrophic effects in skeletal muscle of postabsorptive young and older men.[20,23,29,30] Androgens such as testosterone clearly result in net skeletal muscle protein synthesis, and protein deposition also occurs in other tissues as a result of the androgen-induced positive nitrogen balance. However, as a cautionary note, androgens also possess well-documented androgenic and virilizing effects including stimulation of male secondary sexual characteristics, prostate hyperplasia, hirsutism, and virilization in females. This may limit their clinical use to specific patient populations such as hypogonadal men. Ultimately, if the optimal dose and timing of androgen that gives a maximal anabolic response with minimal side effects can be determined, the expectation is that androgens could be beneficial in modulating age-related sarcopenia in a broader range of patients.

Androgens and Muscle Strength

Physical strength is often compromised with increasing age.[4,24] This functional decrement is accompanied by the loss of lean body mass and increase in body fat and often facilitated by lower circulating levels of serum testosterone. Evidence from a 12-year longitudinal study indicated that a reduction in muscle cross-sectional area was a major contributor to the reduction in muscle strength seen with aging.[31] As noted, the adoption of a more sedentary lifestyle contributes to this loss of muscle mass; however, serum testosterone concentrations also play a regulatory role. In fact, there is a positive correlation between circulating levels of testosterone and the capacity to improve strength in older adults involved in a resistance training program.[16] Further, it is generally accepted that, irrespective of age, muscle strength correlates with muscle mass. However, as an individual ages, strength often decreases to a greater extent than muscle mass. Thus, the possibility of increasing muscle strength and lean body mass via the administration of therapeutic doses of androgen alone would be of considerable benefit.

The primary goal of androgen therapy in the older man is to improve or maintain muscle strength to an extent where functional capability is also maintained or improved.[1,32] However, results thus far have been equivocal. Several studies have demonstrated improved functional capacity following the administration of therapeutic doses of androgens to hypogonadal men.[20,23,33,34] Others have been unable to demonstrate increases in muscle strength with testosterone administration alone.[22,32] In the absence of definitive findings, it is reasonable to consider that a combination of hormone replacement therapy, physical activity, and nutritional intervention would provide the greatest opportunity for increases in muscle strength and function to be realized.

Following an exercise or androgen intervention, a desired outcome is an increase in muscle strength. However, when designing an intervention program (androgen, exercise, and/or diet) other components of fitness such as power, muscular endurance, flexibility, and balance must also be considered and targeted if the primary functional goal is to improve the ability to perform practical tasks and activities of daily living.

Summary

Life expectancy continues to increase as advances in medicine, nutrition, and technology are made. However, the importance of prolonged life expectancy per se is perhaps diminished if the quality of life is less than optimal. Many aging men experience significant functional impairment and a decreased quality of life due in part to alterations in body composition associated with the aging process. Although there is evidence to support the efficacy of androgen therapy in older men afflicted with symptoms of sarcopenia, we are far from understanding many of the underlying mechanisms. This chapter demonstrates the need for additional cross-sectional and longitudinal studies examining the timing, dosing, and mechanisms

responsible for slowing progressive sarcopenia in the aging man.

Androgens and Bone

Age-Associated Osteoporosis in Men

Osteoporosis is a condition characterized by low bone mass and microarchitectural deterioration of bone tissue, with a consequent increase in bone fragility and susceptibility to fracture.[35] Due to the prevalence of osteoporosis in postmenopausal women, very little attention is given to bone health in men. Nevertheless, the risk of fracture in men increases with age and ~1.5 million men >65 years have osteoporosis, with another 3.5 million estimated to be osteopenic.[36] The incidence of hip fractures in both sexes rises exponentially with age, with the majority of fractures occurring in men over the age of 80 years.[37] Although lifetime bone mineral density (BMD) losses are estimated to be 30 to 40% in women, men lose an estimated 20 to 30% in their lifetime. Various estimates of the lifetime risk of osteoporotic-related fractures in men have been suggested, with some reporting as high as 25%[21] and others stating the apparent lifetime risk nationwide to be ~13%.[38] Regardless, this represents a considerable portion of the male population and as such certainly represents a significant public health issue. Compounding the problem, when men do experience hip fractures, they have higher rates of morbidity and mortality than women.[39] Specifically, ~20% of male patients die of thromboembolus or infection-related to hip fractures,[40] and hip fractures are not the only age-related fractures that increase with age in men. Spine, proximal humerus, pelvis, ankle, and distal forearm fractures occur more often in men as they age, although incidence rates remain lower than in women.[37] These data clearly demonstrate the need to evaluate BMD status in men as part of all general health screenings.

To address the growing concern of diagnosing osteoporosis in men, the International Society for Clinical Densitometry met in 2001 for a position conference. Of particular interest was the society's position on the application and use of T-scores in men for the assessment of BMD as currently applied to women. In women, the World Health Organization (WHO) has established a threshold for osteoporosis with bone densities of either the lumbar spine or the proximal femur of at least 2.5 standard deviations (SD) below the young adult mean. However, this threshold has not been validated for men, and recent data indicate that use of a single scale factor for T-scores in men may not be appropriate, and doing so may erroneously identify men as experiencing fractures at a higher relative BMD as compared with age-matched women.[41-43] Thus it appears that a gender-specific scale must be developed to accurately assess BMD in men and women.[41,43,44]

Hypogonadism and Secondary Osteoporosis in Men

The maintenance of bone mass in the aging man is complicated by the decline in androgens associated with andropause. Significant reductions in serum testosterone concentration is a key physiological event in those affected by andropause, with men aged 70 years and older experiencing the lowest androgen levels.[45] As noted, although significant endogenous reductions in testosterone can occur as early as the fifth decade,[3] there is no definitive age of onset of andropause.

The andropause phenomena and the resulting hypogonadism has multiple consequences on BMD and is a primary factor in the development of osteoporosis in men.[21] Hypogonadism, whether it be primary or secondary, is reported in 15 to 20% of men with spinal osteoporosis.[37] Table 10-2 summarizes the common causes of secondary osteoporosis. Further, men with hypogonadism are at 6.5 times the risk of developing an osteoporotic fracture than eugonadal men.[46] The loss of androgens with age is clinically significant and has multiple direct effects on bone. Testosterone has been shown to increase the activity and number of both osteoblasts and osteoclasts.[47] This increased cellular activity in bone translates into increased bone turnover (i.e., an upregulation of both bone synthesis and bone breakdown), which has been shown to be essential for the maintenance of good bone health. In similar studies involving 3 months of testosterone or placebo replacement, those men receiving testosterone were found to have a reduction in markers of bone resorption[32] and an increased level of osteocalcin, a marker of bone formation.[48] It has been suggested that in aging men, estradiol is the dominant sex steroid regulating bone resorption, with bone formation being regulated by both estradiol and testosterone.[49] Although reductions in estradiol levels have been shown to correlate with BMD in both aging women and men,[50] the full impact of estradiol on bone health in men has not been realized.

Perhaps more significant with respect to bone health is that testosterone works to prevent the release of sex hormone–binding globulin (SHBG).[47] Although serum total testosterone concentrations are influenced by SHBG, serum free testosterone concentrations also decrease in healthy aging men,[27,28] indicating that bioavailable testosterone also decreases. Thus the amount of free testosterone available to the tissues is further decreased, ultimately affecting tissue growth and repair. In addition to binding testosterone, SHBG also binds estrogen. This may be one of the most important downstream effects of decreased testosterone

on bone health. Estrogen acts on estrogen receptors in osteoblasts and increases both the activity and numbers of these bone matrix secreting cells. Although the role of estrogen on osteoclasts is controversial, it is thought to increase osteoclast cell death.[47] The combination of these effects is thought to cause the well-recognized phenomenon of estrogen-delayed bone loss in postmenopausal women.

The relationship between male hypogonadism and reduced bone density has been well documented.[46,50,51] However, the importance of estrogen on the male skeleton becomes apparent upon review of two particular case reports.[52,53] In both instances, the men, in their mid-20s, did not experience a pubertal growth spurt, though they had normal secondary sexual characteristics. Further, both men had low BMD with several unfused epiphyses. One of the men was found to have a defective estrogen receptor, so despite high testosterone and estrogen levels, his tissue-level response was impaired.[53] The other case was determined to be an aromatase deficiency resulting in low levels of estrogen,[52] and treatment with estrogen led to an increase in BMD and fusion of the growth plates. These case reports support the premise that a portion of the androgen action on the male skeleton is mediated by estrogens.

Disease States and Osteoporosis in Aging Men

Although 50% of cases of osteoporosis in men appear to be idiopathic, the remaining cases are generally attributed to various diseases and treatment medications (Table 10–2).[21] Chronic glucocorticoid therapy is a common cause of osteoporosis in both men and women.[21] Further, some diseases of the kidney can decrease renal tubule reabsorption of calcium, also affecting BMD. In

TABLE 10–2. Common Causes of Secondary Osteoporosis in Men

Medical Interventions	Secondary Diseases
Medications:	*Endocrine disorders:*
Glucocorticoids	Hyperparathyroidism
Antipsychotics	Hypogonadism
Loop diuretics (furosemide)	Pituitary tumors
Gonadotrophin-releasing hormone agonists	Type 1 diabetes
	Addison's disease
Chemotherapy (methotrexate, cyclosporine)	Cushing's syndrome
Anticoagulants (heparin)	*Other:*
Anticonvulsants	Malnutrition
Antacids with aluminum	Malabsorption
	Hepatic insufficiency
	Vitamin C or D deficiency
Surgeries:	Renal failure
Orchiectomy	Anemia
Gastrectomy	Alcohol

addition, loop diuretics such as furosemide also decrease the reabsorption of calcium. Furthermore, a study of men at a VA hospital found chronic obstructive pulmonary disease (COPD) and antipsychotic medications were associated with lower BMD, independent of smoking status and corticosteroid use.[54] Antipsychotics, which work by decreasing dopamine, can cause an increase in prolactin, which in turn inhibits the release of estrogen and testosterone.[54] Finally, a phenomenon well known to urologists is the negative effect of prostate cancer therapy on BMD.[40] Following orchiectomy, men have an increased incidence of osteoporotic fractures. One study reported that 38% of men 5 years postorchiectomy had at least one osteoporotic fracture.[55] These results clearly demonstrate the need for intervention strategies to prevent osteoporosis in men.

Treatments for Osteoporosis

In patients with no clear cause of osteoporosis, many of the same treatment strategies that apply to women are also applied to men. Current recommendations include the ingestion of 1200 to 1500 mg of calcium per day, along with up to 600 IU of vitamin D.[56,57] However, because some studies have shown a link between overconsumption of calcium and prostate cancer, calcium intake should be monitored in men,[58] and they should consider regular evaluations by their physicians. Though more studies are needed, testosterone supplementation is not recommended at this time as supplementation appears to have only a minimal effect on BMD.[59] Studies with estrogen supplementation have not been conducted in men, but in light of recent discoveries of increased risk of heart disease in estrogen-supplemented postmenopausal women, this is not a likely future direction of research.

Finally, several medications developed for osteoporotic postmenopausal women have also been used experimentally in men. One study in particular has shown that calcitonin can improve BMD in castrated men.[60] However, although calcitonin is moderately effective, bisphosphonates are more widely used, have proven to be an effective treatment strategy in women with osteoporosis, and have shown positive effects on BMD and a decreased rate of fractures in men.[61]

Summary

Although the focus of this topic has been osteoporosis in aging men, there is some evidence that treatment strategies used for women may also prove appropriate for men. Unlike women, however, men more commonly have an underlying pathology contributing to their osteoporosis. Future studies should evaluate the extent to which the normal aging process and the resulting endocrine dysfunction contribute to alterations in bone metabolism, bone loss, and skeletal function.

Androgens and Hair

Hair is a prerequisite to be classified as a mammal, yet has no physiological function in humans.[62] Through evolution and scientific advances, humans no longer requires hair for maintenance of temperature, protection, or camouflage. Although the physiological function of hair is lost, the psychological importance remains.[63] Hair for males and females is a powerful social and sexual communication mechanism, through both visual and olfactory senses. A man's beard has been compared with a lion's mane as a visual sexual attractant for the opposite sex.[62,64] Fullness of scalp, beard, and chest hair has classically been a visual correlate to a man's masculinity. The psychological ramification of thinning hair in these areas, especially the scalp, has been the topic of much research and the target of pharmaceutical investment.

Physiology of Hair Growth

Approximately 5 million hair follicles cover the entire human body, sparing only the palms and soles. A hair follicle is defined as an invagination of the skin with continuously proliferating specialized matrix cells.[65] The hair follicle is responsible for producing the keratinaceous proteins that compose a strand of hair. After birth, hair follicles can produce two different types of hair. Vellus hair is short, soft, nonpigmented, and lacks an arrector pili muscle. At birth, vellus hair covers the entire body except the palms and soles. Terminal hair is large, pigmented, and coarse, and at birth, comprises scalp hair, eyelashes, and eyebrows.[62,64,65]

Each hair follicle undergoes three phases of activity: a growth phase (anagen), a transition phase (catagen), and a resting phase (telogen). Anagen, which varies at different body sites and between individuals, produces hair continuously and lasts for several months to years. The average duration of this phase is 2 to 4 years.[62,64–66] In normal human scalp, the total number of hairs is ~100,000. At any given time, 90% of scalp hair follicles are in the anagen phase. Anagen determines the length of the hair at different body sites and the growth rate of anagen hairs is ~0.35 mm per day.[67] Less than 1% of scalp hair follicles are in the catagen phase, a transition phase characterized by regressive changes of the hair follicle. The average duration for catagen is 2 weeks. Telogen follicles comprise ~8 to 10% of scalp hair. Similar to anagen, the duration of telogen varies by different sites and between individuals. During the telogen phase, the hair remains secured in the base of the follicle, known as the club. The typical length of the telogen phase is approximately 2 to 3 months during which the follicle begins a new growth cycle and the hair is shed with the beginning of another anagen.[68] The average ratio of anagen to telogen hairs on the scalp at any given time is 9:1, with ~30 to 100 hairs shed on a daily basis with moderate seasonal variation.[69,70]

Androgen Effects on Hair

Androgens are the most potent regulator of hair growth.[71,72] Although other hormones including thyroid hormone, growth hormone, and estrogen affect hair growth, androgens appear to control hair growth and distribution to the greatest extent. At the onset of puberty in males, the large increase in androgens cause vellus hair, located in the pubic and axillary regions, to transform to intermediate hair.[73] These changes are graded by the Tanner scale in children and adolescents. This hair eventually becomes long, thick, pigmented terminal hair, which is found in adults. Similar changes occur on the chest, arms, legs, beard, and scalp as puberty continues and adulthood begins. These hairs and follicles cycle through the three phases described previously for the lifetime of the adult male (Fig. **10–1**).

In genetically predisposed males, androgens lead to male pattern baldness or androgenetic alopecia (AGA) by inhibiting terminal hair follicles and transforming them into vellus hair.[74,75] The importance of androgens on hair growth, male pattern baldness, and an inheritance component was described over 60 years ago by Hamilton.[76] Prepubertal castration led to the retention of juvenile hairline, whereas castration after puberty led to normal distribution of hair seen in the general population. Despite the age of the patients or the degree of baldness at the time of castration, none of the patients had progression of hair loss over a 1-year observation period. Testosterone administration to these castrated patients, even at subphysiological doses, produced progressive hair loss and follicular miniaturization in men with a significant family history for balding. Men without a significant family history displayed no further loss of hair.

The effect of androgens on hair growth varies by body site.[77–79] Hair follicles are classified into one of three categories: androgen-insensitive (growth without the influence of androgens), androgen-sensitive (anagen phase shortening and miniaturization of hair follicle leading to transformation of terminal to vellus hair upon exposure to androgens), or androgen-dependent (growth of hair requires presence of androgens). A list of body sites for each category is documented in Table **10–3**. Currently, it is unknown how androgens

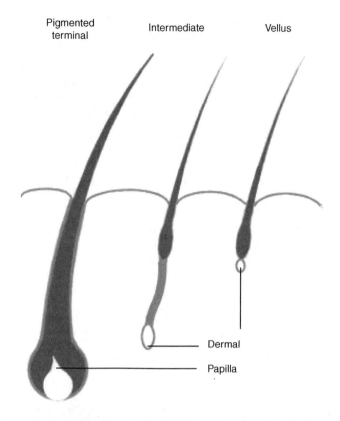

FIGURE 10–1. Different phases of hair growth. A major difference between pigmented terminal and vellus phases of hair growth is the activation of a hexose monophosphate shoot that produces NADPH, which is required by the 5α-reductase pathway to produce PHT. The pigmented terminal phase follicles produce more PHT. The effects of androgens on hair growth varies by body site. Follicles are either androgen insensitive or androgen sensitive.

produce these paradoxical growth effects on various sites and the genes that are involved.

Androgens derived from both the testes and the adrenal glands play a crucial role in follicle stimulation and suppression. Hair follicles not only respond to androgens but are also actively involved in androgen metabolism. Hair follicles at different body sites contain various enzymes.[77] Both steroid sulfatase and 5α-reductase have been localized in the dermal papilla of scalp follicles.[72,77,80] Both of these enzymes are responsible for converting weaker androgens into more potent metabolites. Hair follicles metabolize testosterone to dihydrotestosterone (DHT) through the 5α-reductase enzyme. DHT is considered more potent as it has five times the affinity toward androgen receptors. Androgen-dependent sites (e.g., beard) showed a higher 5α-reductase activity compared with androgen-insensitive body sites (e.g., occipital scalp).[81]

A major biochemical difference between the anagen and telogen phases of hair growth is the activation of a hexose monophosphate shunt that produces

TABLE 10–3. Hair Follicle Classification

Androgen-insensitive
  Eyebrows
  Eyelids
  Occipital scalp

Androgen-sensitive
  Scalp: frontal
  Scalp: vertex

Androgen-dependent
  Pubic
  Axilla
  Chest
  Extremities
  Beard

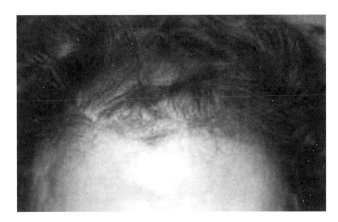

FIGURE 10–2. Patient with androgenetic alopecia, in which androgen-sensitive terminal hair follicles are transformed to vellus hair. See Table 10–3 for effects of androgens on hair by body site.

reduced nicotinamide adenine dinucleotide phosphate (NADPH), which is required by the 5α-reductase pathway to produce DHT.[78] This results in the anagen hair follicles producing more DHT than the resting ones. Scalp follicles have also been found to contain steroid sulfatase within their dermal papilla.[82,83] Dehydroepiandrosterone sulfate (DHEAS) is converted to dehydroepiandrosterone (DHEA) via the enzyme steroid sulfatase found in scalp hair follicles. The DHEA is metabolized to androstenedione and testosterone systemically and to DHT in the hair follicles, which is another mechanism that increases androgen concentration at the hair follicle.

Androgenetic alopecia occurs in genetically predisposed men and women and is thought to be related to the concentration of potent androgens, specifically DHT, at the site of the frontal and vertex scalp follicles[84] (Fig. **10–2**). The dermal papilla of balding hair follicles was found to have higher concentrations of androgen receptors compared with nonbalding scalp follicles.[85] Dermal papilla of hair follicles from balding (AGA) subjects were also smaller and grew less well under normal growth conditions compared with nonbalding papilla.[86] In 1974, the description of patients with incomplete male pseudohermaphroditism established that DHT rather than testosterone was the androgen involved in AGA.[87,88] Male patients were homozygous for mutation of the gene that encodes for steroid 5α-reductase. They were born with ambiguous genitalia and raised initially as girls. At puberty, the boys virilized, and identity was changed from female to male. These patients through puberty and adulthood developed normal male muscle mass and libido, but never experienced prostate enlargement or male pattern hair loss. These patients had markedly decreased levels of DHT with high-normal to elevated levels of testosterone. Confirmation of the importance of DHT over testosterone in the development

of AGA occurred with the use of finasteride, a 5α-reductase inhibitor, in the treatment of this disorder.[89–91] Finasteride was found to increase hair weight when given in low doses to patients with early AGA.[92] In patients with late AGA, patients were found to decrease the amount of hair loss, with some increase in new growth.

Andropause Effects on Hair Growth

After detailing the effects that androgens have on hair follicles, it is simple to elucidate the effects of andropause on hair growth. Andropause, a syndrome of androgen deficiency with aging, leads to a decrease in all androgen production including testosterone, DHEA, and DHT.[93] With the loss of these potent androgens, men experience loss of hair at androgen-dependent body sites. Men complain of decreased hair on their chest, extremities, and beard. Androgen-insensitive sites experience no change in volume or rate of growth, and androgen-sensitive sites experience decreased hair loss in genetically predisposed patients. These findings are similar to the patients described by Hamilton[76] during the castration studies. Androgen replacement by injection, patch or transdermal gel reverses these symptoms and signs causing increased total body hair,[94–96] but will resume the male pattern hair loss in genetically susceptible individuals.

Case Discussion

The 66-year-old man described previously is clearly having symptoms and clinical physical findings consistent with age-associated androgen deficiency. His laboratory data are consistent with an andropause state and do not appear to result from other hypogonadal etiologies. The complaints of muscle weakness, loss of beard growth, and decreased scalp hair loss in a man with the physical exam finding of male pattern baldness are also consistent with andropause. This patient, because of his symptoms, exam findings, and laboratory data, was diagnosed with androgen deficiency most likely secondary to andropause. He began testosterone replacement with Androgel 5 g applied topically daily. Within 4 weeks his testosterone level increased to midnormal ranges and his complaints of depressive symptoms, decreased energy, and erectile dysfunction resolved. After 6 months of treatment he noticed increased muscular strength and increased thickness of his beard, but complained that his scalp hair has begun to thin again. He is currently contemplating starting finasteride and is scheduled for a follow-up BMD scan in 6 months to evaluate his bone health.

REFERENCES

1. Casiano ER, Paddon-Jones D, Ostir GV, Sheffield-Moore M. Assessing functional status measures in older adults: a guide for health care professionals. Phys Ther Rev 2002;7:89–101

2. Deslypere JP, Vermeulen A. Leydig cell function in normal men: effect of age, life-style, residence, diet, and activity. J Clin Endocrinol Metab 1984;59:955–962

3. Moroz EV, Verkhratsky NS. Hypophyseal-gonadal system during male aging. Arch Gerontol Geriatr 1985;4:13–19

4. Dutta C, Hadley EC. The significance of sarcopenia in old age. J Gerontol A Biol Sci Med Sci 1995;50(Spec No):1–4

5. Mastrogiacomo I, Feghali G, Foresta C, Ruzza G. Andropause: incidence and pathogenesis. Arch Androl 1982;9:293–296

6. Simon D, Charles MA, Nahoul K, et al. Association between plasma total testosterone and cardiovascular risk factors in healthy adult men: The Telecom Study. J Clin Endocrinol Metab 1997;82:682–685

7. Short KR, Nair KS. Mechanisms of sarcopenia of aging. J Endocrinol Invest 1999;22:95–105

8. Kenney WL, Buskirk ER. Functional consequences of sarcopenia: effects on thermoregulation. J Gerontol A Biol Sci Med Sci 1995;50(Spec No):78–85

9. Kohrt WM, Holloszy JO. Loss of skeletal muscle mass with aging: effect on glucose tolerance. J Gerontol A Biol Sci Med Sci 1995;50(Spec No):68–72

10. Beaufrere B, Morio B. Fat and protein redistribution with aging: metabolic considerations. Eur J Clin Nutr 2000;54(suppl 3):S48–S53

11. Morio B, Hocquette JF, Montaurier C, et al. Muscle fatty acid oxidative capacity is a determinant of whole body fat oxidation in elderly people. Am J Physiol Endocrinol Metab 2001;280:E143–E149

12. Fiatarone MA, Marks EC, Ryan ND, Meredith CN, Lipsitz LA, Evans WJ. High-intensity strength training in nonagenarians: effects on skeletal muscle. JAMA 1990;263:3029–3034

13. Frontera WR, Meredith CN, O'Reilly KP, Evans WJ. Strength training and determinants of VO2max in older men. J Appl Physiol 1990;68:329–333

14. Frontera WR, Meredith CN, O'Reilly KP, Knuttgen HG, Evans WJ. Strength conditioning in older men: skeletal muscle hypertrophy and improved function. J Appl Physiol 1988;64:1038–1044

15. Parise G, Yarasheski KE. The utility of resistance exercise training and amino acid supplementation for reversing age-associated decrements in muscle protein mass and function. Curr Opin Clin Nutr Metab Care 2000;3:489–495

16. Hakkinen K, Pakarinen A. Serum hormones and strength development during strength training in middle-aged and elderly males and females. Acta Physiol Scand 1994;150:211–219

17. Mazzeo RS, Tanaka H. Exercise prescription for the elderly: current recommendations. Sports Med 2001;31:809–818

18. Morley JE, Kaiser FE, Sih R, Hajjar R, Perry HM III. Testosterone and frailty. Clin Geriatr Med 1997;13:685–695

19. Sih R, Morley JE, Kaiser FE, Perry HM III, Patrick P, Ross C. Testosterone replacement in older hypogonadal men: a 12-month randomized controlled trial. J Clin Endocrinol Metab 1997;82:1661–1667

20. Ferrando AA, Sheffield-Moore M, Paddon-Jones D, Wolfe RR, Urban RJ. Differential anabolic effects of testosterone and amino acid feeding in older men. J Clin Endocrinol Metab 2003;88:358–362

21. Bilezikian P. Osteoporosis in men. J Clin Endo Metab 1999; 84:3431–3434

22. Snyder PJ, Peachey H, Hannoush P, et al. Effect of testosterone treatment on body composition and muscle strength in men over 65 years of age. J Clin Endocrinol Metab 1999;84:2647–2653

23. Ferrando AA, Sheffield-Moore M, Yeckel CW, et al. Testosterone administration to older men improves muscle function: molecular and physiological mechanisms. Am J Physiol Endocrinol Metab 2002;282:E601–E607

24. Reed RL, Pearlmutter L, Yochum K, Meredith KE, Mooradian AD. The relationship between muscle mass and muscle strength in the elderly. J Am Geriatr Soc 1991;39:555–561

25. Mauras N, Hayes V, Welch S, et al. Testosterone deficiency in young men: marked alterations in whole body protein kinetics, strength, and adiposity. J Clin Endocrinol Metab 1998;83:1886–1892

26. Abbasi AA, Drinka PJ, Mattson DE, Rudman D. Low circulating levels of insulin-like growth factors and testosterone in chronically institutionalized elderly men. J Am Geriatr Soc 1993;41:975–982

27. Nankin HR, Caulkins JH. Decreased bioavailable testosterone in aging normal and impotent men. J Clin Endocrinol Metab 1986;63:1418–1420

28. Tenover JS, Matsumoto AM, Plymate SR, Bremner WJ. The effects of aging in normal men on bioavailable testosterone and luteinizing hormone secretion: response of clomiphene citrate. J Clin Endocrinol Metab 1987;65:1118–1125

29. Sheffield-Moore M. Androgens and the control of skeletal muscle protein synthesis. Ann Med 2000;32:181–186

30. Sheffield-Moore M, Urban RJ, Wolf SE, et al. Short-term oxandrolone administration stimulates net muscle protein synthesis in young men. J Clin Endocrinol Metab 1999;84:2705–2711

31. Frontera WR, Hughes VA, Fielding RA, Fiatarone MA, Evans WJ, Roubenoff R. Aging of skeletal muscle: a 12-yr longitudinal study. J Appl Physiol 2000;88:1321–1326

32. Tenover JS. Effects of testosterone supplementation in the aging male. J Clin Endocrinol Metab 1992;75:1092–1098

33. Bhasin S, Storer TW, Berman N, et al. Testosterone replacement increases fat-free mass and muscle size in hypogonadal men. J Clin Endocrinol Metab 1997;82:407–413

34. Urban RJ, Bodenburg YH, Gilkison C, et al. Testosterone administration to elderly men increases skeletal muscle strength and protein synthesis. Am J Physiol 1995;269:E820–E826

35. Sherman S. Preventing and treating osteoporosis: strategies at the millennium. Ann N Y Acad Sci 2001;949:188–197

36. Siddiqui NA, Shetty KR, Duthie EH Jr. Osteoporosis in older men—discovering when and how to treat it. Geriatrics 1999; 54:20–30

37. Boonen S, Vanderschueren D. Bone loss and osteoporotic fracture occurrence in aging men. In: Lundenfeld B, Gooren L, eds. *Textbook of Men's Health*. New York: Parthenon; 2002:455–475

38. Melton LJ 3rd, Atkinson EJ, O'Connor MK, O'Fallon WM, Riggs BL. Bone density and fracture risk in men. J Bone Miner Res 1998; 13:1915–1923

39. Seeman E. The structural basis of bone fragility in men. Bone 1999;25:143–147

40. Smith M. Osteoporosis during androgen deprivation therapy for prostate cancer. Urology 2002;3(suppl 1):79–86

41. de Laet CE, van der Klift M, Hofman A, Pols HA. Osteoporosis in men and women: a story about bone mineral density thresholds and hip fracture risk. J Bone Miner Res 2002;17:2231–2236

42. Leib ES, Lenchik L, Bilezikian JP, Maricic MJ, Watts NB. Position statements of the International Society for Clinical Densitometry: methodology. J Clin Densitom 2002;5(suppl):S5–S10

43. Selby PL, Adams JE. Do men and women fracture bones at similar bone densities? Osteoporosis 2000;11:153–157

44. Binkley NC, Schmeer P, Wasnich RD, Lenchik L. What are the criteria by which a densitometric diagnosis of osteoporosis can be made in males and non-Caucasians? J Clin Densitom 2002; 5(suppl):S19–S27

45. Sheffield-Moore M, Urban RJ. Androgens and lean body mass in the aging male. In: Lundenfeld B, Gooren L, eds. *Textbook of Men's Health*. New York: Parthenon; 2002:241–246

46. Stanley HL, Schmitt BP, Poses RM, Deiss WP. Does hypogonadism contribute to the occurrence of a minimal trauma hip fracture in elderly men? J Am Geriatr Soc 1991;39:766–771

47. Riggs B, Khosla S, Melton LJ 3rd. Sex steroids and the construction and conservation of the adult skeleton. Endocr Rev 2002;23:279–302

48. Morley JE, Perry HM III, Kaiser FE, et al. Effects of testosterone replacement therapy in old hypogonadal males: a preliminary study. J Am Geriatr Soc 1993;41:149–152

49. Falahati-Nini A, Riggs BL, Atkinson EJ, O'Fallon WM, Eastell R, Khosla S. Relative contributions of testosterone and estrogen in regulating bone resorption and formation in normal elderly men. J Clin Invest 2000;106:1553–1560

50. Greendale GA, Edelstein S, Barret-Connor E. Endogenous sex steroids and bone mineral density in older women and men: the Rancho Bernardo Study. J Bone Miner Res 1997;12:1833–1843

51. Murphy S, Khaw KT, Cassidy A, Compston JE. Sex hormones and bone mineral density in elderly men. Bone Miner 1993; 20:133–140

52. Morishima A, Simpson ER, Fischer C, Qin K. Aromatase deficiency in male and female siblings caused by a novel mutation and the physiological role of estrogens. J Clin Endoclinol Met 1995;80:3689–3698

53. Smith EP, Boyd J, Frank GR, et al. Estrogen resistance caused by a mutation in the estrogen receptor gene in a man. N Engl J Med 1994;331:1056–1061

54. Yeh SS, Phanumas D, Hafner A, Schuster MW. Risk factors for osteoporosis in a subgroup of elderly men in a Veterans Administration nursing home. J Invest Med 2002;50:452–457

55. Daniell HW. Osteoporosis after orchiectomy for prostate cancer. J Urol 1997;157:439–444

56. Consensus development conference: diagnosis, prophylaxis, and treatment of osteoporosis. Am J Med 1993;94:646–650

57. NIH Consensus Conference. Optimal calcium intake. NIH Consensus Development Panel on Optimal Calcium Intake. JAMA 1994;272:1942–1948

58. Giovannucci E, Rimm EB, Wolk A, et al. Calcium and fructose intake in relation to risk of prostate cancer. Cancer Res 1998; 58:442–447

59. Shetty KR, Maas DL. Testosterone replacement for hypogonadal elderly men. Clin Geriatr 1998;6:60–68

60. Stepan JJ, Lachman M, Zverina J, et al. Castrated men exhibit bone loss: effect of calcitonin treatment on biochemical indices of bone remodeling. J Clin Endocrinol Metab 1989;69:523–527

61. Orwall E, Weiss S, Miller P, et al. Alendronate for the treatment of osteoporosis in men. N Engl J Med 2000;343:604–610

62. Ebling FJ. Hair. J Invest Dermatol 1976;67:98–105

63. Sternbach H. Age-associated testosterone decline in men: clinical issues for psychiatry. Am J Psychiatry 1998;155:1310–1318

64. Ebling FJ. The biology of hair. Dermatol Clin 1987;5:467–481

65. Mercurio MG, Gogstetter DS. Androgen physiology and the cutaneous pilosebaceous unit. J Gend Specif Med 2000;3:59–64

66. Tosi A, Misciali C, Piraccini BM, Peluso AM, Bardazzi F. Drug-induced hair loss and hair growth: incidence, management and avoidance. Drug Saf 1994;10:310–317

67. Myers R, Hamilton J. Regeneration and rate of growth of hair in man. Ann N Y Acad Sci 1951;53:562–568

68. Orentreich N, Durr NP. Biology of scalp hair growth. Clin Plast Surg 1982;9:197–205

69. Courtois M, Loussouarn G, Hourseau S, Grollier JF. Periodicity in the growth and shedding of hair. Br J Dermatol 1996; 134:47–54

70. Randall VA, Ebling FJ. Seasonal changes in human hair growth. Br J Dermatol 1991;124:146–151

71. Ebling FJ. Hair follicles and associated glands as androgen targets. Clin Endocrinol Metab 1986;15:319–339

72. Randall VA, Thornton MJ, Hamada K, et al. Androgens and the hair follicle. Cultured human dermal papilla cells as a model system. Ann N Y Acad Sci 1991;642:355–375

73. Hiort O, Holterhus PM, Nitsche EM. Physiology and pathophysiology of androgen action. Baillieres Clin Endocrinol Metab 1998;12:115–132

74. Sawaya ME. Biochemical mechanisms regulating human hair growth. Skin Pharmacol 1994;7:5–7

75. Sawaya ME. Clinical updates in hair. Dermatol Clin 1997;15: 37–43

76. Hamilton JB. Male hormone stimulation is prerequisite and an incitement in common baldness. Am J Anat 1942;71:451

77. Hoffmann R. Enzymology of the hair follicle. Eur J Dermatol 2001;11:296–300

78. Mooradian AD, Morley JE, Korenman SG. Biological actions of androgens. Endocr Rev 1987;8:1–28

79. Randall VA, Hibbetrs NA, Thornton MJ, et al. The hair follicle: a paradoxical androgen target organ. Horm Res 2000;54:243–250

80. Drake L, Hordinsky M, Fiedler V, et al. The effects of finasteride on scalp skin and serum androgen levels in men with androgenetic alopecia. J Am Acad Dermatol 1999;41:550–554

81. Itami S, Kurata S, Sonoda T, Takayasu S. Mechanism of action of androgen in dermal papilla cells. Ann N Y Acad Sci 1991; 642:385–395

82. Hoffmann R, Rot A, Niiyama S, Billich A. Steroid sulfatase in the human hair follicle concentrates in the dermal papilla. J Invest Dermatol 2001;117:1342–1348

83. Sawaya ME. Steroid chemistry and hormone controls during the hair follicle cycle. Ann N Y Acad Sci 1991;642:376–383; discussion 383–374.

84. Hoffmann R, Happle R. Current understanding of androgenetic alopecia. Part I: etiopathogenesis. Eur J Dermatol 2000; 10:319–327

85. Hibberts NA, Howell AE, Randall VA. Balding hair follicle dermal papilla cells contain higher levels of androgen receptors than those from non-balding scalp. J Endocrinol 1998; 156:59–65

86. Randall VA, Hibberts NA, Hamada K. A comparison of the culture and growth of dermal papilla cells from hair follicles from non-balding and balding (androgenetic alopecia) scalp. Br J Dermatol 1996;134:437–444

87. Imperato-McGinley J, Guerrero L, Gautier T, Peterson RE. Steroid 5alpha-reductase deficiency in man: an inherited form of male pseudohermaphroditism. Science 1974;186: 1213–1215

88. Walsh PC, Madden JD, Harrod MJ, Goldstein JL, MacDonald PC, Wilson JD. Familial incomplete male pseudohermaphroditism, type 2. Decreased dihydrotestosterone formation in pseudo-vaginal perineoscrotal hypospadias. N Engl J Med 1974; 291:944–949

89. Hoffmann R, Happle R. Current understanding of androgenetic alopecia. Part II: clinical aspects and treatment. Eur J Dermatol 2000;10:410–417

90. Whiting DA. Advances in the treatment of male androgenetic alopecia: a brief review of finasteride studies. Eur J Dermatol 2001;11:332–334

91. Wolff H, Kunte C. Current management of androgenetic alopecia in men. Eur J Dermatol 1999;9:606–609

92. Price VH, Menefee E, Sanchez M, Ruane P, Kaufman KD. Changes in hair weight and hair count in men with androgenetic alopecia after treatment with finasteride, 1 mg, daily. J Am Acad Dermatol 2002;46:517–523

93. Heaton JP. Andropause: coming of age for an old concept? Curr Opin Urol 2001;11:597–601

94. Bhasin S, Bremner WJ. Clinical review 85: emerging issues in androgen replacement therapy. J Clin Endocrinol Metab 1997; 82:3–8

95. Rolf C, Nieschlag E. Potential adverse effects of long-term testosterone therapy. Baillieres Clin Endocrinol Metab 1998; 12:521–534

96. Winters SJ. Current status of testosterone replacement therapy in men. Arch Fam Med 1999;8:257–26

11

Obesity in Aging Men and the Androgen Relationship

Robert S. Tan and Shou-Jin Pu

Case History

Mr. L.K. is a 52-year-old man who had been struggling with weight problems for the past 10 years. He reached a maximum weight of 290 pounds, and decided to be proactive and alter his lifestyle drastically. He went on a strict diet of fish, vegetables, nuts, and minimal carbohydrates and lost ~30 pounds in 6 months. He was also prescribed Orlistat (Xenical) to help with weight loss. He wanted more dramatic results and decided to consult a plastic surgeon for help. On direct questioning, the patient confessed to easy fatigability, depressed moods, loss of libido, and erectile dysfunction. When he was seen, he weighed 260 pounds. His current medications included aspirin 325 mg, multivitamins, and protein powder. On physical examination, the patient had a blood pressure of 150/100 mm Hg, height 5 feet 11 inches. His cardiovascular and respiratory examination was benign. His face was bulging with excessive subcutaneous tissue, the abdominal wall was sagging, and he had multiple spider telangiectasias over the upper chest. Body fat content as measured by electrical impedance methodology was 33.6%. No masses were felt in the abdomen. Laboratory data gathered revealed total testosterone 318 ng/dL (241–827), free testosterone 10.6 pg/mL (7.2–24.0), estradiol 55 pg/mL (0–54), hemoglobin (Hgb) 15.5 g/dL, total cholesterol 240 mg/dL, and prostate-specific antigen (PSA) 0.7 ng/dL. He had consulted with a view toward plastic surgery as well as the possibility of androgen therapy to assist in his weight loss. He also said he wanted a physician to monitor his weight loss and to be a coach for this process.

Introducing a Relationship Between Obesity and Androgens in Men

It must be stated at the onset that andropause is a biochemical-physiological state reflecting a low level of bioavailable testosterone and may or may not be clinically symptomatic. This parallels menopause in a sense, as not all women suffer from transitory symptoms of menopause. Ultimately, however, all women suffer from the long-term effects of estrogen deprivation, such as bone, cognitive, functional, and libido loss, among others, as do andropausal men in most instances. The prevalence of obesity in aging men has been increasing in the past decade. The National Health and Nutrition Examination Survey has found that the prevalence of obesity in aging men has increased from 7.0% to 15.2% during the years 1960–1994 (Fig. 11–1).[1]

Men begin to be obese in middle age, and thus obesity impacts not only on mortality events but also the quality of life in later life. Obesity can be associated with an increased incidence of cardiovascular diseases, type 2 diabetes, hypertension, stroke, hyperlipidemia, osteoarthritis, and some cancers. Mortality can be also increased with obesity.[2] Data on the characteristics of obesity in the aging men are somewhat limited and they vary; however, we will review the work of others and summarize our own findings. Both total and bioavailable testosterone levels decline with advancing age.[3] From our research and others, it can be stated that obesity can be a strong predictor of testosterone deficiency in aging men.[4] In addition to obesity, hypogonadal men may actually have reduced lean body mass and increased visceral fat. The cascade reaction may predispose aging men to an increase in diabetes and cardiovascular diseases.[5] Recently, there has been a growing interest in studying the effect of testosterone replacement in aging men. However, it is prudent to have further studies especially in assessing the risk-to-benefit ratio and quality of life, so as to understand the impact of testosterone on obesity and body composition in men.

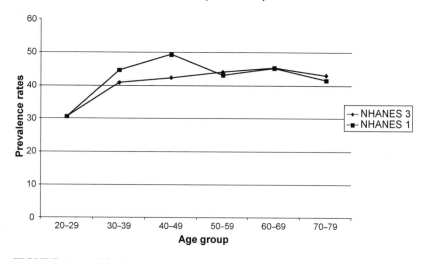

FIGURE 11–1. Obesity prevalence rates in men in the National Health and Nutrition Examination Survey (NHANES) during the years 1960 to 1994. Body mass index of 25–29.9 defines obesity. (Adapted from Flegal KM, Carroll MD, Kuczmarski RJ, Johnson CL. Overweight and obesity in the United States: prevalence and trends, 1960–1994. Int J Obes 1998;22:39–47, with permission.)

Impact of Hypogonadism in Men

Normal aging in men is accompanied by a decline in the serum level of testosterone. Abnormalities at all levels of the hypothalamic-pituitary-testicular axis contribute to a decline in testosterone concentration in older men. Free testosterone declined by 1.2%/year, and albumin-bound testosterone by 1.0%/year. Sex hormone–binding globulin (SHBG) increased by 1.2%/year, with the net effect that total serum testosterone declined more slowly (0.4%/year).[6] Hypogonadism may not correlate directly with andropause-related symptoms. Using total testosterone as a criterion, the incidence of hypogonadism is 20% of men over 60, 30% over 70, and 50% over 80 years of age. The percentage of hypogonadism is even greater when bioavailable testosterone criteria are used.[3] Despite the high prevalence rate, many men with hypogonadism remain undiagnosed and untreated. This is partly because of lack of knowledge of symptoms by both patients and providers.[7,8] Besides aging, factors such as medical illness, diabetes, obesity, sleep disturbance, and medication may contribute to the decline in testosterone with advancing age. Hypogonadism in aging men can lead to symptoms such as decreased libido, erectile dysfunction, fatigue, loss of energy, change in body composition, decrease in muscle mass, increase in fat mass, muscle weakness, osteoporosis, and mood and memory changes—often classified as "andropause symptoms."

Measuring and interpreting testosterone levels can be difficult. As such, it often confounds the results of epidemiological studies. SHBG binds testosterone, and the affinity of the binding is affected with aging process. In addition, obesity is associated with a decrease in SHBG concentration.[9–11] Therefore, men with mild to moderate obesity often have low total testosterone but normal free testosterone concentration because of lower SHBG concentrations. As such, this is not real hypogonadism.[11] Accordingly, androgen deficiency may be misclassified in men with low SHBG.[12] In massively obese men, there is real hypogonadotrophic hypogonadism with decreased free testosterone levels, a consequence of functional alterations at the hypothalamic-pituitary-testicular axis.[11] Although rare, some hypothalamic syndrome, such as Prader-Willi syndrome, can be associated with both obesity and hypogonadotropic hypogonadism. Hyperphagia-induced obesity, hypogonadism, growth hormone deficiency, and insulin and leptin resistance are all hypothalamic dysfunction in origin.[13]

Obesity Is a Global Public Health Concern Today

Obesity has increased steadily over the past several decades in most of the developed and developing world. The World Health Organization in 2000 proposed a classification of being overweight as body mass index (BMI) ranging from 25 to 29.9 kg/m² and obesity set at BMI 30 kg/m² or above (WHO 2000). The prevalence of obesity varies according to age,

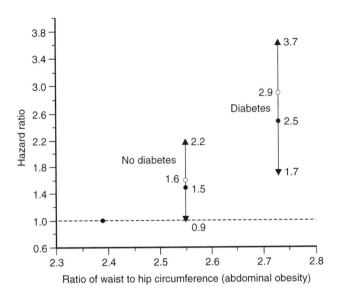

FIGURE 11–2. Relationship of abdominal obesity to risk (hazard ratio) of diabetes. A form of measure of the hazard ratio is risk of death. (Adapted from Cassano PA, Rosner B, Vokonas PS, Weiss ST. Obesity and body fat distribution in relation to the incidence of non-insulin-dependent diabetes mellitus. Am J Epidemiol 1992;136:1474–1486, with permission.)

gender, race, and socioeconomic class in both developed and developing countries. The prevalence of obesity of older men and women has also increased in the last 10 years in the United States, partly because of an increasing aging population. The prevalence of obesity over 70 years of age has increased from 11.4% in 1991 to 14.6% in 1998 in United States. In addition, the prevalence of obesity increased progressively from adolescence (18–29 years: 12.1% in 1998) to age 50 to 59 years (23.8% in 1998) and then decreased progressively with age thereafter (≥70 years: 14.6% in 1998).[14] Studies conducted in other countries point toward similar trends. Geographic and ethnic differences in the prevalence were large. Obesity is common in the sixth and seventh decades of life and then the rates declines. Many obese older persons were obese as middle-aged adults. After middle age, decrease in mean BMI was found in both men and women.[15,16] It is interesting to note that the prevalence of obesity is low in elderly Asian men (Japan: 0.99%, Taiwan: 3.2% in 1991) but high in Middle Eastern and Western elderly men (Kuwait: 32.3%, Mexico: 15.6%, Sweden: 6.9%, United States: 15.2% in 1989).[1,17] Comparative data of the decline of salivary testosterone with age seem to suggest that it is more rapid in Western as compared with Asian or South American men. In general, mean BMI among elderly women is greater for men of a similar age. Elderly men by and large have a lower prevalence of being

overweight and being obese as compared with elderly women.[17,18]

Consequences of Obesity in Aging Men

Obesity is associated with an increased incidence of cardiovascular diseases, type 2 diabetes mellitus, hypertension, stroke, hyperlipidemia, sleep apnea, osteoarthritis, and some cancers.[2] After the age of 60, body weight on average tends to decrease, but fat tends to be redistributed with advancing age toward more visceral fat.[19] BMI has often been used as an indicator of overall obesity. Using BMI as a weight measure can provide an indirect measure of fatness but does not reflect fat distribution, which may reflect the risk of comorbidity independent of BMI. Moreover, BMI does not distinguish between fat mass and lean muscle mass and may sometimes underestimate fatness in older adults who have greater amount of fat at a given BMI than younger adults, due to age-related decline in muscle mass.[20]

Several studies have found patterns of fat distribution, particularly abdominal or central obesity, to be more closely related than BMI for increased incidence of diabetes mellitus and cardiovascular diseases.[21,22]

Central or abdominal obesity is represented by the ratio of waist to hip circumferences (WHR). A WHR in men greater than 1.0 indicates abdominal fat accumulation and is a risk factor for diabetes, hypertension, hyperlipidemia, and cardiovascular diseases (WHO 2000). The middle-age increase in BMI and proportion of body fat typically occurs in men after 40 years of age, and may extend to the fifth and sixth decade of life. A study on aging men confirms the correlation between abdominal obesity and the risk of diabetes (Fig. **11–2**).[21] Several studies have found abdominal obesity to be associated with cardiovascular disease in aging men.[22,23] Greater BMI was associated with higher mortality from all causes and from cardiovascular diseases in men and women up to 75 years of age. However, the relative risk associated with greater BMI declined after 75 years of age in older men and women.[16] Heiat et al[24] reviewed 444 articles of overweight and obesity in elderly persons. These data do not support the BMI range of 25 to 27 as a risk factor for all-cause and cardiovascular mortality among elderly persons. Federal guideline standards for ideal weight (BMI 18.7 to <25) may be overly restrictive as they apply to the elderly. BMI was not correlated with mortality among persons 75 years or older. However, WHR was a significant risk factor for coronary heart disease (CHD) among men 65 years or older. A longitudinal study revealed that mortality was lower in obese men compared with thin and normal older people age 70 and older.[18] However, remaining free of obesity was one of the factors associated with maintenance of good health in older adults.[23] The best denominators for predicting survival and sustained freedom from clinical illness and from physical and cognitive impairment in older Asian men were low blood pressure, low serum glucose, not smoking cigarettes, and not being obese.[25]

Summary of Study on Relationship Between Obesity, Diabetes, and Hypogonadism

As part of an ongoing quality improvement program to improve care to aging men in our geriatric clinic, we assessed and treated 71 consecutive males. Many of the patients had multiple illnesses including hypertension, diabetes, hyperlipidemia, dementia, and osteoarthritis. The routine geriatric assessments were performed including diagnoses of these patients' age, weight, smoking, drinking, BMI, functionality scales [activities of daily living (ADL), instrumental activity of daily living (IADL), nutrition], and cognitive assessment scales [Mini–Mental Status Examination (MMSE), Clock drawing test (CDT)]. Patients had their total and free testosterone and PSA levels determined if they presented with symptomatic andropause such as fatigue, loss of libido, erectile dysfunction, mood, and memory changes. Serum total testosterone was measured by the radioimmunoassay method. Data were entered into Excel, and descriptive and comparative analyses were performed using Statistical Package for Social Scientists (SPSS).

The average age of the patients was 73 years. All 71 patients had at least one of the andropause symptoms. The mean total testosterone was 405 ng/dL (range: 32–877 ng/dL). Hypogonadism was defined as total testosterone <300 ng/dL. Hypertension was defined as systolic blood pressure >140 mm Hg or diastolic pressure >90 mmHg or current use of antihypertensive medication. Diabetes mellitus was defined as fasting blood glucose ≥126 mg/dL or symptoms of diabetes and random plasma glucose ≥200 mg/dL or current taking insulin or oral hypoglycemic agents. Hyperlipidemia was defined as total cholesterol ≥200 mg/dL or low-density lipoprotein (LDL) cholesterol ≥130 mg/dL or triglyceride ≥160 mg/dL or current use of lipid-lowering agents. Thirty-one percent of this group of patients had hypogonadism, 33% had diabetes mellitus, 76% had hypertension, and 18% had hyperlipidemia. Fifty-two percent of men with low testosterone (<300 ng/dL) had diabetes mellitus. Patients with diabetes were significantly more likely to be hypogonadic ($p = .031$). Prevalence of hypogonadism in patients with diabetes was 64%, and in those without diabetes was 38%. However, we did not find any significant association of hyperlipidemia or

hypertension with hypogonadism ($p = .558$ and .729, respectively). On the other hand, BMI > 27 was significantly related to hypogonadism ($p = .026$). Fig. **11–3** summarizes the results of our study, showing an inverse relationship between BMI and total testosterone.

Thirty-one percent of men who presented with andropausal symptoms had low total testosterone levels. Some studies reported that bioavailable testosterone identifies a higher proportion of men as hypogonadal than total testosterone.[26] Obesity and diabetes in our group is associated with hypogonadism. The Massachusetts Male Aging Study also reported that the two strongest single predictors of testosterone deficiency were treated diabetes and obesity.[27] Obese men were more likely to be hypogonadal perhaps because of increased aromatization of testosterone to estradiol in peripheral fat tissue[5] and alteration of the hypothalamic-pituitary-testicular axis.[11] Low testosterone may increase the activity of the hypothalamic-pituitary-adrenal axis and trigger the development of visceral obesity.[28] Low testosterone also decreases the activity of lipolysis of abdominal fat.[29] Some studies suggest that low testosterone plays some role in the development of insulin resistance and type 2 diabetes.[30–32]

The Complex Relationship Between Obesity, Body Composition, Testosterone, Growth Hormone, Insulin, Leptin, and Dehydroepiandrosterone

Aging in men involves a cascade of hormonal, biochemical, and physiological changes. As far as androgens are concerned, it is known that hypogonadal men have reduced lean body mass but an increased BMI and fat mass compared with eugonadal men of a similar age.[33] Vermuellen et al[34] reported in a study of 57 men aged 70 to 80 years that testosterone levels correlated negatively with percentage of body fat, abdominal fat, and plasma insulin levels. Moreover, the increase in fat mass as it occurs in aging men is in itself associated with low levels of free testosterone and growth hormone, which both normalize after weight reduction. A longitudinal male aging study reported that increased obesity may have preceded increased insulin levels or vice versa.[30] Tsai et al[31] established that a strong inverse relationship between the baseline level of total testosterone and accumulation of computed tomography (CT)-measured intraabdominal fat area over time in aging male. The study also showed C-peptide inversely correlated with total testosterone. This would suggest that by predisposing to an increase in visceral adiposity, low levels of testosterone might increase the risk of type 2 diabetes mellitus. Chang et al[35] also showed that elderly men with type 2 diabetes had a higher WHR, BMI, skinfold

Graph of Distribution of Body Mass Index Against Total Testosterone

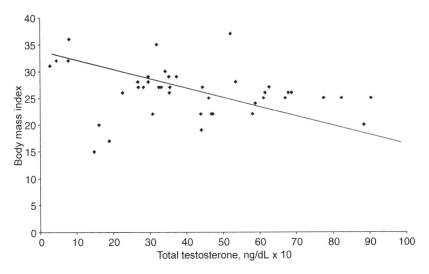

FIGURE 11–3. Relationship between body mass index and total testosterone.

thickness, and lower serum testosterone levels compared with elderly men without type 2 diabetes. A study of men born in 1913 reported that higher BMI, WHR, and lower testosterone are predictors for the development of diabetes. The diabetic state predicts strongly the incidence of myocardial infarction and stroke.[36]

Many studies have shown an association between relative hypogonadism and insulin resistance. Obesity is the most common cause of insulin resistance. Visceral obesity is associated with an increase in free fatty acid levels and decreased hepatic insulin binding and extraction.[37] Intraabdominal fat may be part of the pathway through which lower testosterone is related to insulin resistance.[31] Marin et al[38] demonstrated that testosterone treatment could decrease visceral fat mass, insulin resistance, blood glucose, diastolic pressure, and serum cholesterol in abdominally obese men. These changes might act via the effects of testosterone on visceral fat or via direct effect on muscle insulin sensitivity. Dihydrotestosterone (DHT) treatment increases visceral fat mass and did not change insulin sensitivity and blood glucose.[28] Older diabetic men with lower testosterone level are also associated with diabetic dyslipidemia.[39] However, our understanding of the effect of testosterone and glucose metabolism is incomplete. In human studies, low-dose testosterone (replacement dosage) may increase insulin action, whereas high-dose testosterone (pharmacological dosage) appears to decrease insulin action.[40]

Leptin, the adipocyte-secreted protein product of the *ob* gene, has been strongly linked to obesity and is postulated to regulate weight and the adipose tissue mass by signaling satiety or hunger and energy expenditure or conservation.[41] Several studies reported that leptin was correlated positively with age, BMI, serum insulin, and fat mass, and inversely with testosterone in aging men. The study of Saffele[42] in healthy elderly men showed leptin level correlated positively with age, BMI, and insulin levels. Free testosterone levels were negatively correlated with age and serum leptin. Leptin level is higher in aging men with lower testosterone, and testosterone replacement normalizes elevated serum leptin levels in hypogonadal men.[43–45] Cohen[5] hypothesizes that the increased adipose tissue in aging men is associated with an increase in the enzyme aromatase that converts testosterone to estradiol and leads to diminishing testosterone levels. Low testosterone may contribute to the accumulation of visceral fat.[5,28,46] As total body fat mass increases, hormone resistance develops for leptin and insulin. Increasing leptin fails to prevent weight gain and the hypogonadal-obesity cycle ensues, causing further visceral obesity and insulin resistance. The progressive insulin resistance can lead to diabetes and high triglyceride–low high-density lipoprotein (HDL) pattern dyslipidemia, so-called metabolic syndrome, and increase risk of cardiovascular diseases.[5] It has been reported that hypotestosteronemia may be a risk factors for coronary arteriosclerosis in men.[47]

Testosterone is not the only androgen that declines with age. Dehydroepiandrosterone (DHEA) also declines with age. A well-known study demonstrated that dehydroepiandrosterone sulfate (DHEAS)

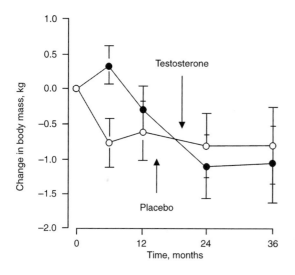

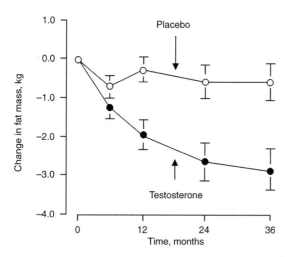

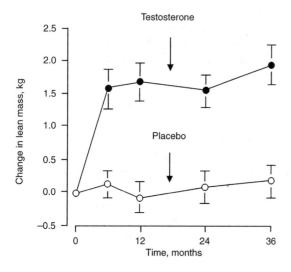

concentration is independently and inversely related to death from any cause and death from cardiovascular disease in men over age 50.[48] The Massachusetts Male Aging Study also supports the hypotheses that low DHEA and DHEAS can predict ischemic heart disease in men.[49] DHEA is available without prescription in the United States. However, there are still insufficiency data to advise routine DHEA supplement in elderly men.[50]

Impact of Testosterone Replacement on Obesity and Body Composition

Total, free, and bioavailable testosterone levels decline progressively with age in men. Lower testosterone levels in men are associated with decreased muscle mass, increased body fat, and visceral obesity.[34] Only a few studies focus on testosterone replacement on obesity and body composition in aging men. Synder et al[51] analyzed 108 men over 65 years of age who were randomized to receive either testosterone patch delivering testosterone 6 mg/day or placebo; 96 completed the entire 3 years of the study. The testosterone-treated subjects experienced a significant decrease in body fat mass and an increase in lean mass. There was no significant difference of BMI between the two groups (Fig. 11–4).

Sih et al[52] reported that testosterone replacement with 200 mg testosterone cypionate biweekly for 12 months improved muscle strength and lowered leptin levels in older hypogonadal men over 65 years of age. Fig. 11–5 shows the effects of testosterone on leptin.

Testosterone did not alter body fat or BMI in this study.[52] A retrospective analysis of testosterone replacement with 200 mg testosterone enanthate or cypionate every 2 weeks in older hypogonadal men showed no significant differences in body weight and BMI in two groups during 2 years follow-up.[53] Morley et al[52] treated eight hypogonadal men, aged 79 to 89 years, with testosterone enanthate 200 mg every 2 weeks for 3 months, and observed no changes in fat mass or body weight, although they did observe an increase in grip strength. Tenover[54] treated 70 hypogonadal men, aged 65 to 83 years, with testosterone enanthate 150 mg twice a week for up to 3 years and observed an increase in lean body mass as well as decrease in body fat. Wang et al[55] treated 227

FIGURE 11–4. Effect of testosterone on body fat and lean mass in men over 65 years of age. (Adapted from Synder PJ, Peachey H, Hannoush P, et al. Effect of testosterone treatment on body composition and muscle strength in men over 65 years of age. J Clin Endocrinol Metab 1999; 84:2647–2653, with permission.)

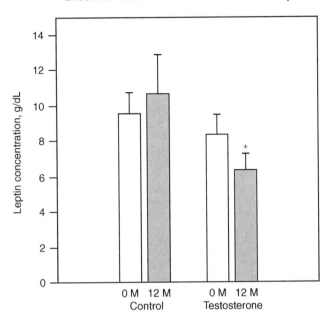

Effects of Testosterone Administration on Leptin

FIGURE 11–5. **Effect of testosterone on leptin, suggesting testosterone can decrease leptin resistance. Leptin is often elevated in obesity and the rise is a measure of leptin resistance. Testosterone has been shown in clinical trials to decrease leptin suggesting that it may control obesity. (Adapted from Sih R, Morley JE, Kaiser FE, Perry HM, Patrick P, Ross C. Testosterone replacement in older hypogonadal men: a 12-month randomized controlled trial. J Clin Endocrinol 1997;82:1661–1667, with permission.)**

hypogonadal men with transdermal testosterone gel (50–100 mg/day) for 180 days and observed an increase in lean mass and a decrease in fat mass. Taken together, these data suggest that variation in responsiveness of fat to testosterone replacement depends on the duration of therapy. Other factors such as pretreatment body composition, the method of body composition assessment, and the age of the subjects might also affect the responsiveness of body fat to testosterone administration. Testosterone replacement is absolutely contraindicated in men with prostate cancer or breast cancer. Thorough urological evaluation should be performed before testosterone replacement and frequent monitoring of PSA in older men during receiving testosterone replacement is indicated. The other potential risks of testosterone therapy in men include leg swelling, gynecomastia, sleep apnea, and polycythemia. Liver toxicity has not been a significant issue with injectable esters, transdermal patches, or gel.

Discussion of the Case History

Mr. L.K. did receive androgen therapy, in the form of compounded testosterone 5 mg with 100 mg of zinc in a topical form. The zinc was added to act as a potential aromatase inhibitor. Oral zinc is not tolerated well in larger doses as it can cause diarrhea and vomiting, and for that reason it was applied topically. The pharmacokinetics and pharmacodynamics of absorbed zinc has been studied and documented. Testosterone was prescribed as his levels were low normal, and he suffered the symptoms. The androgens given boosted his energy level, which allowed him to go the gymnasium 6 days a week. He did a combination of weight training and aerobic exercise in 30 minutes sessions each. His libido returned, although a phosphodiesterase inhibitor had to be prescribed to aid his erectile function. He did undergo a series of plastic surgeries over a 6-month period, including liposuction of his chin and abdomen. He subsequently underwent an abdoplasty as well. At a follow-up visit, he mentioned that he was very happy with himself, and that he is now religiously following his diet and exercise regimen. On examination, he had surgical scars from his surgery, body fat content is down to 23.8%, and blood pressure is 132/88 mm Hg. His laboratory data include total testosterone 560 ng/dL, estradiol 40 pg/dL, and total cholesterol 190 mg/dL. Overall, the patient looks different and he reports he feels different, and he believes he is living life again. This is an example of a dramatic and drastic attempt of an older man wanting to lose weight for an improved quality of life.

Conclusion and Key Points

In most studies mean body weight or BMI increases with age up to about age 60 to 70 and then declines with age. Some studies actually suggest a protective effect of overweight in the oldest age groups. Visceral obesity may be a better indicator of risk of diabetes mellitus and related cardiovascular disorders than BMI in aging men.

Obesity is an issue that is increasingly affecting aging men not only in developed countries but also developing ones. Excess wealth and poor education lead to bad eating habits. With aging, there is a decline in androgens as well. There are implications for the health of men as a result of hypogonadism. Overall, there seems to be an inverse relationship between BMI and testosterone levels. Severe obesity seems to suppress the production of testosterone. It has been hypothesized that there is increased aromatization of testosterone to estradiol and alteration of hypothalamic-pituitary-adrenal axis in obese older men. Some hormones can affect obesity in aging, including leptin, insulin, DHEA, and growth hormone. The relationship of obesity to these hormones in the aging men was reviewed. Testosterone replacement can alter body composition whereby fat may be exchanged for muscle. Further studies in this field are recommended to evaluate long-term benefits and risks.

- Obesity is reaching epidemic portions not only in developed nations but also in the developing world.

- Overall, there seems to be an inverse relationship between BMI and testosterone.

- Obesity not only affects cardiovascular and cerebrovascular morbidity, but also may impact androgens and hence quality of life.

- Low testosterone is associated with insulin resistance and high leptin.

- There is a possibility of a reduction in obesity and the associated metabolic syndrome in some cases with physiological doses of testosterone replacement.

REFERENCES

1. Flegal KM, Carroll MD, Kuczmarski RJ, Johnson CL. Overweigh and obesity in the United States: prevalence and trends, 1960–1994. Int J Obes Relat Metab Disord 1998;22:39–47

2. Must A, Spadano J, Coakley EH, Field AE, Colditz G, Dietz WH. The disease burden associated with overweight and obesity. JAMA 1999;282:1523–1529

3. Harman SM, Metter EJ, Tobin JD, Pearson J, Blackman MR. Longitudinal effects of aging on serum total and free testosterone levels in healthy men. Baltimore Longitudinal Study of Aging. J Clin Endocrinol Metab 2001;86:724–731

4. Tan RS. *The Andropause Mystery.* Houston: AMRED Publishing, 2001

5. Cohen PG. Aromatase, adiposity, aging and disease: the hypogonadal-metabolic-atherogenic-disease and aging connection. Med Hypotheses 2001;56:702–708

6. Gray A, Feldman HA, McKinlay JB, Longcope C. Age, disease, and changing sex hormone levels in middle-aged men: results of the Massachusetts Male Aging Study. J Clin Endocrinol Metab 1991;73:1016–1025

7. Tan RS, Philip PS. Perceptions of and risk factors for andropause. Arch Androl 1999;43:97–103

8. Bhasin S, Buckwalter JG. Testosterone supplementation in older men: a rational idea whose time has not yet come. J Androl 2001;22:718–731

9. Haffner SM, Valdz RA, Stern MP, Katz MS. Obesity, body fat distribution and sex hormones in men. Int J Obes Relat Metab Disord 1993;17:643–649

10. Ukkola O, Gagnon J, Rankinen T, et al. Age, body mass index, race and other determinants of steroid hormone variability: the HERITAGE Family Study. Eur J Endocrinol 2001;145:1–9

11. Vermeulen A. Decreased androgen levels and obesity in men. Ann Med 1996;28:13–15

12. Winters SJ, Kelly DE, Goodpaster B. The analog free testosterone assay: are the results in men clinically useful? Clin Chem 1998;44:2178–2182

13. Nagai T, Mori M. Prader-Willi syndrome, diabetes mellitus and hypogonadism. Biomed Pharmacother 1999;53:452–454

14. Mokdad AH, Serdula MK, Dietz WH, Bowman BA, Marks JS, Koplan JP. The spread of the obesity epidemic in the United States, 1991–1998. JAMA 1999;282:1519–1522

15. Launer LJ, Harris T. Weight, height and body mass index distribution in geographically and ethically diverse samples of older persons. Age Ageing 1996;25:300–306

16. Steven J, Cai J, Pamuk ER, Williamson DF, Thun MJ, Wood JL. The effect of age on the association between body mass index and mortality. N Engl J Med 1998;338:1–7

17. Chiu HC, Chang HY, Mau LW, Lee TK, Liu HW. Height, weight and body mass index of elderly persons in Taiwan. J Gerontol A Biol Sci Med Sci 2000;55:M684–M690

18. Lerman-Garber I, Villa AR, Martinez CL, et al. The prevalence of obesity and its determinants in urban and rural aging Mexican populations. Obes Res 1999;7:402–406

19. Siedell JC, Visscher TL. Body weight and weight change and their health implications for the elderly. Eur J Clin Nutr 2000;54:S33–S39

20. Teh BH, Pan WH, Chen CJ. The relocation of body fat toward the abdomen persists to very old age, while body mass index declines after middle age in Chinese. Int J Obes Relat Metab Disord 1996;20:683–687

21. Cassano PA, Rosner B, Vokonas PS, Weiss ST. Obesity and body fat distribution in relation to the incidence of non-insulin-dependent diabetes mellitus. Am J Epidemiol 1992; 136:1474–1486

22. Grinker JA, Tucker KL, Vokonas PS, Rush D. Changes in patterns of fatness in adult men in relation to serum indices of cardiovascular risk: the normative aging study. Int J Obes Relat Metab Disord 2000;24:1369–1378

23. Burke GL, Arnold AM, Bild DE, et al. Factors associated with healthy aging: the cardiovascular health study. J Am Geriatr Soc 2001;49:254–262

24. Heiat A, Vaccarino V, Krumholz HM. An evidence-based assessment of federal guidelines for overweight and obesity as they apply to elderly persons. Arch Intern Med 2001;161:1194–1203

25. Reed DM, Foley DJ, White LR, Heimovitz H, Burchfiel CM, Masaki K. Predictors of healthy aging in men with high life expectancy. Am J Public Health 1998;88:1463–1468

26. Morley JE, Charlton E, Patrick P, et al. Validation of a screening questionnaire for androgen deficiency in aging males. Metabolism 2000;49:1239–1242

27. Smith KW, Feldmen HA, McKinlay JB. Construction and field validation of a self-administered screener for testosterone deficiency (hypogonadism) in aging men. Clin Endocrinol (Oxford) 2000;53:709–711

28. Marin P, Arver S. Androgen and abdominal obesity. Baillieres Clin Endocrinol Metab 1998;12:441–451

29. Marin P, Oden B, Bjorntorp P. Assimilation and mobilization of triglycerides in subcutaneous abdominal and femoral adipose tissue in vivo in men: effect of androgens. J Clin Endocrinol Metab 1995;80:239–243

30. Lazarus R, Sparrow D, Weiss S. Temporal relation between obesity and insulin: longitudinal data from the Normative Aging Study. Am J Epidemiol 1998;147:173–179

31. Tsai EC, Boyko EJ, Leonetti DL, Fujimoto WY. Low serum testosterone level as a predictor of increased visceral fat in Japanese-American men. Int J Obes Relat Metab Disord 2000;24:485–491

32. Goodman-Gruen D, Barrett-Connor E. Sex differences in the association of endogenous sex hormone levels and glucose tolerance status in older men and women. Diabetes Care 2000;23:912–918

33. Katznelson L, Finkelstein JS, Schoenfeld DA, Rosenthal DI, Anderson EJ, Klibanski A. Increase in bone density and lean body mass during testosterone administration in men with acquired hypogonadism. J Clin Endocrinol Metab 1996;81:4358–4365

34. Vermeulen A, Goemaere S, Kaufman JM. Testosterone, body composition and aging. J Endocrinol Invest 1999;22:110–116

35. Chang TC, Tung CC, Hsiao YL. Hormonal changes in elderly men with non-insulin-dependent diabetes mellitus and the

hormonal relationship to abdominal adiposity. Gerontology 1994; 40:260–267

36. Tibblin G, Adlerberth A, Lindstedt G, Bjorntorp P. The pituitary-gonadal axis and health in elderly men. A study of men born in 1913. Diabetes 1996;45:1605–1609

37. Lemieux S. Contribution of visceral obesity to insulin resistance syndrome. Can J Appl Physiol 2001;26:273–290

38. Marin P, Holmang S, Jonsson L, et al. The effects of testosterone treatment on body composition and metabolism in middle-aged obese men. Int J Obes Relat Metab Disord 1992;16:991–997

39. Barrett-Connor E. Lower endogenous androgen levels and dyslipidemia in men with non-insulin-dependent diabetes mellitus. Ann Intern Med 1992;117:807–811

40. Rizza R. A. Androgen effect on insulin action and glucose metabolism. Mayo Clin Proc 2000;75(suppl):S61–S64

41. Bray GA, York DA. Clinical review 90: leptin and clinical medicine: a new piece in the puzzle of obesity. J Clin Endocrinol Metab 1997;82:2771–2776

42. Saffele JKV, Goemaere S, Bacquer DD. Serum leptin levels in healthy ageing men: are decreased serum testosterone and increased adiposity in elderly men the consequence of leptin deficiency? Clin Endocrinol (Oxford) 1999;55:81–88

43. Perry HM III, Miller DK, Morley PP. Testosterone and leptin in older African-American men: relationship to age, strength, function, and season. Metabolism 2000;49:1085–1091

44. Jockenhovel F, Blum WF, Vogel E, et al. Testosterone substitution normalizes elevated serum leptin levels in hypogonadal men. J Clin Endocrinol Metab 1997;82:2510–2513

45. Luukkaa V, Pesonen U, Huhtaniemi I, et al. Inverse correlation between serum testosterone and leptin in men. J Clin Endocrinol Metab 1998;83:3243–3246

46. Marin P. Testosterone and regional fat distribution. Obes Res 1995;(Suppl 4):609S–612S

47. Phillips GB, Pinkernell BH, Jing TY. The association of hypotestosteronemia with coronary artery disease in men. Arterioscler Thromb 1994;14:701–706

48. Barrett-Connor E, Khaw KT, Yen SS. A prospective study of dehydroepiandrosterone sulfate, mortality, and cardiovascular disease. N Engl J Med 1986;315:1519–1524

49. Feldman HA, Johannes CB, Araujo AB, Mohr BA, Longcope C, McKinlay JB. Low dehydroepiandrosterone and ischemic heart disease in middle-aged men: perspective results from the Massachusetts Male Aging Study. Am J Epidemiol 2001;153: 79–89

50. Porsova-Dutoit I, Sulcova J, Starka L. Do DHEA/DHEAS play a protective role in coronary heart disease? Physiol Res 2000;49(suppl 1):S43–S56

51. Synder PJ, Peachey H, Hannoush P, et al. Effect of testosterone treatment on body composition and muscle strength in men over 65 years of age. J Clin Endocrinol Metab 1999;84: 2647–2653

52. Sih R, Morley JE, Kaiser FE, Perry HM, Patrick P, Ross C. Testosterone replacement in older hypogonadal men: a 12-month randomized controlled trial. J Clin Endocrinol Metab 1997;82:1661–1667

53. Hajjar RR, Kaiser FE, Morle JE. Outcomes of long-term testosterone replacement in older hypogonadal males: a retrospective analysis. J Clin Endocrinol Metab, 1997;82: 3793–3796

54. Tenover JS. Effects of testosterone supplementation in aging male. In: 1st World Congress, "The Aging Male" Geneva: 1997

55. Wang C, Swerdloff RS, Iranmanesh A, et al. Transdermal testosterone gel improves sexual function, mood, muscle strength, and body composition parameters in hypogonadal men. Testosterone Gel Study Group. J Clin Metab 2000;85: 2839–2853

12

Arthritis in Aging Men

CARLOS A. PLATA AND ROBERT S. TAN

Case History

Mr. L.C. is a 75-year-old man who presented to the Geriatrics Evaluation and Management Clinic after being referred by his primary care physician. His main complaint was that he was not able to cope at home. His wife died 6 months ago, and he confesses to being sad. His main concern was that his mobility was limited partly because of "weakness of his knees." He said that he is unable to walk because of pain and that he was dependent on a walker. His past medical history included insulin-dependent diabetes mellitus, hypertension, and osteoarthritis. He was also recently diagnosed by his primary physician to have rheumatoid arthritis (RA) of his left knee after he presented with knee swelling and had synovial fluid aspirated. Physical examination revealed a frail gentleman, weighing 187 pounds. His blood pressure was 176/86 mm Hg and cardiopulmonary examination was normal. He had clinical suggestion of hypogonadism because of loss of axillary hair and testicular atrophy. Muscle strength in the lower extremities was 4/5 bilaterally and reflexes were diminished. There was a mild degree of genu varus of the knee, but the right knee was noticeably swollen and felt doughy. There was presence of Heberden's nodes in both hands, but his metacarpophalangeal joints were normal. His Folstein Mini–Mental State Examination was 27/30, suggesting some degree of cognitive impairment. His Geriatric Depression Score was 3/15, suggesting that he was not clinically depressed at this stage. Blood work revealed that his blood chemistry, thyroid-stimulating hormone (TSH), and prostate-specific antigen (PSA) were within normal range. Of note was the hemoglobin at 11.5 g/dL, positive rheumatoid factor, erythrocyte sedimentation rate (ESR) 66 mm/hour, total testosterone 190 ng/dL (low), free testosterone 24.4 ng/dL (low), follicle-stimulating hormone (FSH) 9.6 mIU/mL (elevated), and HbA_{1c} 8 (elevated).

To summarize, Mr. L.C. was diagnosed to have a fall risk from arthritis and muscle wasting. In addition, he has suggestions of early dementia but was not clinically depressed. The rise in FSH is consistent with Leydig cell failure, leading to andropause. He needed better control of his blood pressure and diabetes. Incidentally too, he was hypogonadal as evidenced by examination and laboratory testing. The treatment plan included increasing the dose of his antihypertensive and diabetic medications. He was given testosterone for his hypogonadal state. His arthritis medications were unchanged.

Arthritic Problems in Aging Men

"Arthritis" can mean more than a hundred different diseases with different etiologies, manifestations, and prognoses. Further, arthritis has always been associated with aging. We have heard our parents and grandparents complain of pain, and relate it to their age. However, the aging process does not imply the presence of arthritis. There are physiological changes in bone, joint, and muscle related to age, but none of those cause symptoms that can be diagnosed as arthritis. Nevertheless, the prevalence of the different types of arthritis in general increases with age, making this group of diseases the leading cause of disability in the elderly.[1,2]

The elderly man can be affected by any of the different diseases we know as "arthritis." From osteoarthritis, the most common type, to systemic lupus erythematosus, all can cause morbidity in this population group. This chapter discusses aspects of rheumatology that affect aging men such as osteoarthritis, RA, and polymyalgia rheumatica/giant cell arteritis. Ankylosing spondylitis and other spondyloarthropathies and collagen vascular diseases are also discussed.

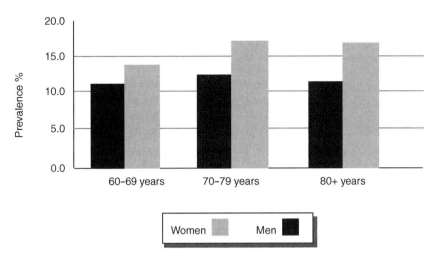

FIGURE 12–1. Prevalence of significant hip pain on most days in older adults, stratified by age and sex. Osteoarthritis and symptoms are reported more frequently in females. (Adapted from Christmas C, Crespo CJ, Franckowiak SC, et al. How common is hip pain among older adults? Results from the Third National Health and Nutrition Examination Survey. J Fam Pract 2002;51(4), with permission.)

Osteoarthritis is the most common of the diseases of the joints.[3,4] It affects a large percentage of the population causing significant disability.[5] Its incidence and prevalence increases progressively with age, with persons being affected starting at variable age depending on the cause of the disease. Fig. **12–1** shows the rates of reporting of hip pain in both men and women. Osteoarthritis is *not* a normal consequence of aging. Actually, the changes noted in aging cartilage are in many ways the *opposite* of the changes noted in osteoarthritic cartilage.[4,6] For example, the aging cartilage has a decrease in the amount of fluid and the number of chondrocytes, and an increase in the amount of collagen and in hyaluronate content. Osteoarthritic cartilage has an increase of the amount of fluid, due to the fibrillation of the surfaces, changes in the amount in hyaluronate, and an initial increase in the amount of chondrocytes to try and repair the damage. Eventually the cartilage cannot repair itself anymore, and a loss of volume occurs.

The etiology of osteoarthritis is unclear.[4,6] Theories suggest the presence of poor cartilage submitted to normal stress, the presence of normal cartilage submitted to excessive stress, or a combination of both. The *poor-cartilage theory* explains the nodular osteoarthritis of the hands, which is clearly inherited, affecting mostly females, and the osteoarthritis associated with diseases like hemochromatosis or hypothyroidism. The *excessive stress theory* explains the osteoarthritis seen after direct joint trauma, the one noted in persons in specific occupations or in overweight individuals. However, primary osteoarthritis cannot be in many cases clearly explained by one of

the mechanisms, and is probably a combination of both. Osteoarthritis affects more females than males,[7] and it is probably due to the *predominance of estrogen receptors* in the bone and cartilage cells that modify the rate of metabolism.[8]

Osteoarthritis affects a large percentage of the population.[9] It usually manifests with pain that occurs *with activity*. With time, pain occurs with less activity and eventually starts to bother the individual at rest. Patients usually feel a short-lived sensation of stiffness in the joints, like the joints are "glued together," that lasts less than half an hour. Joint inflammation is usually not a component of the disease. However, "flares" may occur that can cause synovial fluid accumulation and effusions. These are commonly present after episodes of overactivity. Some patients may have calcium deposits in the cartilages, causing chondrocalcinosis. These patients are exposed to acute joint swelling episodes and are diagnosed with pseudogout. *Primary* osteoarthritis affects knees, hips, and digits, mostly at the distal interphalangeal joints. *Secondary* osteoarthritis may affect other joints, which may help to diagnose the origin of the disease.[10] For example, involvement of the metacarpophalangeal joints suggests the presence of hemochromatosis, or damage to wrists suggests the presence of chondrocalcinosis. Heberden and Bouchard nodules are typical of osteoarthritis of the hands, and are significantly more common in females. These changes are usually familial in origin.

Regrettably we still do *not* have a specific treatment for osteoarthritis. However, that does not mean that there is not much to do, as many patients are told.

There are many ways to keep patients with osteoarthritis from suffering, and perhaps, to slow down the progression of disease.[11-13] The initial part of the therapy is education. Many patients feel that the disease is not treatable and that they will be disabled in a short time. That is not the case for most patients. An *exercise program*, directed to muscle strengthening and range of motion decreases pain in many cases by decreasing the load inside the joint, and improves results in case the patient eventually requires surgery. Use of a cane and weight reduction in patients who are overweight also decrease load and slow progression of disease. For patients with mild pain, the use of analgesics like acetaminophen may be enough. However, when pain becomes more severe, use of nonsteroidal antiinflammatory drugs (NSAIDs) is recommended. Both the American College of Rheumatology and the American Pain Society recommend the use of Cox-2–specific medications in this setting, because the elderly are at significant risk of gastrointestinal complications or nonspecific ones.[14] Use of a proton pump inhibitor or misoprostol with nonspecific NSAIDs are other options.

Local steroid injections are indicated for patients with "flares" or patients who have pseudogout episodes. However, repeated use of the intraarticular steroids may accelerate the progression of disease. When effusions are present, the muscles around a joint will be inhibited, causing significant loss of strength and muscle atrophy. Therefore, effusion should in general be drained. *Hyaluronic acid derivatives* have been approved for the use in osteoarthritis of the knee.[15] It is not clear what the mechanism of action is, but patients may have long-lasting pain control after the series of injections. These are usually prescribed for patients who cannot or will not take antiinflammatory medications, or for patients who still have pain with adequate medication use.

Chondroprotective medications, drugs that can improve or protect the cartilage from further damage, have been under investigation for a long time. *Glucosamine sulfate*, an over-the-counter product, suggests in limited studies a chondroprotective effect.[16] Doxycycline has also been found to have some chondroprotective effect in vitro, but clinical studies have not been conclusive.

When pain does not respond to nonpharmacological and pharmacological treatment or when deformity is progressive, patients are considered surgical candidates. Techniques have evolved significantly, and unicompartmental procedures are becoming more frequent. Total replacements of the knee and the hip have become common practice and have in general excellent results. Arthroscopic procedures are indicated in patients that have mechanical derangements in the joints, but have not been very helpful for patients with only osteoarthritic changes. Osteoarthritis will continue to be a major public health problem with the graying of our society. We are far from a cure, but adequate treatment will decrease significantly morbidity in our patients.

Rheumatoid arthritis is the most common of the inflammatory arthritides. It affects about 1% of the general population. However, its distribution varies depending on sex and age. It is more common in females and increases progressively with age. If untreated, it can cause significant disability and increased mortality. *Seropositive RA*, the disease that has a positive rheumatoid factor (~80% of patients), has not such a significant female preponderance. There is an unsolved question about the difference between RA presenting early in life and late-onset RA. Initial studies suggested that the elderly had a lower incidence of rheumatoid factor and that the disease seemed to be milder. However, later studies indicate that the disease seems to be *as aggressive in the elderly population* as in the younger one, requiring a similar approach to diagnosis and therapy.[17]

The cause of RA is not known. There is a genetic predisposition to the disease related to the presence of human leukocyte antigen (HLA)-DR4 or similar histocompatibility antigens. Theories have included infections with different *viruses*, like the Epstein-Barr virus, or bacteria, like mycoplasma or other intracellular organisms. These have been found by polymerase chain reaction (PCR) in joints of affected individuals. However, whatever the cause, it seems to be only a trigger for a faulty immunological response that arms a cell-mediated attack to the synovium and other tissues. The cytokine balance is abnormal in patients with RA, presenting increased amount of tumor necrosis factor-α and other cytokines with decrease of interferon-γ and other cytokines. Regardless of the origin, the inflammatory infiltrate in the joint creates significant synovial growth and erosion of the bone. If left unchecked, that disorderly growth can destroy the cartilage and the periarticular bone, causing complete joint destruction and in some cases ankylosis.[18] RA is a *systemic disease* that may affect other organs like the pleura, the pericardium, the heart, and blood vessels. Large effusions may be the presenting characteristic of the disease, and nodules may affect the heart rhythm. Systemic vasculitis and leucopenia with splenomegaly (Felty's syndrome) are very serious complications of the extraarticular disease.

Rheumatoid arthritis usually presents with slowly progressive symptoms that become more prevalent over the course of months.[19] Patients initially feel frequent tiredness and morning stiffness, usually associated with inflammatory changes in the joints of the hands and occasionally the feet. With time the pain and stiffness affects other joints, including shoulders, elbows, knees, ankles, and hips. Temporomandibular and neck joints are affected later, and the thoracic and lumbar spine are rarely affected. A minority of patients may present with an explosive onset of disease, or with an affection of only one or two joints. There is a belief among many rheumatologists that the disease tends to start *more acutely in the elderly*. However, it is difficult sometimes to differentiate polymyalgia rheumatica, which usually has an acute onset, from RA. Extraarticular manifestations are rare early in the disease, but large pleural effusions may be the heralds of RA.

The treatment of RA has changed significantly in the last 10 years, and especially in the last 3 or 4. By the last decade of the twentieth century it was clear that RA caused not only very significant morbidity, with loss of income and independence for the affected individuals, but also a significant increase in mortality. Also it was noted that damage to the joints occurred early in the disease, and that most of the impact of the disease took place within 3 years of diagnosis.[20] Therefore, it was important to treat the disease early and aggressively to avoid that initial joint damage. It was clear that NSAIDs, although they decreased pain and somewhat decreased inflammation, did not change the progression of disease. The pyramid of treatment, which allowed for initial nonpharmacological treatment, then use of NSAIDs, and then the slow introduction of drugs considered at that time modifiers of disease (gold salts, penicillamine), was abandoned. The introduction of methotrexate gave rheumatologists a drug that was easier to use and that improved significantly the joint inflammation.[21,22] Combination therapy with hydroxychloroquine and sulfasalazine and the use of low-dose steroid therapy were later introduced for patients who did not respond adequately to methotrexate.[23] Some rheumatologists, following the lead of oncologists, decided to use a very aggressive "remission" treatment with combination therapy at the beginning of disease.[24] Leflunomide was a new option for treatment to replace methotrexate, or to complement treatment of patients who failed methotrexate.[25] The introduction of the biologicals, tumor necrosis factor-α inhibitors (etanercept, infliximab, adalimumab), and interleukin-1 inhibitors (anakinra), was a revolution in the treatment of the disease.[26]

At this time the treatment of RA should be considered an emergency. It can be parallel to the treatment of hypertension or hyperlipidemia. The goal is to control the disease before it can cause any significant anatomical damage or any major disability. There has been a question of whether the treatment in the elderly should be different, but studies suggest that the impact of the disease is similar in all age groups and all should be treated aggressively.[27] Therefore, when a diagnosis of RA is made, treatment should be started with a disease-modifying agent. Because so many are available, with different toxicity and activity profiles, a rheumatologist or a physician with experience in the use of these medications should be consulted. The prevalence of rheumatoid arthritis has decreased in our society, but the disease still causes significant morbidity and increases mortality. New discoveries have revolutionized treatment, and although we still are far from knowing the specific cause of the disease, we can now improve the quality of life and slow or stop the progression of disease in many, if not most, patients.

Polymyalgia rheumatica is a disease that causes significant morbidity in the elderly. Its causes are not known. Theories, similar to those of RA, have included viral and bacterial infections. It affects only the elderly, being extremely rare before age 55. It has a significant predominance in females over males. However, most affected females are postmenopausal, making a hormonal cause unlikely. There are no adequate pathology reports because the disease is usually self-limited and does not cause any deformity. The disease is clearly inflammatory, because its hallmark is elevation of the sedimentation rate. Whole-body bone scan has shown increased uptake in the shoulders and hips. The disease presents acutely—the patient usually goes to sleep feeling well and wakes up with very severe stiffness that affects mostly the shoulders and the hips. Some patients have such severe disease that they may not be able to get out of bed without help. Low-grade temperature, loss of appetite, general malaise, and loss of weight are frequent symptoms.[28] Some patients may respond to treatment with NSAIDs, but most of them will require a low dose of glucocorticoids. The disease is frequently self-limited, but may last 1 year or longer. Therefore, patients require usually long-term glucocorticoids.[29] These patients should be educated about the risks of the medications, and required bone density tests and osteoporosis prophylaxis with calcium, vitamin D, and a diphosphonate.

Temporal arteritis or giant cell arteritis is a disease that is usually associated with polymyalgia rheumatica. It is unknown if those two diseases are different manifestations of the same pathological process or two independent entities that have similar trigger mechanisms. Giant cell arteritis affects *mostly elderly women*, though not exclusively. It presents mostly in patients who have symptoms of polymyalgia rheumatica. It causes inflammatory changes in the walls of the branches of the carotid arteries or the ascending aorta. It affects mostly the extracranial vessels.[30] It is a very inflammatory disease causing frequently fever, malaise, weight loss, and extreme fatigue. Patients present with severe headache, usually located in the temporal areas, and with claudication of the areas irrigated by the extracranial branches of the ascending aorta, especially the temporal muscles. The most dreaded consequence of the disease is blindness. It is initially unilateral, but if not treated, can rapidly affect the contralateral side. Diagnosis of temporal arteritis can be difficult. It should be suspected in patients with polymyalgia rheumatica who present also with headaches or jaw claudication. However, if it does not present in the company of polymyalgia rheumatica, it can be confused with migraines, cardiovascular complications, or cerebrovascular incidents, with awful results. In some cases a fever of unknown origin in an elderly individual may be the only manifestation of temporal arteritis. The sedimentation rate and the C-reactive protein (CRP) are usually elevated, which helps in making the diagnosis. When the disease is suspected, treatment with high-dose glucocorticoids should be started without waiting for a pathology report. However, the only way to confirm the diagnosis is by temporal artery biopsy. In general, the surgeon should obtain a long segment, and be ready to do a bilateral procedure if the preliminary report is negative. The treatment is with high-dose glucocorticoids, for example, 60 to 80 mg of prednisone in divided doses. The dose is decreased slowly during the next few months, and most patients need treatment for at least 1 year.

Ankylosing spondylitis is an inflammatory disease that affects mostly *men in the third and fourth decades of life*. It presents in about 0.5% of the general population. Its etiology is not clearly known, but most patients with the disease present with histocompatibility antigen HLA-B27 or a similar one.[31] The prevalence of the disease in a population parallels the frequency of that antigen in the same group. It seems to be triggered by an infectious agent, and *Shigella, Salmonella,* and *Chlamydia* have been blamed. The disease starts with alternating pain and stiffness in the gluteal area that slowly ascends through the spine, causing significant pain and limitation of range of motion. If the disease is not treated, patients may lose range of motion of the spine, and of the large peripheral joints. The disease is frequently associated with uveitis, fibrotic lung changes, and nonrheumatic aortic insufficiency. Fig. **12–2** depicts a patient with uveitis, which is often associated with ankylosing spondylitis. Elderly patients usually have had the disease for a long time, resulting in significant limitations. The disability caused by the disease can be severe. However, even after this long disease duration, it may remain active, and many patients require aggressive treatment to preserve use of peripheral joints like the hips or knees.

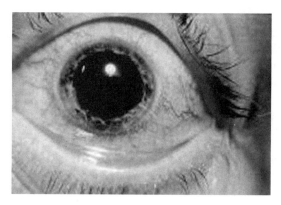

FIGURE 12–2. Anterior uveitis commonly associated with ankylosing spondylitis.

The diagnosis of ankylosing spondylitis is clinical. There are sets of criteria for diagnosis, including the New York[32] and the European[33] sets. These criteria are important for clinical research, but in clinical settings a more empirical approach is required. Patients have the previously indicated symptoms and x-rays or bone scan suggesting inflammation of the sacroiliac joints. Computed tomography (CT) scan is very sensitive, but its cost is generally not justified. The use of HLA-B27 as a diagnostic test is *not* recommended in general because the prevalence of that gene in the general population is significantly higher than the one for the disease. In the elderly the diagnosis is usually easy because the disease has progressed for many years and patients present with significant loss of range of motion of the lumbar spine in all planes. X-rays show syndesmophytes and in many cases complete obliteration of the sacroiliac joints.

Treatment of the disease has changed significantly. Until recently the NSAIDs, and especially older drugs like indomethacin, were the only accepted treatment. Sulfasalazine and methotrexate have been used in doses similar to the ones used in RA with mixed results.[34,35] The introduction of the *tumor necrosis factor-α inhibitors* has made a big difference for these patients.[36] The inflammatory component of the disease responds very well, preventing in many cases progression of disease. In the elderly, the treatment should remain aggressive if the disease is active and the patient has severe pain, or progression can impact activities of daily life. These patients may need a combined approach that may include aggressive rehabilitation and in many instances surgery. Therefore, these patients benefit from a referral to a rheumatologist or to a physician with experience in the management of these diseases.

Collagen-vascular diseases are a group of systemic autoimmune diseases that include systemic lupus erythematosus, Sjogren's syndrome, polymyositis/dermatomyositis, and scleroderma; there are overlaps between these diseases.[37] The etiology of these diseases is unknown, but all share the presence of autoantibodies directed to the nucleus of the cell. They have similar genetic predisposition. The target of the inflammatory response and the symptoms help us differentiate among them. They are diseases more commonly diagnosed in females, and in younger groups. However, new onset of these diseases has been described in all ages and all sexes. Therefore, an elderly man who presents with fever, sun-sensitive rash, arthralgia and arthritis, and a positive antinuclear antibody will have lupus, and will require an aggressive approach to diagnosis of affected major organs, and treatment depending on the findings.[38] Lupus in the elderly tends to be a less aggressive disease, with less glomerular compromise. However, treatment of lupus should be similar in all age groups and should be directed at the seriousness of the disease. Once again, the use of glucocorticosteroids has higher risks, especially for osteoporosis and worsening of arteriosclerotic lesions.

Dermatomyositis differs from the other autoimmune diseases. It has a significant association with malignancy. Therefore, patients who present with this disease should undergo an extensive evaluation. The disease, however, is not a paraneoplastic syndrome, and it progresses independently of the malignancy. Diagnosis of polymyositis/dermatomyositis is made in patients who present with significant muscle weakness and elevated muscle enzymes. Electrophysiological studies and muscle biopsy are important to confirm the diagnosis and rule out other myopathies. This is more important in the younger population that may have muscle dystrophies.

Scleroderma is usually *advanced* when seen in the elderly. In those cases making the diagnosis is easy, if it has not been made earlier. However, the treatment options are more limited, and are directed more to the complications of the disease, such as lung fibrosis, gastroesophageal reflux disease, esophageal stenosis, loss of digits, and chronic extremity pain.

Sjogren's syndrome is more frequent in postmenopausal women and is rare in men. However, it can cause fever of unknown origin, blindness, oral abnormalities, rashes, lymphadenopathy, and other symptoms that can be difficult to classify. Suspicion of Sjogren's syndrome can help explain many mystery diseases in the elderly.[39]

A comment is necessary on fibromyalgia.[40] Again, it is a disease more frequent in women. Many men have chronic pain, but have never been evaluated or treated for the disease. It has no specific diagnostic test. Patients present with morning stiffness, easy fatigability, and pain in areas around the joints. A diagnosis of RA is usually considered, but because no swelling is noted

and the rheumatoid factor and the acute reactants are negative, it is ruled out. On examination, the presence of tenderness in the fibromyalgia points is noted. It is important to make this diagnosis, because this disease is usually not progressive. Patients can be reassured, and treated with low-dose tricyclic antidepressants. An exercise program can yield great improvement. New medications are being introduced to treat pain specifically in fibromyalgia.

Rheumatoid Arthritis and Relationship to Androgens

Rheumatological diseases are less common in men than in women. It has been argued that this may be because of the protective effect of androgens, as men have a substantial larger amount of testosterone as compared with women. Indeed, women have only ~10% of the testosterone that men have. This may be a simplistic view; but studies in both men and women have found that there is association of low testosterone and dehydroepiandrosterone (DHEA) with rheumatoid disease. For example, in a study at Karolinska Institute in Sweden, it was found that men with RA had lower levels of bioavailable testosterone, and a large proportion were considered hypogonadal. The low levels of luteinizing hormone (LH) suggested a central origin of the relative hypoandrogenicity.[41] In this study basal serum concentrations of total testosterone (T), sex hormone–binding globulin (SHBG), and LH were measured in 104 men with RA, and the quotient T/SHBG were calculated. The data were compared with those of 99 age-matched healthy men. The results were analyzed separately for the age groups 30 to 49, 50 to 59, and 60 to 69 years. The RA men had lower biovailable testosterone levels than the healthy men in all age groups. Testosterone levels and the T/SHBG ratio were lower only in the 50 to 59 age group, but SHBG did not differ significantly. LH was significantly lower in the patients than in the controls. Thirty-three of the 104 patients were considered to have hypogonadism compared with 7 of the 99 healthy men. The only clinical variable apart from age that had a significant impact on biovailable testosterone was the Stanford Health Assessment Questionnaire (HAQ) score.

Biological Role of Androgens in Rheumatoid Arthritis and Molecular Basis for Testosterone Adjuvant Therapy

It is generally believed that androgens exert antiinflammatory properties, whereas estrogens are immunoenhancing.[42] Certain chemicals found in the environment, such as those in plastics, pesticides, plants, and agricultural products mimic estrogens or "phytoestrogens" and can also block androgen action. Exposure to these antiandrogens may cause changes similar to those associated with estrogen exposure.[43] To exert direct immunoregulatory effects, neurohormones and steroid hormones need to diffuse passively into target cells, interact with intracellular receptors, and then be translocated into the genome. Human and murine macrophages exhibit functional cytoplasmic and nuclear androgen and estrogen receptors. They can also metabolize gonadal and adrenal androgen precursors [testosterone and dehydroepiandrosterone sulfate (DHEAS)] into their active metabolites [dihydrotestosterone (DHT) and DHEA]. As such, chronic inflammation such as in RA may lead to a hypogonadal state. In addition, studies of the effects of intraarticular testosterone and DHT on cartilage breakdown and inflammation in animal models of RA indicate significant inhibitory activity of these steroids on synovial hyperplasia and cartilage erosion.[44] In a laboratory experiment by Steward, testosterone and its metabolite 5α-dihydrotestosterone (DHT) were compared with dexamethasone 21-acetate in two different animal models of arthritis. The researchers found interesting effects on cartilage breakdown and inflammation. In the mouse air pouch, at the three dose levels used, significant effects were obtained with DHT and were more pronounced on cartilage breakdown than on inflammation. Even at low doses, there was a 64% inhibition of collagen breakdown and 18% inhibition of glycosaminoglycans (GAGs) breakdown with androgens. In the antigen-induced arthritis mouse model, testosterone had significant inhibitory effects on synovial hyperplasia and cartilage erosion.[45]

Genetic polymorphism may also contribute to RA etiology by endocrine interactions. For example, the estrogen synthase (CYP19) locus is the cytochrome p450 that catalyzes the conversion of C19 androgens to C18 estrogens. In RA, a linkage to this locus has been described in sibling pair families having an older age at onset of the disease. CYP19 polymorphisms that lead to higher levels of CYP19 or higher enzyme activity lead to reduced levels of androgens and hence may put an older individual at risk for RA.[46] The rise in incidence in RA with age is associated with a decline in androgen production declines in both older males and females.

Stress may influence neuroendocrine and immune mechanisms adversely resulting in a reduction of testosterone production. A subgroup of RA patients who were found to have high stress at onset of the disease showed a worse disease prognosis.[47] Even minor life events might precipitate further RA flares. Studies have shown a greater occurrence of increased disease activity, joint tenderness, and pain in RA even with minor stress.

Clinical Trials of Testosterone in the Treatment of Rheumatoid Arthritis

In clinical practice, patients placed on testosterone therapy for hypogonadism-related symptoms, such as loss of libido, report a surprising improvement in their pain threshold from their associated arthritis. However, the few prospective trials of the use of testosterone as an adjuvant have been equivocal. For instance, in Hall et al's[48] 1996 trial reported in the *British Journal of Rheumatology*, there was no suggestion of a positive effect of testosterone on disease activity in men with RA (Fig. **12–3**). In that study, 35 men, aged 34 to 79 years, with definite RA were recruited from outpatient clinics and randomized to receive monthly injections of testosterone enanthate 250 mg or placebo

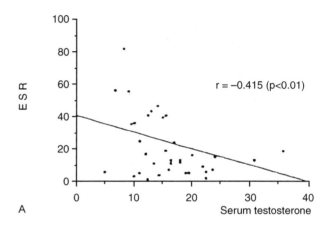

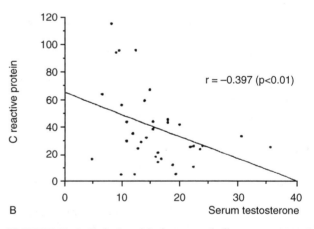

FIGURE 12–3. Relationship between inflammatory markers such as (A) erythrocyte sedimentation rate (ESR) and (B) C-reactive protein (CRP) with serum testosterone levels. The lower the testosterone levels, the higher the marker of inflammation. (Adapted from Hall GM, Larbre JP, Spector TD, et al. A randomized trial of testosterone therapy in males with rheumatoid arthritis. Br J Rheumatol 1996;35: 568–573, with permission.)

as an adjunct therapy for 9 months. End points included disease activity parameters and bone mineral density (BMD). At baseline, there were negative correlations between the ESR and serum testosterone ($r = -0.42$, $p < .01$) and BMD (hip, $r = -0.65$, $p < .01$). Thirty patients completed the trial, 15 receiving testosterone and 15 receiving placebo. There were significant rises in serum testosterone, dihydrotestosterone, and estradiol in the treatment group. There was no significant effect of treatment on disease activity overall, and five patients receiving testosterone underwent a "flare." Differences in mean BMD following testosterone or placebo were nonsignificant.

In another study from Italy, however, Cutolo and his group[49] seemed to have proved that testosterone was useful in those men with active RA. In that trial seven men were treated daily for 6 months with oral testosterone undecanoate plus an NSAID in an attempt to evaluate the immunological response, the overall clinical response, and the sex hormone response to such replacement therapy. At the end of the 6 months, there was a significant increase in serum testosterone levels ($p < .05$), an increase in the number of CD8+ T cells, and a decrease in the CD4+:CD8+ T-cell ratio. The immunoglobulin M (IgM) rheumatoid factor concentration decreased significantly ($p < .05$). There was a concurrent significant reduction in the number of affected joints ($p < .05$) and in the daily intake of NSAIDs ($p < .05$). The authors report that the immunosuppressive action of androgens probably contributed to the findings in these RA patients.

In our opinion, both these trials are limited by sample size, and hence testosterone's role in RA remains to be determined. The negative result of the first trial could be the result of an inadequate dosing of testosterone at 250 mg monthly. In contrast, the positive result of the second trial was achieved because of a daily dosing of testosterone undecanoate, which resulted in eugonadal levels. It was interesting that NSAID usage was decreased in the second trial. This is significant especially for patients who are at risk for peptic ulcer disease. In a separate study, Martens and his colleagues[50] reported that men with RA who are *not* taking prednisone have significantly elevated levels of FSH and LH with normal testosterone levels, suggesting a state of compensated partial gonadal failure. However, men with RA *taking low doses of prednisone have lower testosterone and gonadotropin levels, suggesting that prednisone may suppress the hypothalmic-pituitary-testicular axis.* Because testosterone affects immune function as well as bone and muscle metabolism, androgen deficiency in some men with RA may predispose them to more severe disease and to increased complications of steroid therapy such as myopathy and osteoporosis. Overall, there exists an interesting interplay of androgens with RA, but the relationship may be multifactorial. As such, there is no blanket recommendation to use adjuvant testosterone therapy in RA; it should be individualized. Testosterone levels should be measured to assist in making a clinical decision.

Discussion of the Case History

Mr. L.C. was diagnosed to have a fall risk from his arthritis and muscle wasting. It is an important geriatric medicine principle to prevent falls, as they lead to further morbidity such as pneumonia and incontinence, and even to mortality. As such, physical therapy assessment and treatment is very important. In this case, testosterone enanthate 200 mg IM every 2 weeks was added to the patient's treatment. The patient returned to the clinic after 6 weeks and said that the pain in his knees had completely disappeared. Testosterone was used with the intent to regain muscle strength, rather than treat his arthritis. Although he was already on Cox-2 inhibitors, the testosterone given could have helped reduced the rheumatoid inflammation in his knee. He surprised the nurses in the clinic by turning up without his walker and needing no assistance in ambulation. He claims that his memory improved, but this was not measurable on the Folstein. His blood pressure and diabetes were also better controlled. Overall, what is most important is that Mr. L.C. achieved a better quality of life. He even reported a return of early morning erections!

Conclusion and Key Points

Many presume that age and rheumatic diseases are linked. Although osteoarthritis is frequent in the elderly, it is *not* a consequence of aging. Also, elderly men can present with multiple other causes of muscle and joint pain, and suspicion and purposeful evaluation is important to make an adequate diagnosis and establish a treatment. If the diagnosis is not clear or the treatment seems complex or difficult, a consult to a rheumatologist or geriatrician may be needed.

- Osteoarthritis is not a function of aging per se. It may be more common among women because of the predominance of estrogen receptors in bone and cartilage cells that modify the rate of metabolism.

- Although there are no specific treatments for osteoarthritis, exercise plays an important role. Cox-2 inhibitors are recommended for older individuals.

- Seropositive RA does not have a significant predominance among women, as compared with seronegative RA.

- Current thinking is that RA may be more acute in older individuals, and the impact similar to that in younger individuals, and as such should be treated aggressively.

- Ankylosing spondylitis occurs predominantly in men in the third and fourth decades of life.

- Collagen vascular disease and fibromyalgia can occur in men as well.

- There is an association between low testosterone and RA in men.

- Androgens have antiinflammatory properties, whereas estrogens are immunoenhancing. There is complex interplay between androgens and RA.

REFERENCES

1. Demlow L, Liang MH, Eaton HM. Impact of chronic arthritis in the elderly Clin Rheum Dis 1986;12:329–335

2. Hootman JM, Helmick CG, Schappert SM. Magnitude and characteristics of arthritis and other rheumatic conditions on ambulatory medical care visits, United States, 1997. Arthritis Rheum 2002;47:571–581

3. Christmas C, Crespo CJ, Franckowiak SC, et al. How common is hip pain among older adults? Results from the Third National Health and Nutrition Examination Survey. J Fam Pract 2002;51

4. Creamer P, Hochberg MC. Osteoarthritis. Lancet 1997;350: 503–509

5. Hamerman D. Clinical implications of osteoarthritis and ageing. Ann Rheum Dis 1995;54:82–85

6. Berenbaum F, Osteoarthritis A. Epidemiology, pathology and pathogenesis. In: Klippel JH, ed. *Primer of Rheumatic Diseases.* Atlanta: Arthritis Foundation; 2001

7. Belanger A, Martel L, Berhelot JM, et al. Gender differences in disability-free life expectancy for selected risk factors and chronic conditions in Canada. J Women Aging 2002;14:61–83

8. Hawker GA, Wright JG, Coytre PC, et al. Differences between men and women in the rate of use of hip and knee arthroplasty. N Engl J Med 2000;342:1016–1022

9. Creamer P, Lethbridge-Cejku M, Hochberg MC. Factors associated with functional impairment in symptomatic knee osteoarthritis. Rheumatology (Oxford) 2000;39:490–496

10. Schumacher HR Jr. Secondary osteoarthritis. In: Moskowitz Rw, Howell DS, Golberg VM, et al, eds. *Osteoarthritis: Diagnosis and Surgical Management.* Philadelphia: WB Saunders; 1993:367–398.

11. Hochberg MC, Altman RD, Brant KD, et al. Guidelines for the medical management of osteoarthritis, Part 1: osteoarthritis of the hip. Arthritis Rheum 1995;38:1535–1540

12. Hochberg MC, Alman RD, Brant KD, et al. Guidelines for the medical management of osteoarthritis, Part II: osteoarthritis of the knee. Arthritis Rheum 1995;38:1541–1546

13. American College of Rheumatology Subcommittee on Osteoarthritis Guidelines. Recommendation for the medical management of osteoarthritis of the hip and knee: 2000 update. Arthritis Rheum 2000;43:1905–1915

14. AGS Panel on Chronic Pain on Older Persons. Clinical Practice Guidelines: the management of chronic pain in older persons. J Am Geriatr Soc 1998;46:635–651

15. Brandt KD, Smith GN Jr, Simon LS. Intraarticular injection of hyaluronan as treatment for knee osteoarthritis: what is the evidence? Arthritis Rheum 2000;43:1192–1203

16. McAlindon TE, LaValley MP, Guilin JP, Felson DT. Glucosamine and chondroitin for the treatment of osteoarthritis: a systematic quality assessment and meta-analysis. JAMA 2000;283:1469–1475

17. Mavragani CP, Moutsopoulos HM. Rheumatoid arthritis in the elderly. Exp Gerontol 1999;34:463–471

18. Goronzy JJ, Weyand CM. Rheumatoid arthritis: A: epidemiology, pathology and pathogenesis. In: Klippel JH, Crofford JH, Weyand CM, eds. *Primer on the Rheumatic Diseases.* Atlanta: Arthritis Foundation; 2001

19. Anderson RJ. Rheumatoid arthritis: B: clinical and laboratory features. In: Klippel JH, Crofford JH, Weyand CM, eds. *Primer on the Rheumatic Diseases.* Atlanta: Arthritis Foundation; 2001

20. Van der Heijde DM, van Reil PL, van Leeuwen MA, et al. Prognostic factors for radiographic damage and physical disability in early rheumatoid arthritis: a prospective study of 147 patients. Br J Rheumatol 1992;31:519–525

21. Krause D, Schleusser B, Herberon G, et al. Response to methotrexate treatment is associated with reduced mortality in patients with severe rheumatoid arthritis. Arthritis Rheum 2000;43:14–21

22. Wluka A, Buchbinder R, Mylvaganam A, et al. Long-term methotrexate use in rheumatoid arthritis: 12 year follow-up of 460 patients treated in community practice. J Rheumatol 2000;27:1864–1871

23. O'Dell JR, Haire CE, Erikson N, et al. Treatment of rheumatoid arthritis with methotrexate alone, sulfasalazine and hydroxychloroquine, or a combination of all three medications. N Engl J Med 1996;334:1287–1291

24. Boers M, Verhoeven AC, Markusse HM, et al. Randomised comparison of combined step-down prednisolone, methotrexate and sulphasalazine alone in early rheumatoid arthritis. Lancet 1997;350:309–318

25. Stand V, Tugwell P, Bombardier C, et al. Function and health-related quality of life: results from a randomized controlled trial of leflunomide vs. methotrexate or placebo in patients with active rheumatoid arthritis. Arthritis Rheum 1999;42:1870–1878

26. Weinblatt ME, Keystone EC, Furst DE, et al. Adalimumab, a fully human anti-tumor necrosis factor alpha monoclonal antibody, for the treatment of rheumatoid arthritis in patients taking concomitant methotrexate: the ARMADA trial. Arthritis Rheum 2003;48:35–45

27. Pease CT, Bhakta BB, Devlin J, Emery P. Does the age of onset of rheumatoid arthritis influence phenotype?: a prospective study of outcome and prognostic factors. Rheumatology (Oxford) 1999;38:228–234

28. Labbe P, Hardouin P. Epidemiology and optimal management of polymyalgia rheumatica. Drugs Aging 1998;13:109–118

29. Weyand CM, Fulbright JW, Evans JM, et al. Corticosteroid requirements in polymyalgia rheumatica. Arch Intern Med 1999;159:577–584

30. Brack A, Martinez-Taboada V, Stanson A, et al. Disease pattern in cranial and large vessel giant cell arteritis. Arthritis Rheum 1999;42:311–317

31. Wordsworth P. Genes in the spondyloarthropathies. Rheum Dis Clin North Am 1998;24:845–863

32. Khan MA. Ankylosing spondylitis: clinical features. In: Klippel JH, Dieppe PA, eds. *Rheumatology.* 2nd ed. London: Mosby; 1998

33. Amor B, Dougados M, Mijiyama M. Criteres de classification des spondyloarthropaties. Rev Rhum 1990;57:85–89

34. Clegg DO, Reda DJ, Abdellatif M, et al. Comparison of sulfasalazine and placebo for the treatment of axial and peripheral articular manifestations of the seronegative spondyloarthropathies: a Department of Veterans Affairs cooperative study. Arthritis Rheum 1999;42:2325–2329

35. Biasi D, Carletto A, Caramaschi P, et al. Efficacy of methotrexate in the treatment of ankylosing spondylitis: a three-year open study. Clin Rheumatol 2000;19:114–117

36. Brandt J, Haibel H, Cornely D, et al. Successful treatment of active ankylosing spondylitis with the anti-tumor necrosis factor alpha monoclonal antibody infliximab. Arthritis Rheum 2000; 43:1346–1352

37. Kimberley RP. Connective-tissue diseases. In: Klippel JH, Crofford JH, Weyand CM, eds. *Primer on the Rheumatic Diseases.* Atlanta: Arthritis Foundation; 2001

38. Baer AN, Pincus T. Occult systemic lupus erythematosus in elderly men. JAMA 1983;249:3350–3352

39. Tishler M, Yaron I, Shirazi I, Yaron M. Clinical and immunological characteristics of elderly onset Sjogren's syndrome: a comparison with younger onset disease. J Rheumatol 2001; 28:795–797

40. Clauw DJ. Fibromyalgia and diffuse pain syndromes. In: Klippel JH, Crofford JH, Weyand CM, eds. *Primer on the Rheumatic Diseases.* Atlanta: Arthritis Foundation; 2001

41. Tengstrand B, Carsltrom K, Hafstrom I. Biovailable testosterone in men with rheumatoid arthritis- high frequency of hypogonadism. Rheumatology (Oxford) 2002; 41: 285–289

42. Cutolo M. Sex hormone adjuvant therapy in rheumatoid arthritis. Rheum Dis Clin North Am 2000;26:881–895

43. Kelce WR, Stone CR, Laws SC, et al. Persistent DDT metabolite p,p-DDE is a potent androgen receptor antagonist. Nature 1995;375:581–585

44. Da Silva JA, Larbre JP, Seed M, et al. Sex differences in inflammation induced cartilage damage in rodents. J Rheumatol 1994;21:330–337

45. Steward A, Bayley DI. Effects of androgens in models of rheumatoid arthritis. Agents Actions 1992;35:268–272

46. John S, Myerscough A, Eyre S, et al. Linkage of a marker in intron D of the estrogen synthase locus to rheumatoid arthritis. Arthritis Rheum 1999;42:1617–1620

47. Huyser B, Parker JC. Stress and rheumatoid arthritis: an integrative review. Arthritis Care Res 1998;11:135–145

48. Hall GM, Larbre JP, Spector TD, et al. A randomized trial of testosterone therapy in males with rheumatoid arthritis. Br J Rheumatol 1996;35:568–573

49. Cutolo M, Balleari E, Giusti M, et al. Androgen replacement therapy in male patients with rheumatoid arthritis. Arthritis Rheum 1991;34:1–5

50. Martens HF, Sheets PK, Tenover JS, et al. Decreased testosterone levels in men with rheumatoid arthritis: effect of low dose prednisone therapy. J Rheumatol 1994; 21:1427–1431

13

Cardiovascular Heart Disease in Older Men and the Role of Androgens

GRANT C. FOWLER, EMMA CID, AND ROBERT S. TAN

Case History

Mr. C.M. is a 50-year-old man who consulted with a view toward testosterone replacement therapy. He is very proactive about his health, and has visited many health centers throughout the United States. He has an extensive cardiac history including hypertension and hypercholesterolemia. He had a heart attack 5 years ago, and underwent angioplasty and a stent placement. He feels well, and has no chest pains at the moment. He states that he does get chest pains off and on, however. He confesses to be short of breath on extreme exertion like running a mile, but has no orthopnea. He has a positive family history of hypertension and coronary heart disease. His current medications include atorvastatin, atenolol, and felodipine. He also is a chronic drinker and consumes two or three beers each day, but does not smoke. On physical examination, the patient was noted to be centrally obese, and his weight was 250 pounds. His blood pressure was 135/100 mm Hg while on medication. He had some signs of hypogonadism, including testicular atrophy but no problems with his sense of smell. Muscle strength was normal and there were no focal neurological deficits. After confirming biochemical hypogonadism, he was put on testosterone gel. He was also prescribed a course of aerobic and strength training exercise. The patient was followed up regularly to ascertain his progress.

Aging and the Cardiovascular System

Due to adequate compensatory mechanisms, the changes affecting the cardiovascular system due to normal aging do *not* necessarily result in dramatic changes in the heart rate, left ventricular systolic function, or cardiac output at rest. When cardiovascular disease develops, however, its impact is superimposed on the already stressed cardiovascular system, resulting in increased functional impairment, morbidity, and mortality in the elderly population. Both atherosclerotic cardiovascular disease and hypertensive cardiovascular disease have a very high prevalence in the United States and other Western societies. The increase in prevalence is dramatic with age: statistics show that the majority of all cardiovascular deaths in the United States occur in patients older than 65 years of age. More than half of the patients hospitalized annually for acute myocardial infarctions (MIs) are more than 65 years old. Autopsy data from the 1950s and 1960s showed significant coronary artery stenoses in approximately three fourths of men over age 50 years. Congestive heart failure is the most common medical cause for hospital admissions in the elderly. Although recent trends in diet, smoking, physical activity, management of hypertension, and hyperlipidemia may have a significant impact on the prevalence of cardiovascular disease, the magnitude of the change is not known at this time.

Coronary Artery Disease in Men

Although the incidence may have reached a plateau in our Western population, coronary artery disease (CAD) in men continues to be the largest killer by far. By the age of 40, a man has greater than a 3% chance of a coronary event over the next 5 years. Even younger men with multiple risk factors may have this high of a risk. By the age of 50, a man has a 10% risk of a coronary event over the next 10 years. After the age of 40, if stroke is considered along with coronary events, more men die from cardiovascular events than all cancers combined.

Atherosclerotic CAD begins very early in life. At autopsy, 20% of 30- to 34-year-olds have been noted to have 40% stenosis in the left anterior descending coronary artery.[1] In fact, fatty streaks have been noted in the aortas of newborns whose mother had severe hyperlipidemia.[2] And this is despite the fact that among children and adolescents, the mean total cholesterol is only 165 mg/dL. Only ~10% of adolescents, ages 12 to 19 years, have total cholesterol > 200 mg/dL.

At this age, obesity, hypertension, and inadequate levels of high-density lipoprotein (HDL) cholesterol are more commonly associated with atherosclerotic streaking. There is not a gender difference, and this is interesting because girls have significantly higher levels of both total cholesterol levels and low-density lipoprotein (LDL) cholesterol from ages 4 to 19 years. Perhaps part of the explanation is that before puberty, there is little gender difference in HDL cholesterol levels.

Following puberty, the synthesis of testosterone in males suppresses hepatic production of apolipoprotein A (ApoA), the precursor of the HDL particle. This leads to a 20% reduction in HDL cholesterol in men compared with women. This difference persists through adulthood and is now thought to explain the gender difference in early heart disease.

Interestingly, beginning at age 50, women catch up with men from the perspective of total cholesterol levels. In fact, at this age a higher percentage of women than men have a total blood cholesterol of 200 mg/dL or higher. Granted total cholesterol is not the only risk factor, the American Heart Association (AHA) reports that a 10% decrease in cholesterol levels may result in a 30% decrease in the incidence of CAD. A reduction of total cholesterol in both sexes is therefore universally beneficial in our Western population.

Another impact upon HDL cholesterol and a risk factor for CAD that is exploding in prevalence is the cardiovascular metabolic syndrome. This syndrome affects ~25% of men by the age of 40 and ~45% of men by the age of 65. The diagnosis is made if a patient has three of the five factors listed in Table 13–1.

Being overweight is the most common attribute of those patients with metabolic syndrome. With over half of individuals in Western countries being overweight, the incidence of the metabolic syndrome is sure to

increase. What is becoming apparent is that adipose tissue is an endocrine organ, especially adipose tissue located at the waist in men, the so-called central adiposity. Although adipose tissue is supposed to produce leptin when there is sufficient energy stored, which is intended to decrease the appetite, this mechanism does not function properly in many individuals. Another functional problem with central adipose tissue it that it is exquisitely sensitive to catecholamines. With the slightest exposure to catecholamines, it releases free fatty acids, which travel to the already overworked liver to be stored and unfortunately contribute further to insulin resistance. In addition, central adipose tissue has been found to release inflammatory mediators, possibly contributing to the atherosclerotic process. This process is possibly monitored with one of the new markers for risk of CAD, the high-sensitivity C-reactive protein (hs-CRP or cardio-CRP).

One method of quantifying insulin resistance is in the form of the hemoglobin A_{1C}. From the EPIC-Norfolk Study, it was noted that for nondiabetic men, the difference between an A_{1C} of 5.0 and 5.4 doubled their risk of a cardiovascular event.[3] Perhaps these data explain why insulin-resistant men's risk of a cardiovascular event increases 10 years before they become diabetic.

Regarding other risk factors, smoking is more common among men, and it is the *most preventable* risk factor for CAD. More than a quarter of men still smoke. Hypertension is slightly more common in men, with 26% being affected. Although men are less likely than women to be physically inactive, 30 to 50% are sedentary. Obesity and a sedentary lifestyle are directly associated with CAD in addition to contributing to most of the risk factors already mentioned.

Hypertension in Older Men

In the United States, hypertension is defined by the National Institutes of Health (in 1997) as systolic blood pressure (SBP) 140 mm Hg or greater, diastolic blood pressure (DBP) 90 mm Hg or greater, or taking antihypertensive medication. The World Health Organization (WHO) defines hypertension as a blood pressure higher than 160/95 mm Hg. Isolated systolic hypertension is defined as a SBP above 150 mm Hg with a DBP below 90 mm Hg. The estimates for the prevalence of hypertension in persons 65 years or older vary widely (10 to 50%) depending on the study and definition of hypertension used. Consistently, however, the prevalence of hypertension increases with age and the risk of cardiovascular events increases with the severity of the hypertension. In addition, the prevalence of hypertension in elderly black patients is higher than in elderly white patients. It is estimated that about one fifth

TABLE 13–1. Metabolic Syndrome Criteria: Patients Are Diagnosed If They Have Three of Five

Waist over 40 inches
Triglycerides > 150 mg/dL
HDL cholesterol < 40 mg/dL
Hypertension (BP > 130/85)
Insulin resistance (fasting plasma glucose > 110 mg/dL)

of elderly men in the United States have isolated systolic hypertension (ISH).

The Framingham data has shown that ISH is more predictive for future stroke, myocardial infarction (MI), and congestive heart failure (CHF) than the presence of diastolic hypertension.[4] Both reduced arterial compliance and increased cardiac output are implicated in the pathogenesis of ISH. Significant morbidity is associated with hypertension: the Framingham Heart Study showed a twofold increase in CHF in men in the presence of hypertension; in the Veterans Administration Cooperative Study on Antihypertensive Agents, half of all morbidity, heart failure, and stroke occurred in men older than 60 years even though they only made up 20% of the subjects in the trial. Fortunately, control of hypertension decreases the risk of cardiovascular death, heart failure, and stroke in elderly patients. Control of isolated systolic hypertension in the Systolic Hypertension in the Elderly Program (SHEP) using low-dose chlorthalidone as initial therapy and low-dose β-blockade added as needed reduced stroke by 36%, heart failure by 50%, and all cardiovascular events by 32%. Similar outcomes occurred in the Swedish Trial in Old Patients with Hypertension-2 (STOP-2) with therapies using β-blockade, diuretics, angiotensin-converting enzyme (ACE) inhibitors, and calcium channel blocking agents. Notably, elderly patients receiving ACE inhibitors had fewer MI and heart failure episodes compared with those patients treated with calcium channel blocking agents.

In elderly patients with mild hypertension, non-pharmacological therapies including reduction in salt intake, weight reduction, and regular physical activity should be considered. The goals of treatment should include lowering the blood pressure to 140/90 mm Hg without causing postural hypotension, minimizing side effects, and using affordable drugs. More aggressive lowering of blood pressure should be considered in the hypertensive elderly patient who is also a diabetic or has chronic kidney disease. Initiating drug therapy at half the usual adult dose and measuring the blood pressure in the sitting and standing positions may minimize symptomatic side effects. Methods to promote long-term adherence should be considered, for example, providing written information describing the medical therapy and the blood pressure goal, and using simplified dosing regimens such as once-daily regimens, when possible, and pill-box systems.

Diagnosis of Heart Disease in Men

Compared with women, symptoms in men are much more specific for CAD. Typical angina is described as substernal chest discomfort relieved by rest or nitroglycerine. If the chest discomfort is characterized by two of these three factors, it is labeled atypical chest pain, and by one of the three factors, nonanginal chest pain. Tables and figures are available that delineate the likelihood of CAD by correlating these symptoms with age and gender. As it turns out, these tables alone are very accurate for predicting CAD in men. In fact, for men with typical angina, few studies need to be performed to confirm the diagnosis. From the tables, a 55-year-old man with typical angina has greater than a 90% chance of having CAD. In this case, a cardiac exercise test is not needed to make the diagnosis; an exercise test is usually more helpful for predicting the *prognosis* and managing the disease.

In addition, for men with a pretest likelihood of disease greater than 70%, the accuracy of the test is insufficient to exclude disease even if the test is negative. A positive exercise test is not much help in this range, such as in the previous example. An exercise test is also not useful in the lower pretest likelihood range (<20%) unless the test result is profoundly positive (>3 mm ST segment deviation), which is rare. A patient with <20% pretest likelihood of disease with a positive exercise test is much more likely to have a false-positive test unless it is profoundly positive. Therefore, it may be safer to follow patients frequently until they develop a higher pretest likelihood of disease.

Fortunately, false-positive results are much less common for men undergoing exercise testing. In women, estrogen is apparently the cause of many false positives. With most experts now declaring only a flat or down-sloping ST-segment change as a positive, even false-positive results in women are minimized. The number of millimeters of ST-segment deviation should be measured at the j-point.

For those with a positive exercise test, the Duke nomogram is probably the most important development in the past 20 years for managing CAD. Use of the Duke nomogram allows the clinician to provide counseling that is very specific for the patient's condition. In fact, patients can make informed decisions knowing their annual risk of a cardiovascular event if they choose medical as opposed to surgical management. In most centers, the risk of serious life-threatening complications from coronary artery bypass grafting (CABG) is ~2.5%, so patients can compare their risk of complications with surgery against medical management or even watchful waiting.

When evaluating options for treatment, the exercise test is again helpful.[5] There are study results that support medical management for men able to achieve 10 METS (metabolic equivalents of a task; 1 MET = amount of oxygen consumed at rest) with exercise, even if they have a positive result. However, this evidence needs to be considered in light of the age and activity level of the individual. For a man younger than 50 years of age, the diagnosis of CAD indicates aggressive CAD. It may be helpful to know the architecture of the coronary arteries; therefore, cardiac catheterization may be reasonable and helpful. For very active men older than 50 years, there may also be benefit of knowing the coronary architecture.

Because such a cardiac catheterization would be performed after having obtained objective evidence of ischemia (e.g., a positive exercise test or perfusion study), angioplasty or stenting may be reasonable at the time of the original catheterization. The current literature suggests that stenting may have surpassed angioplasty from the perspective of durability.

For those men with objective evidence of coronary ischemia at low work loads(<5 METS), revascularization is prudent, for which there are now options. CABG has been shown to improve outcomes of individuals with left main equivalent CAD (significant left main obstruction, or three-vessel obstruction with resultant left ventricular ischemic dysfunction). For all other men, angioplasty and/or stenting may be an option. The goal for most men undergoing angioplasty or stenting is to manage the symptoms or to buy time until their eventual CABG. Prior to our ability to aggressively manage cholesterol with medications, the average durability of a CABG was only 10 to 12 years. At that point, it usually had to be repeated. Therefore, the best candidate for a CABG is an elderly man, so that it will have to be performed only once or possibly twice. Durability data for angioplasty and stenting is still evolving and not fully available; therefore, the goal of these procedures is to buy time until the eventual CABG.

Primary Prevention

Primary prevention should be targeted toward risk factor reduction. Patients who smoke should be counseled to stop and should be enrolled in a smoking cessation program. Hypertension should be treated aggressively. Patients should be screened with a lipid profile at 21 years of age, and every 5 years thereafter if the profile is normal. For patients with multiple risk factors, or a strong family history of dyslipidemia, a lipid profile should be checked earlier. For those patients with multiple risk factors, it may also be prudent to measure a cardio-CRP level and possibly a homocysteine level. If homocysteine levels are elevated, there may be benefit of attempting to normalize them with high doses of folic acid (2–3 mg/day). Although we have few data proving the benefit of such treatment, the studies are currently ongoing and there is almost no detriment to folic acid supplementation. Folic acid is so safe that it was added to the commercially produced bread in the United States a few years ago.

Current evidence for medical management for the prevention of cardiovascular events indicates that aspirin is the most potent agent for prevention. As a primary preventive measure, 80 to 160 mg a day has been proven cardioprotective. The United States Preventive Task Force (USPSTF) recommends men take a low-dose aspirin daily starting at 40 years of age, and recently the AHA has recommended this dose beginning at the age of 50. For patients with multiple risk factors for CAD or an abnormal cardio-CRP, aspirin therapy should be started earlier.

For patients with multiple risk factors under the age of 50 years, there may be benefit of obtaining an electron beam computed tomography (EBCT) coronary calcium score. If the EBCT is abnormal, a cardiac exercise test should be performed. After the age of 50 years, a large percentage of patients have calcium on their EBCT so there is little value of an EBCT over a cardiac exercise test.

Interestingly, primary prevention goals in diabetics may be slightly different. And primary prevention is crucial in diabetics, because 8 out of 10 diabetics succumb to a cardiovascular event. From the United Kingdom Prospective Diabetes Study (UKPDS), more value was found in antihypertensives for preventing cardiovascular events than any other class of medications.[6] Next most valuable were the lipid-lowering agents. Glycemic control also prevented events. Aspirin was also valuable for preventing cardiovascular events, but the number of patients needed to treat to prevent a cardiovascular event with aspirin was higher than with any of the other three classes of medications.

In nondiabetics, lipid-lowering agents clearly prevent cardiovascular events in high-risk individuals. Guidelines suggest treating anyone with greater than a 20% risk of a cardiovascular event over the next 10 years to obtain an LDL cholesterol level of less than 100 mg/dL. Lipid-lowering agents will also usually decrease cardio-CRP levels.

Patients should also be counseled to eat a low-fat, low-cholesterol diet or at least a Mediterranean-style diet. They should be given a prescription for exercise in the aerobic range at least 3 days a week for 30 minutes each day. Clinicians should spend some time learning about the topic of adherence to improve the likelihood that their patients will follow recommended treatments and lifestyle changes. Less than half of most medications are taken appropriately and even fewer lifestyle changes are maintained. As a result, the art and science of adherence are currently evolving.

Treatment and Secondary Prevention

As mentioned previously, for primary prevention clinicians must become experts in issues related to adherence if they want to be successful at treating and preventing events in patients with CAD. Adherence becomes even more of a challenge as the cost and number of medications increases and the lifestyle changes become more important. Such is the case in patients with CAD.

Antiplatelet therapy, including aspirin, is invaluable for secondary prevention following the initial diagnosis, a revascularization procedure, or a cardiovascular event. It may be the most effective medication available for preventing future cardiovascular events. Lifelong therapy in those individuals in whom it is not contraindicated is probably beneficial. For those in whom aspirin therapy is contraindicated, other antiplatelet agents should be considered, such as clopidogrel.

Lipid-lowering therapy is also indicated in most patients with CAD, with the ultimate goals still being determined. Most guidelines utilize LDL cholesterol as the treatment variable, and indicate a desirable LDL cholesterol goal is <100 mg/dL in those patients with known CAD or cerebrovascular disease, or those with what is now determined to be a cardiovascular risk equivalent [e.g., diabetes, peripheral arterial disease (PAD), abdominal aortic aneurysm (AAA), symptomatic carotid disease]. Recent data from the Heart Protection Study and other studies, including several in progress, indicate that a goal for LDL cholesterol of <100 mg/dL may not be low enough. Lipid-lowering therapy not only prevents progression of disease but also has been shown to decrease cardiovascular events and mortality. It may even hasten regression of CAD.

For patients with impaired left ventricular function, ACE inhibitors are indicated. In addition, post-MI patients benefit from lifelong β-blocker therapy. Cardiac rehabilitation programs have also been proven to prevent cardiovascular events. Every necessary effort should be made to support the patient in an exercise program, including peripheral revascularization as needed. An exercise program will harm no patients with CAD, even those following an anterior MI, if they have been screened carefully by their clinician or through a cardiac rehabilitation program. In fact, most patients with known CAD should be advised to exercise for up to an hour a day, four or five times a week.

The patient should also be advised to follow a low-fat diet or a traditional Mediterranean diet. As it turns out, neither diet may affect the overall measured cholesterol levels; however, they have been shown to prevent cardiovascular events. Diets high in omega-3 fish oils have also been shown to prevent cardiovascular events.

Androgens and the Cardiovascular System

Heart disease is a very broad category of diseases, and this section addresses the impact of androgens on arteriosclerosis, blood pressure, and vascular disease. These three areas do have common etiological factors and are often interrelated. Hormones, including reproductive hormones, can regulate normal physiological factors and are discussed in more detail.

Sex steroids play a complex role in the vascular vessel wall system. In the past, many clinical trials have supported the protective effect of estrogens in arteriosclerosis. However, the recent Women's Health Initiative estrogen replacement therapy failed to demonstrate a reduction of cardiovascular mortality.[7] It seems that the effects of androgens on cardiovascular diseases is even more contentious, partly because of the lack of large clinical trials looking at this aspect of androgen influence in cardiovascular health. In the past, androgens were believed to be adverse to the cardiovascular system. However, recent studies in men have documented several beneficial actions of testosterone in the arterial vascular system. In short, androgens have been demonstrated to affect lipid metabolism including LDL, HDL, lipoprotein a [Lp(a)], and hemostasis through platelet aggregation and fibrinolytic activity. In addition, testosterone seems to affect coronary vessel dilatation. These effects are examined in the following subsections.

Arteriosclerosis and Androgens

In recent years there has been exciting evidence that androgens may indeed influence cardiovascular risk through the influence on arteriosclerosis. Some older men on androgen replacement seem to be protected from arteriosclerosis and hence arterial disease, but these trials are small and in no way conclusive. However, animal studies seem to suggest the protection of some species but not all species from arteriosclerosis.

One of the paradoxes of androgens on cardiovascular health is that they tend to reduce HDL, and HDL has been viewed to be cardioprotective. The reduction of HDL and elevation of LDL is not universal in all cases of androgen supplementation, and the issue is less important in older men.[8] On the positive side, the administration of androgens also results in a reduction of triglyceride and Lp(a), which in turn are also independent risk factors for arteriosclerosis.[9] Androgen supplementation can also increase the hematocrit, and in some older men with heart failure this could be seen as a positive effect. Increases in the hematocrit can result in relief of symptoms of shortness of breath seen in advanced cases of heart failure. Quality of life may thus be enhanced with androgen supplementation in selected patients.

One of the larger studies to determine the association between hypotestosteronemia and arteriosclerosis has been the population-based Rotterdam study (Fig. 13–1),[10] which investigated the association of levels of dehydroepiandrosterone sulfate (DHEAS) and total and bioavailable testosterone with aortic arteriosclerosis among 1032 nonsmoking men and women over 55 years. Aortic arteriosclerosis was assessed by radiographic detection of calcified deposits in the abdominal aorta. Relative to men with levels of total and bioavailable testosterone in the lowest tertile, men with levels of these hormones in the highest tertile had age-adjusted relative risks of 0.4 [95% confidence interval (CI), 0.2–0.9] and 0.2 (CI, 0.1–0.7), respectively, for the presence of severe aortic arteriosclerosis. The corresponding relative risks for women were 3.7 (CI, 1.2–11.6) and 2.3 (CI, 0.7–7.8). In this longitudinal study, men with levels of total and bioavailable testosterone in subsequent tertiles were also protected against progression of aortic arteriosclerosis measured after 6.5 year (SD/ ± 0.5 year) of follow-up (p for trend = .02). No clear association between levels of DHEAS and presence of severe aortic arteriosclerosis was found, either in men or in women. In men, a protective effect of higher levels of DHEAS against progression of aortic arteriosclerosis was suggested, but the corresponding test for trend did not reach statistical significance.

This study also found an independent inverse association between levels of testosterone and aortic arteriosclerosis in men. In women, however, positive associations between levels of testosterone and aortic arteriosclerosis were largely due to adverse cardiovascular disease risk factors. Epidemiological studies point to a protective effect of endogenous testosterone against arteriosclerosis. It is instinctive to suggest that administration of testosterone to older men would protect against arteriosclerosis, but long-term prospective trials looking at this issue are not yet available. Fig. 13–2 shows increasing risk of arteriosclerosis with decreasing quartile of testosterone in men but not women.

Blood Clotting and Androgens

As seen in the previous discussion, hypotestosteronemia is associated with an increased risk of cardiovascular disease. In the literature, there is some evidence that *low baseline fibrinolytic activity* in hypogonadal men results in thromboembolic disease and perhaps even myocardial infarction. We know from studies that hypogonadism in men is associated with an enhancement of fibrinolytic inhibition via *increased synthesis of the plasminogen activator inhibitor-1* (PAI-1).[11]

It has been demonstrated that synthetic androgens such as stanozolol and danazol reduce PAI-1 and are associated with increased fibrinolytic activity.[12] However, in men who abuse anabolic steroids and who typically use supraphysiological levels of androgens, thrombosis has been reported.[13] A prothrombotic state appears when the threshold dose is exceeded. There have been numerous reports on weightlifters dying of atherothrombotic and ischemic heart disease while abusing anabolic steroids.

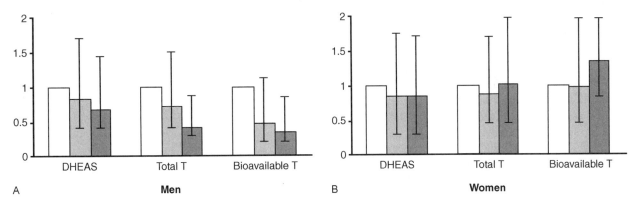

FIGURE 13–1. Results of the population-based Rotterdam study for (A) men and (B) women to determine the association between DHEAS, total testosterone, and bioavailable testosterone and severe aortic arteriosclerosis in 1032 nonsmoking men and women above the age of 55. There was a decrease in severe aortic arteriosclerosis with increases in total and bioavailable testosterone (T). No statistical effect of increases in DHEAS was found. (Adapted from Hak AE, Witteman JC, de Jong FH, et al. Low levels of endogenous androgens increase the risk of atherosclerosis in elderly men: the Rotterdam study. J Clin Endocrinol Metab 2002;87(8):3632–3639, with permission.)

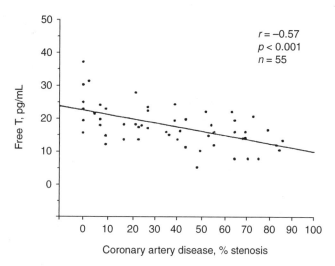

FIGURE 13–2. Effect of free testosterone (T) on the degree of coronary artery disease in a group of patients. Free testosterone correlated with a decrease in degree of coronary artery disease. (Adapted from Phillips GB, Pinkernell BH, Jing TY. The association of hypotestosteronemia with coronary artery disease in men. Arterioscler Thromb 1994; 14:701–706, with permission.)

Besides the activity of androgens on blood clotting, there are also concomitant androgen effects on carbohydrate and lipid metabolism. Some of the individual inconsistency of the effects of androgens on fibrinolytic and hemostatic activity appears to be based on the interrelationship to these metabolic systems. Androgens may have unfavorable effects on the HDL/LDL cholesterol ratio, but may have favorable effects on triglycerides and insulin resistance. Hypertriglyceridemia and insulin resistance are both associated with low fibrinolytic activity and increased PAI-1 levels, and hence androgens may influence clotting through indirect means.

Lipoprotein(a), a recently acknowledged independent risk factor of cardiovascular disease (CVD), has been shown to respond favorably to androgen treatment in not only men but also women. A synthetic androgen medication such as danazol used in premenopausal women for endometriosis was found to reduce PAI-1, suggesting an improvement of the fibrinolytic activity.[14] In addition, hormone replacement therapy (HRT) in women with androgenic progestins are associated with a marked reduction of PAI-1 and thus an improvement of fibrinolytic activity. Further improvement of fibrinolytic activity may be associated with the marked decrease of lipoprotein(a) in women who are on HRT with androgens in combination.

In summary, the current information on the effects of androgens on blood clotting is limited. Part of the difficulty is due to the interaction of androgens with other sex steroids in influencing blood clotting. Synthetic androgens have been shown to influence the synthesis and release of hemostatic factors. These include an increase of the inhibitors of coagulation and a decrease of the inhibitor of the fibrinolytic system. However, the use of androgens in patients with congenital deficiencies of these coagulation factors or previous events of cardiovascular disease has yielded disappointing results.[12] It is with optimism that androgen effects on the reduction of fibrinolytic inhibition (PAI-1) and Lp(a) can lead to the reduction of risk of cardiovascular disease. The profibrinolytic effects of androgens may be of particular interest, and clinical trials are needed in men to determine if clinical outcomes of increased thrombotic states like strokes and myocardial infarctions can indeed be reduced with bioidentical androgen replacement.

Blood Pressure and Androgens

It seems that gender may play an important role in blood pressure. For instance, premenopausal women have lower arterial blood pressure than age-matched men. Postmenopausal women have higher blood pressures, however, suggesting that ovarian hormones can possibly alter blood pressure. Again, it is intuitive to suggest that androgens may be responsible for higher blood pressure in men as much higher levels of androgens are found in men. However, with aging, hypotestosteronemia results and older men have higher blood pressures, which suggests that testosterone may protect against rises in blood pressure. There is a danger of overinterpreting these associations, which may not be causal relationships.

For instance, animals administered testosterone have shown induction of hypertension and hypertrophy of the left ventricle. Crofton and Share[15] demonstrated that in rats with deoxycorticosterone (DOC)-salt hypertension, arterial blood pressure rises more rapidly and reaches a higher level in male than in female rats. However, the course of the hypertension was removed by gonadectomy in male rats but exacerbated by gonadectomy in female rats. The group also found that testosterone aggravates the development of the hypertension in gonadectomized male rats but not in intact females. Progesterone, which has androgenic effects, had no effect on hypertension in ovariectomized rats but when given to ovariectomized rats in combination with estradiol transiently prevented the protective effect of the estradiol. The findings suggest that gonadal steroid hormones may play an important role in modulating the pathogenesis of DOC-salt hypertension in rats. It was suggested that the effects of the gonadal hormones on the course of the hypertension might be due to modulation of the cardiovascular and renal actions of *vasopressin*, because vasopressin is required for this model of hypertension.

In contrast, the effect of testosterone on blood pressure in humans has been positive rather than negative as described previous in animal models. For example, in the Rancho Bernardo study in San Diego, there was a negative correlation between plasma testosterone and testosterone levels (Table **13–2**).[16]

In that study of 1132 men aged 30 to 79 years, those with hypertension, categorically defined as SBP greater than 160 mm Hg and/or DBP greater than 95 mm Hg had significantly lower testosterone levels than non-hypertensives. SBP and DBP are inversely correlated with testosterone levels ($r = 0.17$, $p < .001$ for systolic; $r = -0.15$, $p < .001$ for diastolic) in the whole cohort. The association was present over the whole range of blood pressures and sex hormone levels, with a stepwise decrease in mean SBP and DBP per increasing quartile of testosterone. Obesity accounted for some, but not all, of this relationship, which was reduced, but still apparent after adjusting for age and body mass index. It was of interest that no other hormone (androstenedione, estrone, estradiol) or even sex hormone–binding globulin showed a consistent relationship with blood pressure. However, the association does not prove causality.

To take this association further to prove causality, exogenous testosterone can be administered to men and their blood pressure can then be observed.

However, these intervention trials generally use small numbers of subjects, and thus it is difficult to come to a consensus. In three separate small studies, investigators found no negative influences on blood pressure either with physiological or supraphysiological levels of exogenous testosterone.[17–19] In contrast to testosterone, however, anabolic steroids used for bodybuilding may cause a *slight rise in blood pressure*. In one such study, the effects of anabolic steroids on body composition, blood pressure, lipid profile, and liver functions were studied in male body builders who received a weekly intramuscular injection of nandrolone-decanoate (100 mg) or placebo for 8 weeks in a double-blind study.[20] An increase in diastolic blood pressure was observed, which returned to pre-anabolic values ~6 weeks after cessation of drug administration.

In conclusion, animal models suggest that testosterone administration leads to a rise in blood pressure. However, in human observations, testosterone appears to behave as a coronary and possibly peripheral vasodilator, acting primarily through nitric oxide release and the modulation of endothelial function.[21] Testosterone introduced into the coronary arteries during angiography results in a dose-dependent increase in vessel diameter and blood flow.[22] The administration of physiological doses of testosterone to older men who are hypogonadal appears not to influence blood pressures adversely, and may possibly

TABLE 13–2. The Rancho Bernardo Study Showed Negative Correlation of Plasma Testosterone Level to Systolic Blood Pressure

Hormone	Quartile				
	1	2	3	4	p value
SBP (mm Hg)					
Androstenedione					
Age-adjusted	140.9	139.9	138.7	139.8	0.56
Age- and BMI-adjusted	140.0	139.9	138.8	139.6	0.89
Testosterone					
Age-adjusted	142.8	140.6	139.1	135.3	< 0.001
Age- and BMI-adjusted	141.8	140.5	139.2	136.6	0.01
Estrone					
Age-adjusted	140.0	137.8	140.6	140.2	0.35
Age- and BMI-adjusted	140.5	137.5	140.3	140.0	0.28
Estradiol					
Age-adjusted	139.3	139.2	140.3	139.6	0.90
Age- and BMI-adjusted	139.3	139.7	139.6	139.6	0.99
Sex hormone–binding globulin					
Age-adjusted	141.9	140.2	139.4	137.1	0.05
Age- and BMI-adjusted	141.0	140.2	139.5	137.8	0.36

BMI, body mass index.

Adapted from Hak AE, Witteman JC, de Jong FH, et al. Low levels of endogenous androgens increase the risk of atherosclerosis in elderly men: the Rotterdam study. J Clin Endocrinol Metab 2002;87(8):3632–3639, with permission.

even be beneficial. As explained, the use of anabolic steroids for bodybuilding is associated with a slight rise in blood pressure, which in turn is reversible. The effect of DHEA and androstenedione on blood pressure is unclear at this point.

Influence of Androgens on Coronary and Peripheral Circulation

Testosterone had been suspected of being a coronary vasodilator as far back as the 1940s. Four separate trials in that decade demonstrated the relief of angina pectoris with testosterone injections.[23–26] Several cross-sectional studies also suggest a positive correlation of hypotestosteronemia to poor arterial circulation. For instance, Phillips and his group[27] demonstrated that testosterone correlated negatively with the risk factors fibrinogen, PAI-1, and insulin, and positively with high-density lipoprotein cholesterol. The correlations found in their study between testosterone and the degree of coronary artery disease and between testosterone and other risk factors for MI raise the possibility that in men hypotestosteronemia may be a risk factor for coronary arteriosclerosis (Fig. 13–2).

Models of androgen deprivation in men, as seen in chemical castration for prostate cancer, has demonstrated that androgen withdrawal is associated with a reduction in central arterial compliance.[28] Arterial compliance or "stiffness" is increasingly regarded as a modifiable risk factor for cardiovascular disease. In a study by Webb and her colleagues,[22] the hypothesis that testosterone has direct relaxing effects on coronary arteries was furthered. Thirteen older men with coronary artery disease underwent measurement of coronary artery diameter and blood flow after a 3-minute intracoronary infusion of vehicle control (ethanol) followed by 2-minute intracoronary infusions of acetylcholine. A dose–response curve to 3-minute infusions of testosterone was then determined, and the acetylcholine infusions were repeated. Finally, an intracoronary bolus of isosorbide dinitrate was given. Coronary blood flow was calculated from measurements of blood flow velocity, using intracoronary Doppler, and coronary artery diameter, using quantitative coronary angiography. Testosterone significantly increased coronary artery diameter compared with baseline. A *significant increase* in coronary blood flow also occurred at all concentrations of testosterone compared with baseline. Fig. 13–3 shows the effect of testosterone on coronary diameter and blood flow.

Unfortunately, the Cochrane systematic review of testosterone replacement research concluded that there is no evidence to date that short-term testosterone treatment is beneficial in subjects with lower limb arteriosclerosis. However, the authors admitted that

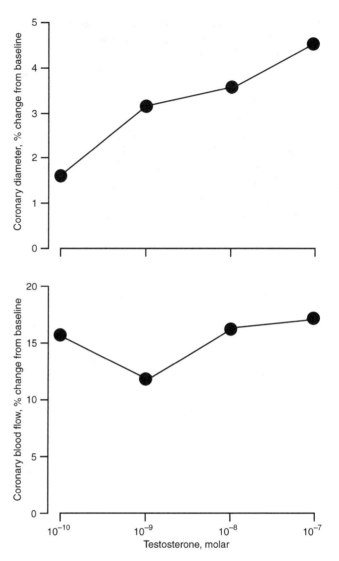

FIGURE 13–3. Effect of testosterone on coronary artery diameter and blood flow. Testosterone correlated significantly with increased coronary artery diameter and blood flow at all levels of administration. (Adapted from Webb C, McNeill J, Hayward C, et al. Effects of testosterone on coronary vasomotor regulation in men with coronary heart disease. Circulation 1998;100:1960–1966, with permission.)

their conclusion was based on limited data rather than the lack of a real effect.[29] The clinical significance of the collective data are that because testosterone may act as a vasodilator, hypogonal patients on nitrate therapy may attain better coronary blood flow if they are replaced to physiological levels of testosterone. Testosterone should not replace conventional therapies for coronary artery disease including nitrates, calcium channel blockers, β-blockers, and others, but could be considered as an adjunct in difficult selected cases if the patient is hypogonadal and there are no contraindications.

Discussion of the Case History

The patient immediately reported improved quality of life after being on testosterone. Partly because of his increased energy from the testosterone, he began to exercise more regularly. Subsequent follow-ups reveal that he lost 40 pounds in total after a year of supervision. His blood pressure went down to 100/80 mm Hg without adjusting his hypertension medications. Felodipine was discontinued, and his pressure normalized to 120/80 mm Hg. The drop in blood pressure is a reflection of his weight loss, increased aerobic activity, and possibly the vasodilatory effects of testosterone. His chest pains totally disappeared. His cholesterol levels did not alter much, and he was maintained on the same dose of atorvastatin.

Conclusion and Key Points

Much has been studied about estrogens and heart disease in women. Androgens certainly play a role in the cardiovascular well-being of men. Trials of testosterone therapy in men are limited, and much more research needs to be done before the effects of testosterone therapy are fully understood. However, the impact of androgens on the cardiovascular system is generally positive, rather than negative, especially when physiological doses are used. Supraphysiological doses as administered by some athletes can be harmful.

- There is still no consensus of the impact of androgens on arteriosclerosis.

- However, it has been demonstrated that androgen therapy decreases fibrinogen, lipoprotein, and PAI-1, all of which are risk factors for arteriosclerosis.

- Hypotestosteronemia in itself is associated with coronary arteriosclerosis.

- Hypotestosteronemia in itself is also associated with hypertension.

- Testosterone replacement does not induce hypertension, but may even be useful in lowering pressure.

- There is increasing evidence that physiological, but not supraphysiological, doses of testosterone result in vasodilatation of coronary and peripheral vasculature.

- Angina is not worsened by testosterone, but may even be improved with it.

REFERENCES

1. McGill HC Jr, Mcmahan CA, Zieske AW, et al. Association of coronary heart disease risk factors with the intermediate lesion of atherosclerosis in youth: the Pathobiological Determinants of Atherosclerosis in Youth (PDAY) Research Group. Arterioscler Thromb Vasc Biol 2000;20:1998–2004

2. Berenson GS, Wattigney WA, Tracey RE, et al. Atherosclerosis of the aorta and coronary arteries and cardiovascular risk factors in persons aged 6 to 30 years and studied at necropsy: the Bogalusa Heart Study. Am J Cardiol 1992;70:851–858

3. Khaw KT, Wareham N, Luben R, et al. Glycated hemoglobin, diabetes, and mortality in men in Norfolk cohort of European prospective investigation of cancer and nutrition (EPIC-Norfolk). BMJ 2001;322:15–18

4. Witteman JC, D'Agostino RB, Stijnen T, et al. G-estimation of causal effects: isolated systolic hypertension and cardiovascular death in the Framingham Heart Study. Am J Epidemiol 1998;148:390–401

5. Fowler GC, Evans CH, Altman MA. Office procedures: exercise testing. Prim Care 1997;24:375–406

6. Stratton IM, Adler AI, Neil HA, et al. Association of glycemia with macrovascular and microvascular complications of type 2 diabetes (UKPDS 35): prospective observational study. BMJ 2000;321:405–412

7. Rossouw JE, Anderson GL, Prentice RL, et al. Risks and benefits of estrogen plus progestin in healthy postmenopausal women: principal results From the Women's Health Initiative randomized controlled trial. JAMA 2002;288:321–333

8. Bhasin S, Buckwalter JG. Testosterone supplementation in older men: a rational idea whose time has not yet come. J Androl 2001;22:718–731

9. Barud W, Palusinski R, Beltowski J, Wojcicka G. Inverse relationship between total testosterone and anti-oxidized low-density lipoprotein antibody levels in ageing males. Atherosclerosis 2002;164:283–288

10. Hak AE, Witteman JC, de Jong FH, et al. Low levels of endogenous androgens increase the risk of atherosclerosis in elderly men: the Rotterdam study. J Clin Endocrinol Metab 2002;87:3632–3639

11. Zollner TM, Veraart JC, Wolter M, et al. Leg ulcers in Klinefelter's syndrome–further evidence for an involvement of plasminogen activator inhibitor-1. Br J Dermatol 1997;136:341–344

12. Winkler UH. Effects of androgens on haemostasis. Maturitas 1996;24:147–155

13. Ferenchick GS, Hirokawa S, Mammen EF, et al. Anabolic-androgenic steroid abuse in weight lifters: evidence for activation of the hemostatic system. Am J Hematol 1995;49:282–288

14. Ledford MR, Horton A, Wang G, et al. Efficacy of danazol in a patient with congenital protein-S deficiency: paradoxical evidence for decreased platelet activation with increased thrombin generation. Thromb Res 1997;87:473–482

15. Crofton JT, Share L. Gonadal hormones modulate deoxycorticosterone-salt hypertension in male and female rats. Hypertension 1997;29:494–499

16. Khaw KT, Barrett-Connor E. Blood pressure and endogenous testosterone in men: an inverse relationship. J Hypertens 1988;6:329–332

17. Whitworth JA, Scoggins BA, Andrews J, et al. Haemodynamic and metabolic effects of short term administration of synthetic sex steroids in humans. Clin Exp Hypertens A 1992;14:905–922

18. White CM, Ferraro-Borgida MJ, Moyna NM, et al. The effect of pharmacokinetically guided acute intravenous testosterone administration on electrocardiographic and blood pressure variables. J Clin Pharmacol 1999;39:1038–1043

19. Marin P, Holmang S, Jonsson L, et al. The effects of testosterone treatment on body composition and metabolism in middle-aged obese men. Int J Obes Relat Metab Disord 1992;16:991–997

20. Kuipers H, Wijnen JA, Hartgens F, et al. Influence of anabolic steroids on body composition, blood pressure, lipid profile

and liver functions in body builders. Int J Sports Med 1991; 12:413–418

21. Lloyd G. Androgens and blood pressure in men. In: *Textbook of Men's Health*. London: Parthenon; 2002:351–359

22. Webb C, McNeill J, Hayward C, et al. Effects of testosterone on coronary vasomotor regulation in men with coronary heart disease. Circulation 1999;100:1690–1696

23. Hamm L. Testosterone propionate in the treatment of angina pectoris. J Clin Endocrinol 1942;2:325–328

24. Siegler LH, Tuglan J. Treatment of angina pectoris by testosterone propionate. NY State J Med 1943;43:1424–1428

25. Walker TC. The use of testosterone propionate and estrogenic substance in the treatment of essential hypertension, angina pectoris and peripheral vascular disease. J Clin Endocrinol 1942;2:560–568

26. Lesser MA. Testosterone propionate therapy in one hundred cases of angina pectoris. J Clin Endocrinol 1946; 6:547–549

27. Phillips GB, Pinkernell BH, Jing TY. The association of hypotestosteronemia with coronary artery disease in men. Arterioscler Thromb 1994;14:701–706

28. Dockery F, Bulpitt CJ, Agarwal S, et al. Testosterone suppression in men with prostate cancer is associated with increased arterial stiffness. Aging Male 2002;5:216–222

29. Price JF, Leng GC. Steroid sex hormones for lower limb arteriosclerosis. Cochrane Database Syst Rev 2002;1:CD000188

14

An Overview of Erectile Dysfunction in Aging Men

PETER HUAT-CHYE LIM, PERIANAN MOORTHY, AND ROBERT S. TAN

Case Histories

Case 1

A 55-year-old man with mild hypertension and coronary heart disease presented with erectile dysfunction (ED). Current medications included antihypertensives and nitrates. He had been counseled not to take sildenafil (Viagra) because of a positive chemical stress test. He was therefore tried on apomorphine (Uprima), unfortunately without a satisfactory outcome. Self-injection of prostaglandin E_1 (PGE_1) worked for him but he wanted to be rid of frequent injections and was opposed to getting a formal penile prosthetic implant. Finally, he had a Brindley penile autoinjector implant, which works by delivering a metered quantity of sodium nitroprusside into the corpora cavernosum after executing the tiny pump embedded into the central part of the scrotum. Each refill done percutaneously via a hypodermic needle permits 50 erections before the next visit to the urologist. This "implant" does not damage the corpus cavernosum like the classic implant. If so desired it can be explanted and the penis can be returned to the original preimplant status. It also does not preclude use of the other methods after it is removed.

Case 2

A 67-year-old man came for ED therapy in the andrology clinic. The use of Viagra restored his erections, but his wife refused to have intercourse with him, saying that she was menopausal, that her sex life was over, and that he must likewise accept the inevitable gracefully and not seek artificial rejuvenation. He gave up trying to reason with her, and sought gratification from prostitutes. His wife found out and came to the doctor's office demanding the doctor refrain from giving her husband Viagra.

Case 3

A 30-year-old man was involved in a motor vehicle accident. He fractured his pelvis and could not produce a good erection. Neurological examination showed no focal deficits. Investigations included a color Doppler ultrasound, which showed an arteriogenic cause. Arteriography revealed damage to the artery in Alcock's canal and no flow into the right deep cavernosal artery. A revascularization procedure using the inferior epigastric artery with microvascular anastomosis of the latter to the deep dorsal vein of the penis (arterialization of the deep dorsal vein of the penis) restored his erections to premorbid level of performance.

Introduction

An erection is a complex, involuntary, neuropsychological, hormone-mediated vascular event that occurs when blood rapidly flows into the penis and becomes trapped in its spongy chambers. Erectile dysfunction, the preferred term for impotence, was defined at the National Institutes of Health Consensus Conference, December 1992, as "the inability to achieve or maintain an erection satisfactory for sexual intercourse." Satisfaction is determined by both patient and partner, making erectile dysfunction a "couple's disease."

For a man, however, sexual performance carries an identity and the sense of self-esteem in his society and the world. Thus sexual performance in the man has unprecedented importance, depending on the erectile function of the male sex organ. Potentially, ED plays an adverse role in human procreation, as male infertility may also be a consequence. In daily life, it is very easy for men to admit having a sore throat or hemorrhoids. However, admitting to having ED is difficult for the male ego, especially if the dysfunction occurs at a younger age.

The Physiology and Mechanism of Penile Erection

Coordinated control of psychological, hormonal, neurological, vascular, and cavernosal events results in an erection. An erection of the penis is a hemodynamic event, influenced by relaxation of smooth muscle cells in the corpora cavernosa and the arteries of the penis, coincident with restriction of penile venous outflow.[1,2] The arterial blood flow into the penis is therefore increased, by way of the cavernosal arteries and the helicine arteries, which deliver blood directly to the cavernosal spaces.[1] There is consequently a sustained elevation of the intracavernosal arterial pressure with relaxation of smooth muscle trabeculae and pooling of blood within the corpora cavernosa, leading to engorgement of cavernosal tissue and finally to penile erection.[1]

The relaxation of smooth muscle cells in the corpora cavernosa is mediated by 3'5'-cyclic guanosine monophosphate (cGMP). This critical event is brought about by nitric oxide (NO).[2,3] The response to a sexual stimulus is the release of NO at the nonadrenergic, noncholinergic nerve endings innervating the penile arterioles, leading to the relaxation of smooth muscle cells.[3] Subsequently, cGMP is hydrolyzed by a cGMP specific phosphodiesterase-5 (PDE-5) isoenzyme, which is richly present in the cavernosal tissues of the penis.[1] Besides the NO system, relaxation of cavernosal smooth muscle is also initiated by other neurotransmitters including acetylcholine, vasoactive intestinal polypeptide (VIP) via the VIP-ergic system, prostaglandin via the prostacyclin system, and NO via the nitrergic system.

Definition and Prevalence of Erectile Dysfunction

Erectile dysfunction (ED) is one form of the sexual dysfunction that leads to the patient's experience of inadequate libido, inefficient orgasm, and retarded or premature ejaculation. In recent times, ED has been labeled as the most common sexual problem among pleasure-seeking males and a complaint of all men irrespective of their age, race, and culture. It is reported that nearly 100 million men around the world are experiencing ED, but only 10 million are opting for treatment, despite enormous advancements and treatment facilities in all parts of the world. To cite a few countries, in China and Korea only 9% and 30% of men, respectively, voluntarily admit to having ED, and in most of the other countries in Asia it is still considered a very sensitive issue with considerable social stigma.

Age-Specific Incidence

The *incidence* of ED is reported to be associated with increasing age. In the United States, the Massachusetts Male Aging Study (MMAS) reported that 5% of men at 40 years, and 15% of men at 70 years are suffering from complete ED. The *prevalence* of some degree of ED is reported as 52% of men for the age range 40 to 70 years, of which 25% are of moderate extent and 10% complete.[4] The risk of ED in this cohort of men was 26 cases per 1000 men annually. The annual incidence rate increased with each decade: from 12.4 cases per 1000 man-years in men between 40 and 50 years to 46.4 cases in men between 50 and 60 years. As a result, some men consider this is as an evil, whereas others consider it as a *necessary evil* (unavoidable aging process). More recently many young men were also found to have ED; however, the problem is more significant after the age of 50 (Fig. **14–1**).[5] Marital status also influences the incidence of ED. Symptoms of ED are significantly higher in unmarried men than in married men.[6]

What Causes Erectile Dysfunction?

Many men believe that ED is the result of stress or anxiety or a direct impact of aging. However, it is also caused by various other comorbid conditions such as diabetes mellitus (DM), heart disease, hypertension, and penile diseases. In addition, other causes such as levels of testosterone, high-density lipoprotein, peptic ulcer, arthritis, alcohol abuse, drug abuse, smoking, and allergy are also associated with ED. Stroke, spinal cord disorders, temporal lobe epilepsy, urinary tract symptoms, and multiple sclerosis might also lead to erectile dysfunction.[6,7] Further, medications for other common problems could be causative and cannot be neglected as an important contribution to this malady, for example in outpatient clinics, around 25% of ED was diagnosed as drug-mediated. Among these drugs, the prime agent is thiazide diuretics followed by β-blockers.[8] These drugs either reduce blood pressure or act on the smooth muscles of the corpus cavernosa. Other drugs such as benzodiazepines, serotonin inhibitors, cimetidine, digoxin, and metoclopramide would also increase ED.[9] In untreated depressions, 20 to 40% prevalence of ED has been reported, and its severity was found to increase with the magnitude of depression. In men over age 50, *myocardial infarction* and *vascular disorders* such as insufficient arterial inflow and excessive venous outflow account for ~50 and 64% of ED, respectively.[10,11] As diabetes and cardiovascular diseases are common in men with ED, we will briefly concentrate on the association of ED with these two diseases. The etiology of ED is summarized in Table **14–1**.

Erectile Dysfunction in Diabetes

In diabetic patients ED is very common.[11] Of late, the number of diabetic patients is on the rise and as a

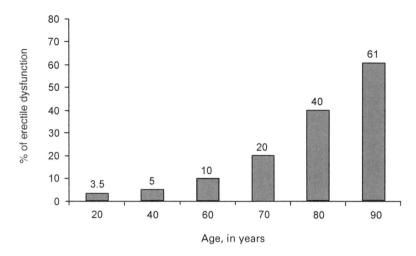

FIGURE 14–1. Incidence of erectile dysfunction by age.

result 30% of ED in Singapore is attributed to diabetes.[12] In the Western world the situation is not so conspicuous with diabetes, where only 25% of ED is associated with diabetes. Moreover, with an increase in age and diabetes the problem of ED becomes compounded (Fig. **14–2**).[13] Because of this potential effect of diabetes on ED, diabetes can also be regarded as a potential diagnostic indicator of ED in men.

Hypogonadism and diabetes has a cumulative effect on the erectile function through their role in changing the ability of endothelial cells to produce neurotransmitters and thus cause ED. An Italian study assessed the prevalence of ED in 1383 men with type 1 DM and in 8373 men with type 2 DM, between the ages of 20 and 69 years.[14] Not unexpectedly, age was a factor in the prevalence, but taking age into account 37/100 men with type 2, and 51/100 with type 1 DM had erectile problems. Romeo et al[15] established that rather than age, glycemic control and peripheral neuropathy were independent predictors of ED.

Erectile Dysfunction in Cardiac Patients

Erectile dysfunction is a common condition in men with cardiovascular disease. Some men realize that there is a degree of cardiac risk associated with sexual activity, which subsequently becomes a concern for many patients with cardiovascular disease, particularly those who have experienced acute cardiac events in the past. The relative risk of an incidence of myocardial infarction following sexual activity was no different in patients with a previous history of myocardial infarction or angina pectoris than in patients with no history of cardiovascular disease.[16,17] ED is also often associated with cardiovascular diseases requiring drug treatment. The medications for these conditions are reported to cause or contribute to ED at a rate of 25%.[8] Also, medications such as antidepressants and luteinizing-releasing hormone analogs may be responsible for ED, too. Hormonal changes such as age-related decline in testosterone and even more so of free testosterone level and other hormonal disorders,

TABLE 14–1. Summary of Etiology Factors in Erectile Dysfunction

Vascular disease affecting large or small vessels
Endocrine disorders (e.g., pituitary problems, gonadal failure, adrenal disorders, thyroid disorders, diabetes mellitus)
Local factors (e.g., cavernous veno-occlusive dysfunction)
Poor overall health (e.g., severe angina or shortness of breath that limits or prevents the physical act of intercourse)
Smoking and consumption of alcohol
Medication use (see text for examples)
Inflammatory conditions of the prostate (e.g., prostatitis), urethra, or seminal vesicles
Surgical procedures (e.g., perineal surgery, radical prostatectomy, cystoprostatectomy, abdominal-perineal resection, vascular surgery)
Pelvic fractures with or without urethral distraction injuries
Lumbar neurological injuries
Neurological conditions (e.g., Parkinson's disease, amyotrophic lateral sclerosis, multiple sclerosis, tabes, peripheral neuropathies)

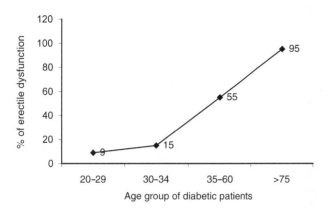

FIGURE 14–2. Incidence of erectile dysfunction in diabetic patients by age group. (Adapted from McCulloch DK, Campbell IW, Wu FC, Prescott RJ, Clarke BF. The prevalence of diabetic impotence. Diabetologia 1980;18:279–283, with permission.)

such as hypo- and hyperthyroidism, are reported to result in ED.[18,19]

Despite several organic causes of ED, potential problems due to psychogenic causes cannot be overlooked. Of these, sexual fear, unsatisfactory interpersonal relationships, depression, identity problems and lifestyle, miseducation toward sex orientation, anxiety, shame or guilt, and past sexual history play an important role in men's perception of sexual prowess and vigor.

Classification and Quantification of Erectile Dysfunction

Based on severity, etiology, and onset, ED is classified into three major categories: mild, moderate, and severe. From a clinical standpoint, the patient's physical and biochemical assessment variables should be considered for proper classification.

Mild ED is defined as a decreased ability to attain and or maintain an erection with intermittent satisfactory sexual performance.

Moderate ED is regarded as a decreased ability to attain and or maintain an erection with infrequent satisfactory sexual performance.

Severe ED is meant as a decreased ability to attain and or maintain an erection with rare or absent satisfactory sexual performance.

The 15-item questionnaire of the International Index of Erectile Function (IIEF) is used to quantify the degree of erectile function in men. It addresses and measures various attributes of sex, such as activity, intercourse, relationship, desire, sexual satisfaction, and stimulation. An abbreviated five-item questionnaire is also available (Table 14–2).

How Can We Investigate ED?

In most instances, a careful history and physical examination is sufficient to evaluate a patient for ED. However, in certain unclear cases investigations are needed. These tests are often performed in clinical trials, so as to have objective outcomes, rather than those subjectively reported by the patient.

Pharmacological Testing

Office pharmacological testing is useful as it is offers information to the physician. The test involves intracavernosal injection of a small amount of an active agent [e.g., $10\,\mu g$ of alprostadil (PGE_1)] that, theoretically, would produce a normal or priapic erection in a patient with normal erectile function but a poor response in a patient with ED. The problem with this test is that unless the patient's sympathetic nerve impulses are completely overcome by the injected agent, he may have a poor erection even though his erectile function is normal.

Hormonal Assays

Measuring serum levels of bioavailable, total and free testosterone, gonadotropins, and gonadotropin-releasing hormone may be helpful in a patient in whom history taking or physical examination suggests lack of androgen stimulation. Androgen insufficiency is not a direct cause of ED, but is often associated with ED. Features of hypogonadism include poor libido, a disproportionately small prostate gland relative to the patient's age, small or soft testicles, and noticeable thinning or diminished growth of the beard. If the patient's total testosterone level is low, the prolactin level should be checked to exclude a prolactinoma.

Nocturnal Studies

Nocturnal studies present perhaps a true picture of ED due to organic causes. The most complete evaluation of nocturnal erectile function is obtained in a sleep laboratory, where patients are monitored for rapid-eye-movement (REM) sleep. Under normal conditions, an erection would be expected to occur with each REM episode. The erection can be described in terms of tumescence and buckling force, which is a measure of rigidity.

Vascular Erectile Testing

Duplex Doppler ultrasonography has been used extensively. As the clinical evaluation is often sufficient, evaluation of erectile function with Doppler ultrasound may be useful for evaluating resistant cases of ED and treatment selection. Doppler studies

TABLE 14–2. Five-Item International Index of Erectile Function (IIEF)

Question	Response	Points	Score	Interpretation
(1) How do you rate your confidence that you could get and keep an erection?	Very low or none at all	1	22–25	No erectile dysfunction
	Low	2	17–21	Mild erectile dysfunction
	Moderate	3	12–16	Mild to moderate ED
	High	4	8–11	Moderate erectile dysfunction
	Very high	5	5–7	Severe erectile dysfunction
(2) When you had erections with sexual stimulation how often were your erections hard enough for penetration?	Almost never or never	1		
	A few times (much less than half the time)	2		
	Sometimes (about half the time)	3		
	Most times (much more than half the time)	4		
	Almost always or always	5		
(3) During sexual intercourse, how often were you able to maintain your erection after you had penetrated (entered) your partner?	Almost never or never	1		
	A few times (much less than half the time)	2		
	Sometimes (about half the time)	3		
	Most times (much more than half the time)	4		
	Almost always or always	5		
(4) During sexual intercourse, how difficult was it to maintain your erection to completion of intercourse?	Extremely difficult	1		
	Very difficult	2		
	Difficult	3		
	Slightly difficult	4		
	Not difficult	5		
(5) When you attempted sexual intercourse, how often was it satisfactory for you?	Almost never or never	1		
	A few times (much less than half the time)	2		
	Sometimes (about half the time)	3		
	Most times (much more than half the time)	4		
	Almost always or always	5		

Score	Interpretation
22–25	No erectile dysfunction (ED)
17–21	Mild ED
12–16	Mild to moderate ED
8–11	Moderate ED
5–7	Severe ED

provide information about both arterial and venous flow. Dynamic infusion caversometry and cavernosography (DICC) can also provide detailed data about pressure related to erectile function, but this information often is more extensive than is required for treatment of ED.

What Are the Treatment Modalities for Erectile Dysfunction?

Because ED is a multifactorial disorder, identifying the exact cause is of more importance than therapeu-

tic intervention at the initial visit. Involving the female partner would also be helpful in circumventing ED. Moreover, a holistic approach using modern medications, surgery, or technology with a personalized touch could mitigate ED when considering the available options to achieve the desired goal. Several therapeutic medications and techniques have evolved to mitigate ED, of which, oral medications has become the first-line nonsurgical treatment option for many patients.[20] Similarly, intracavernosal injections, microsurgical revascularization, and the implantation of penile prostheses[21] have also gained significant attention among a broader group of patients who are affluent in society and nonresponsive to alternative treatment modalities.

The penile prosthesis is a rigid, hinged, mechanical, inflatable, or hydraulic structure used to mechanically overcome ED. Although penile prosthetic implants have been the treatment choice for many nonresponsive patients to alternative medications, proper diagnosis should be made to identify the cause of the ED in these patients. The clinical success of penile implant therapy depends on postoperative infection, the smooth and uncomplicated execution of the operation, avoidance of technical errors, the durability of the prosthetic material, and anatomic features of the penis,[21] as complications arising from these factors often warrant removal of the prosthesis with the sequela of complete ED.

Some of the oral compounds used for ED are beraprost, pinacidil, alprostadil, papaverine, phentolamine, yohimbine, sildenafil, tadalafil, vardenafil, and apomorphine. At times, these treatments are unsatisfactory because of patient reluctance, unpredictable efficacy, inconvenience, and nonavailability, and inevitably they all lack spontaneity. When ED is remedied by pharmacological modalities, an adequate column of penile blood gives an erection of sufficient rigidity and duration for satisfactory sexual intercourse.[4] We will briefly concentrate on widely prescribed and promising oral medicines for ED.

Sildenafil citrate (Viagra) is a selective phosphodiesterase-5 inhibitor and has been shown to be highly efficacious in men with ED originating from a wide variety of etiologies.[20,22] It arrests the metabolic breakdown of cGMP. Compared with other oral agents Viagra has been identified as promising first-line treatment for ED because of its combined high efficacy with a good safety profile. Fig. **14–3** illustrates the clinical efficacy of Viagra compared with placebo in men with ED.

Viagra also has been proven to be effective in diabetic patients. Though many therapeutic agents are available for ED, proper diagnosis and medications are needed for sustained efficacy and durability of these treatments. Despite an increase in erection with Viagra, some of the side effects, such as arrhythmias, transient ischemic attacks, stroke, hypertension, headache, flushing, dyspepsia, and visual abnormalities, should be expected in those at risk and should be predetermined before prescription. One possible disadvantage of Viagra is the interaction with food and shorter half-life. However, the clinical experience with Viagra is unparalleled, with several million men worldwide using this drug.

Tadalafil (Cialis) or IC351 is a new representative compound of the *second generation* of selective phosphodiesterase-5 (PDE-5) inhibitors.[23] The selectivity ratio versus PDE-5 is more than 10,000 for PDE-1 through PDE-4 and PDE-7 through PDE-10 and 780 for PDE-6. In the European daily-dosing trial, the efficacy rates were up to 93% for successful intercourses with completion in the 50-mg dose in patients with mild to moderate ED. In two different dose-ranging studies with 2 to 25 mg taken as needed, efficacy rates of up to 88% improvement in erections and up to 73% successful intercourses with completion were achieved. In a placebo-controlled, fixed-dose (10- and 20-mg) trial in diabetic patients, improved erections of 56% and 64% were reported compared with 25% after placebo. Drug-related adverse effects, with headache in up to 23% of patients (placebo, up to 17%), dyspepsia in up to 11% (placebo, up to 7%), back pain in up to 4.7% (placebo, 0%), and myalgia in up to 4.1% (placebo, up to 2.4%), were mostly mild to moderate. Neither drug-related serious cardiovascular adverse events nor color vision disturbances were encountered. The *long half-life* (>17 hour), with a comfortably long window of opportunity, releases couples from the need to plan

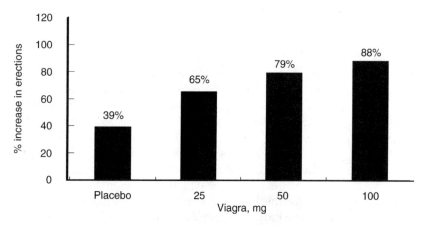

FIGURE 14–3. Increase in erections as a response to increasing dosages of sildenafil (Viagra). (Adapted from Goldstein I, Lue TF, Padma-Nathan H, Rosen RC, Steers WD, Wicker PA. Oral sildenafil in the treatment of erectile dysfunction. N Engl J Med 1998;338:1397–1404, with permission.)

sexual activities and therefore provides the *highest amount of spontaneity* for sexual activities.

Vardenafil (Levitra) selectively inhibits PDE-5, an enzyme that hydrolyzes cyclic guanosine monophosphate in the cavernosum tissue of the penis.[24] Inhibition of PDE-5 results in increased arterial blood flow leading to enlargement of the corpus cavernosum. Because of the increased tumescence, veins are compressed between the corpus cavernosum and the tunica albuginea, resulting in an erection. Vardenafil has a *high bioavailability* and is rapidly absorbed. An erection of >60% rigidity was maintained for approximately twice as long following visual stimulation in patients treated with vardenafil 10 or 20 mg than in recipients of placebo. In a large, placebo-controlled trial in patients with mild to severe ED, vardenafil 5, 10, or 20 mg taken as needed over a 12-week period significantly improved the scores in questions 3 and 4 of the International Index of Erectile Function (IIEF). The rate of successful attempts at intercourse with ejaculation was also significantly higher with vardenafil (71 to 75%) than in the placebo group (39.5%), and significantly more patients treated with vardenafil than placebo responded "yes" to a Global Assessment Questionnaire (GAQ) asking if treatment had improved erections. In a 26-week trial in 736 men with ED of varied etiologies and severity, patients receiving vardenafil 5, 10, or 20 mg experienced significantly improved erections, with 85% of the vardenafil 20 mg recipients reporting improved erectile function (assessed using the GAQ) compared with 28% of placebo recipients. Treatment with vardenafil also significantly improved scores in response to questions 3 and 4 of the IIEF compared with placebo. A 12-week trial in 452 men with ED associated with diabetes mellitus demonstrated that treatment with vardenafil 20 mg compared with placebo significantly improved IIEF erectile function domain scores and the rate of positive responders to the erectile improvement GAQ. Similar results were reported in a placebo-controlled trial of vardenafil 10 to 20 mg involving 440 patients with ED after radical prostatectomy. Adverse events associated with vardenafil were those commonly associated with PDE-5 inhibitors: headache, flushing, dyspepsia, and rhinitis. These were mostly dose-dependent and mild to moderate in intensity.

Apomorphine SL (Ixense; Uprima) is a new oral medication shown to be effective in the treatment of ED.[25] This compound is a dopaminergic agonist with affinity for dopamine receptor sites—mostly D(2)—within the brain known to be involved in sexual function. Apomorphine induces selective activation in the nucleus paraventricularis leading to erectogenic signals. More than 5000 men with ED participated in phase II/III clinical trials assessing the safety and efficacy of doses ranging from 2 to 6 mg. The most favorable risk/benefit ratio is seen with a dose-optimization regimen of 2 to 3 mg: the 3-mg dose provides efficacy comparable to that of 4 mg but with fewer side effects. Consequently, review of clinical studies focuses on data with the 2- to 3-mg dose, the registered dose for use in clinical practice. The primary efficacy end point in most clinical trials with apomorphine SL was the percentage of attempts resulting in erections firm enough for intercourse—one of the most rigorous end points used in ED trials to date. These data were collected from both patients and their partners by reviewing entries in patient diaries and partner Brief Sexual Function Inventory (BSFI) questionnaires. Secondary end points included percentage of attempts resulting in intercourse and improvement in ED severity based on the IIEF. The proportion of attempts resulting in erections firm enough for intercourse was 49.4% with 3 mg compared with the baseline value of 24.3%. Partner evaluations corresponded with those of the patients. Erections occurred between 18 and 19 minutes after taking apomorphine SL 2 or 3 mg. The most common side effect was nausea, which declined with continued use. Vasovagal syncope was reported in <0.2% of men, and was preceded by clear prodromal symptoms. Apomorphine SL is also an effective, well-tolerated drug for ED.

Discussion of the Case Histories

Case 1

The desperate patient fearful of the traditional implant can now use the latest prosthesis invented by Prof. Giles Brindley from the University of London. It is excellent for neurogenic ED, mild to moderate venogenic ED, some cases of arteriogenic ED, and most cases of psychogenic ED. If PGE$_1$ works, then this method will most certainly be effective.

Case 2

Couple counseling is vital as exemplified by the events of this case. The practitioner should try at all costs to get the couple to see how the two parties must interact, and to educate both on sexual health concerns that apply today even for the old and the elderly. In this era of the availability of countless varieties of erectogenic drugs and increasing awareness of female sexuality and female sexual dysfunction, couple concerns and couple therapy have an increasingly vital role to play in sexual medicine.

Case 3

Acute vascular injuries respond well to revascularization procedures in the younger individual. Middle-aged and older men will have a fairly rapid relapse with arterializations procedures because of higher thrombosis rates occurring at the site of microvascular anastomosis in this cohort. Hence we recommend this procedure only to younger men.

Conclusion and Key Points

Erectile dysfunction has become more widely accepted as a disease entity, and treatment should be targeted not only on improving the quality of life, but also on the associated disease states.[26] In terms of numbers, prevalence studies suggest that one in two men over 50 years have an associated ED.

- ED is associated with comorbid states including diabetes mellitus, heart disease, hypertension, etc.

- ED is often associated with hypogonadism and the andropause syndrome. Although hypogonadism does not directly lead to ED, treatment with testosterone is supplementary as well as complementary.

- Some medications can lead to ED, and the practitioner should adjust or remove these medications before adding PDE-5 inhibitors, for instance.

- It is important to objectively classify the stage of ED in the assessment of treatment success or failure.

- Newer oral medication for treatment of ED will mean more patients appearing for treatment, and hence an opportunity to screen for comorbid disorders.

REFERENCES

1. Andersen KE, Wagner G. Physiology of penile erection. Physiol Rev 1998;75:191–236

2. Naylor AM. Endogenous neurotransmitters mediating penile erection. Br J Urol 1998;81:424–431

3. Rajfer J, Aronson WJ, Bush PA, Dorey FJ, Ignarro LJ. Nitric oxide as a mediator of relaxation of the corpus cavernosum in response to nonadrenergic, noncholinergic neurotransmission. N Engl J Med 1992;326:90–94

4. Feldman HA, Goldstein I, Hatzichristou DG, Krane RJ, McKinlay JB. Impotence and its medical and psychosexual correlates: results of the Massachusetts Male Aging study. J Urol 1994;151:54–61

5. Kaiser FE. Erectile dysfunction in the aging man. In: Kaiser FE, ed. *The Medical Clinics of North America: The Aging Male Patient.* Philadelphia: WB Saunders; 1999:1267–1278

6. Laumann EO, Anthony Paik MA, Raymond RC. Sexual dysfunction in the United States: prevalence and predictors. JAMA 1999;281:537–544

7. Kaiser FE. Sexuality and impotence in the aging man. Clin Geriatr Med 1991;7:63–72

8. Buffum J. Prescription drugs and sexual function. Psychiatr Med 1992;10:181–198

9. Burchardt M, Burchardt T, Baer L, et al. Hypertension is associated with severe erectile dysfunction. J Urol 2000;164:1188–1191

10. Lue TF, Hricak H, Schmidt RA, et al. Functional evaluation of penile veins by cavernosography in papaverine induced erection. J Urol 1986;135:479–482

11. Kaiser FE, Korenmann SG. Impotence in diabetic men. Am J Med 1988;85:147–152

12. Lim PH, Li MK, Ng FC, et al. Clinical efficacy and safety of sildenafil citrate (Viagra) in a multiracial population of Singapore: a retrospective study of 1520 patients. Int J Urol 2002;9:308–315

13. McCulloch DK, Campbell IW, Wu FC, Prescott RJ, Clarke BF. The prevalence of diabetic impotence. Diabetologia 1980;18:279–283

14. Fedele D, Bortlotti A, Coscelli C, et al. Erectile dysfunction in type 1 and 2 diabetics in Italy. Int J Epidemiol 2000;29:524–531

15. Romeo JH, Seftel AD, Madhun ZT, Aron DC. Sexual function in men with diabetes type 2: association with glycemic control. J Urol 2000;163:788–791

16. Muller JE, Mittleman A, Maclure M, Sherwood TB, Tofler GH. Triggering myocardial infarction by sexual activity. Low absolute risk and prevention by regular physical exertion. JAMA 1996;275:1405–1409

17. De Busk RF. Sexual activity triggering myocardial infarction. One less thing to worry about. JAMA 1996;275:1447–1448

18. Blackman MR, Kowatch MA, Wehmann RE, Harmann SM. Basal serum prolactin levels and prolactin responses to constant infusions of thyrotropin releasing hormone in healthy aging men. J Gerontol 1986;41:699–705

19. Foster RS, Mulcahy JJ, Callagher JT, Crabtree R, Brashear D. Role of serum prolactin determination in evaluation of impotent patient. Urology 1990;36:499–501

20. Goldstein I, Lue TF, Padma-Nathan H, Rosen RC, Steers WD, Wicker PA. Oral sildenafil in the treatment of erectile dysfunction. N Engl J Med 1998;338:1397–1404

21. Rosen JC. Erectile dysfunction: the medicalization of male sexuality. Clin Psychol Rev 1996;16:497–519

22. Perrin J. And then came Viagra. Pharmaceu J 1998;260:707

23. Porst H. IC351 (tadalafil, Cialis): update on clinical experience. Int J Impot Res 2002;(suppl 1):S57–S64

24. Ormrod D, Easthope SE, Figgitt DP. Vardenafil. Drugs Aging 2002;19:217–227

25. Altwein JE, Keuler FU. Oral treatment of erectile dysfunction with apomorphine SL. Urol Int 2001;67:257–263

26. Jordan GH. Erectile function and dysfunction. Postgrad Med 1999;105:131–134

Depression, Erectile Dysfunction, and Coronary Heart Disease: An Interlinked Syndrome in Men

Robert S. Tan

Case History

Mr. S. is a 54-year-old man who came for a consultation for his loss of libido and erectile dysfunction. He gives a 5-year history of fatigue, and was diagnosed to be hypothyroid and is on Levoxyl. He has consulted many physicians across the country seeking help for his sexual dysfunction symptoms. He admitted to being depressed about his situation, and was prescribed Celexa 40 mg by his psychiatrist. He has a history of angina, and recently underwent angioplasty and had stents introduced by his cardiologist. His urologist has been managing his erectile dysfunction, and the patient was prescribed Viagra. He felt dizzy after taking Viagra, and was thought to have an idiosyncratic reaction to the medication. Subsequently, he was switched to injection Caverject. The response was inadequate and he was subsequently put on an intramuscular compounded mixture of Trimix (prostaglandin, phentolamine, and papaverine). Other medications include Tenormin, Ativan, Lipitor, and a host of vitamins.

Past surgical history include liposuction, arthroscopy, cervical spine decompression, and angioplastic stent insertion. Physical examination findings include blood pressure (BP) 130/86, weight 225 pounds, body fat content 27.8%, and normal testicular and phallic examination. Laboratory results include total testosterone 230 ng/dL, free testosterone 48.4 ng/dL (hypogonadal range), dihydrotestosterone 57 ng/dL (normal), insulin-like growth factor-I (IGF-I) 264 ng/dL (normal: 90–360), prolactin 10 ng/dL (normal: 2–18), prostate-specific antigen (PSA) 0.6 (normal). Baseline electrocardiogram shows poor R wave progression, with S-T depression in V5,6. Review of angiographic studies shows two-vessel disease. Depression screening with the Geriatric Depression Scale (GDS) reveals a score of 10/15.

It is well known from epidemiology studies that the prevalence of depression, erectile dysfunction (ED), and coronary heart disease (CHD) increases with age.[1,2] Physicians tend to consider them as separate entities, partly because physicians specialize in their respective fields. However, men presenting with problems in one organ system may already be aware of comorbid states in other organ systems, but may choose not to discuss them with their physicians. The National Institutes of Health recognized ED as a medical identity following a Consensus Development Conference in 1992,[3] but it was not until the introduction of specific type 5 phosphodiesterase inhibitors that men began seeking help in large numbers. A similar situation occurred with depression before safe and effective treatments were available, and patients shunned treatments or were treated inappropriately. The symptoms related to depression, ED, and CHD are closely interlinked. An older patient presenting with symptoms of depression may have comorbid ED and/or comorbid CHD. Likewise, an older patient who presents with ED may have underlying CHD as well as depression.

This triad of depression, ED, and CHD in aging men is referred to as the *DEC syndrome,* and this chapter discusses the considerable overlap and interlinks among these three components. When a patient presents with one component of the DEC syndrome, physicians should also screen for the other two components (Fig. 15–1). The root cause of the syndrome may be a combination of arteriosclerosis, neurochemical imbalance, and hypogonadism that occurs with aging. The timing of symptoms coincides with hypogonadism in the andropause.[4] Although testosterone replacement may improve depression,[5] it should not be the first line of treatment. Likewise, although testosterone may improve coronary flow,[6] patients with coronary problems should not be offered testosterone as first-line therapy, but may be offered

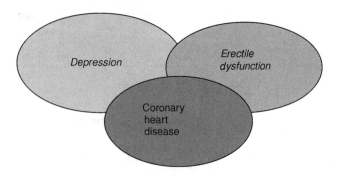

FIGURE 15–1. The interlinked DEC (depression, erectile dysfunction, coronary heart disease) syndrome. The presence of one of these may be a harbinger of a larger, multifactorial problem.

supplementary testosterone if hypogonadism is present and if there are no contraindications.

Older Men with Depression Are More Likely to Develop Coronary Heart Disease

In general, the prevalence of depression in older women is higher than that in older men.[7,8] There may be several reasons for this, including the possibility that men are less likely to report symptoms. However, it should be noted that rates of completed suicide are much higher in older men than in older women.[9] Epidemiology studies reveal that in older men the prevalence of symptomatic depression is ~15 to 25%, and the prevalence of major depression is ~1.5 to 2%.[7] CHD is the leading cause of death worldwide. The lifetime risk of developing CHD at age 40 is 50% for men and 33% for women. The cardioprotective role of estrogens in women is fairly well established.[10] For several years, researchers have been arguing that testosterone provides cardioprotection for men.[6] Thus, a complex relationship exists between depression in later life and CHD. Several prospective studies have suggested that depression may predispose an individual to an increased risk of developing CHD. For example, the Normative Aging Study, which included 1305 older men, demonstrated that depression is positively associated with the risk of CHD.[11] In another longitudinal study, after 40 years of follow-up, investigators have found that depression appears to be an independent risk factor for incident CHD in male medical students.[12] Ariyo and coworkers[13] also found that depressive symptoms constitute an independent risk factor for the development of CHD and total mortality in older Americans. Figure 15–2 shows the association of increasing CHD and death with depression.

Depression can even increase the risk for cardiac mortality in subjects with and without cardiac disease at baseline.[14] Some community surveys have observed an increase of ischemic heart disease among those with depression who were initially free of disease.[15] Investigators have also found after long-term follow-up that history of depression in itself increases the risk of myocardial infarction and mortality.[16,17] On the other hand, a multicenter study reported that baseline depressive symptoms were not related to subsequent cardiovascular events, but rather to prognosis.[18] Table 15–1 is an example of a screening tool, the GDS, commonly used to screen for depression. This tool is particularly useful for older individuals.

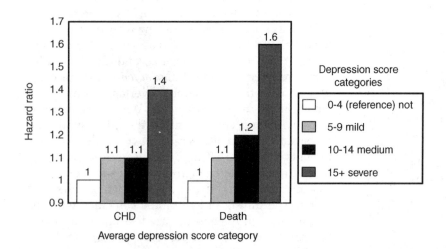

FIGURE 15–2 The association of coronary heart disease (CHD) and death rates with depression. (Adapted from Ariyo AA, Haan M, Tangen CM, et al. Depressive symptoms and risks of coronary heart disease and mortality in elderly Americans: Cardiovascular Health Study Collaborative Research Group. Circulation 2000;102:1773–1779, with permission.)

TABLE 15–1. Short-Form Geriatric Depression Scale

Choose the best answer for how you have felt over the past week ("Yes" or "No" for each):

Are you basically satisfied with your life?
Have you dropped many of your activities and interests?
Do you feel that your life is empty?
Do you often get bored?
Are you in good spirits most of the time?
Are you afraid that something bad is going to happen to you?
Do you feel happy most of the time?
Do you often feel helpless?
Do you prefer to stay at home, rather than going out and doing new things?
Do you feel you have more problems with memory than most?
Do you think it is wonderful to be alive now?
Do you feel pretty worthless the way you are now?
Do you feel full of energy?
Do you feel that your situation is hopeless?
Do you think that most people are better off than you are?

Scoring: A "Yes" answer to items 2–6, 8–10, 12, 14, 15 is one point each. A "No" answer to items 1, 7, 11, 13 is one point each. A score of 0–5 is normal; a score of 6–15 suggests depression.

Older Men with Coronary Heart Disease Are More Likely to Suffer from Depression

It seems natural to think that a person with a chronic disorder is more likely to be depressed; thus, older men with coronary heart disease are more likely to be depressed. Studies have found that ~45% of patients have been found to have either major or minor depression after a myocardial infarction (MI).[19]

Depression in older men with CHD can affect not only morbidity but also mortality. For instance, older post-MI patients with depression have nearly four times the risk of dying within the first 4 months after discharge.[20] Therefore, there is an independent link between depression and cardiac mortality after MI. Bush et al[21] reported that even minimal symptoms of depression could increase mortality risk after acute MI. Conversely, the national survey from 1971 to 1992 reported that depression was not a significant risk factor for CHD mortality.[22] We hypothesize several biological reasons for the increased vulnerability of depressed patients to CHD and increased mortality: hypothalamic-pituitary-adrenocortical and sympathomedullary hyperactivity, diminished heart rate variability, ventricular instability and myocardial ischemia in reaction to mental stress, and alteration in platelet receptor and reactivity.[23]

Depression is underrecognized and undertreated in older men with CHD despite its consequences for prognosis. The clinical implication is that older men recovering from an MI should be evaluated for depression.[19] The selective serotonin reuptake inhibitors (SSRIs) remain the most commonly prescribed class of antidepressants, but they may contribute to ED. As a result they may paradoxically exacerbate existing depression because the patient might not have expected the side effect of ED from SSRI, and the further loss of libido from the medication can be devastating. The mechanism of SSRIs on decreasing sexual functioning is dose related and may vary among patients based on serotonin and dopamine reuptake mechanisms, induction of prolactin release, anticholinergic effects, inhibition of nitric oxide synthase, and the propensity of the SSRI to accumulate over time.[24] The use of tricyclic antidepressants such as amitriptyline, nortriptyline, and desipramine have particularly diminished in older men due to adverse anticholinergic and cardiovascular effects.[25,26] Finally, treating depression may actually lower cardiovascular mortality, but studies are lacking at this point.[25,26]

Association of Erectile Dysfunction with Depression

Erectile dysfunction is a common condition among older men. The Massachusetts Male Aging Study, a cross-sectional, community-based, random survey of men in the Boston area, determined that the combined prevalence of minimal, moderate, and complete ED is ~52%.[27] According to the study, the age group of greatest concern for developing ED is between 40 and 70 years. Indeed, in that group, the prevalence of complete ED tripled, from 5 to 15%. The study also found that ED was directly correlated with hypertension, diabetes, alcoholism, anger, and depression. The probability of ED with treated hypertension was 0.15, with treated diabetes was 0.29, with excessive alcohol was 0.29, with anger was 0.35, and with severe depression was 0.9. This contrasts with the general risk of ED of 0.09 ($p < .01$).

Patients with ED are more likely to be clinically depressed, and patients with clinical depression often have ED. The Massachusetts Male Aging Study determined that ED was associated with depression. After controlling for potential confounders, moderate-to-complete ED was 1.82 times more likely in those who exhibited depressive symptoms as compared with those who did not display symptoms ($p < .01$).[28]

The study from the Johns Hopkins Sexual Behaviors Consultation Unit also revealed that approximately one in three men presenting with sexual dysfunction had comorbid psychiatric problems, including alcoholism and psychosocial stress.[29] Men with ED had high levels of depressive, somatic, and anxious symptoms and scored very high on measures of overall psychological distress.

One possible explanation of how clinical depression can lead to the development of ED may be that depression causes overactivity of the autonomic nevous system. The overactive autonomic nervous system can then produce the inability to initiate or maintain relaxation of the corporal smooth muscle tissue because of increased adrenergic tone in the penile organ.[30] Certain antidepressants such as fluoxetine, paroxetine, and sertraline can also result in side effects that can include ED. Many other substances and medications in addition to antidepressants can cause ED, including tobacco, antihypertensives, alcohol, anxiolytics, mood stabilizers, and antipsychotics.[27,31] It is difficult to interpret the significance of antidepressant-induced ED because there are many confounders. In general, SSRIs can adversely influence orgasm, libido, and arousal.[32] According to a study, bupropion may have the lowest overall rate of sexual dysfunction as compared with other antidepressants.[33] The sustained-release formulation of bupropion had a sexual dysfunction rate of 25%. In contrast, the SSRIs Prozac (fluoxetine), Paxil (paroxetine), Zoloft (sertraline), and Celexa (citalopram) as well as Effexor (venlafaxine) and Remeron (mirtazapine) averaged ~40%. When the different classes of antidepressants are compared, it appears that men report the least problems with sexual dysfunction while taking tricyclic antidepressants.[31]

Depressive symptoms in men with ED may improve with successful ED treatment.[34] Sildenafil is efficacious for ED in men with mild-to-moderate depression. Improvement of ED is associated with marked improvement in depressive symptoms and quality of life (Fig. 15–3).[35]

Correlation of Erectile Dysfunction with Coronary Heart Disease

Erectile dysfunction may be due to compromised blood flow. An erection occurs as a result of increased blood flow to the corpus spongiosum of the penis upon vasodilatation. This is also mediated through neuronal control systems that are dependent on the autonomic nervous system. Arteriosclerosis can affect the penis as it does any other organ, and accounts for ED in approximately half of all affected men over the age of 50 years.[36] If blood flow is inadequate, it usually signals vascular problems, especially if neurogenic controls are intact. The narrow vessels of the penis may be more prone to arteriosclerotic blockage than the larger vessels of the heart.[37]

Patients presenting with ED often have hypertension, and thus have a significantly higher prevalence

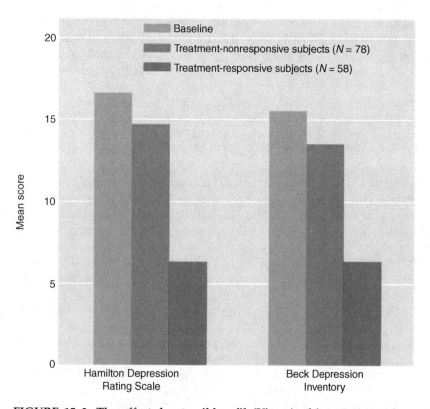

FIGURE 15–3. The effect due to sildenafil (Viagra) of improvement in erectile dysfunction on depressive symptoms. Sildenafil decreases both the Hamilton Depression Rotrip Scale as well as the Beck Depression Inventory. (Adapted from Seidman SN, Roose SP, Menza MA, et al. Treatment of erectile dysfunction in men with depressive symptoms: results of a placebo-controlled trial with sildenafil citrate. Am J Psychiatry 2001;158:1623–1630, with permission.)

Coronary Heart Disease Risk Factors and Erectile Dysfunction

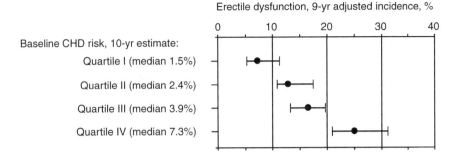

FIGURE 15–4. The association of erectile dysfunction and coronary heart disease risk factors. (Adapted from Feldman HA, Johannes CB, Derby CA, et al. Erectile dysfunction and coronary risk factors: prospective results from the Massachusetts Male Aging Study. Prev Med 2000;30:328–338, with permission.)

of cardiovascular complications.[38] Antihypertensive medication such as β-blockers can impair circulation, mask hypoglycemic symptoms, worsen asthma, and induce ED in older individuals. A study by Greenstein et al[39] demonstrated a significant correlation between ED and the number of coronary vessels involved by arteriosclerosis. ED and CHD share some modifiable risk factors (e.g., smoking and overweight) in men free of ED and heart disease after 6 to 10 years of follow-up.[40] Fig. **15–4** shows the relationship between CHD and ED.

Many managed care formularies have put restrictions on the availability of sildenafil. Unrestricted availability of licensed treatments for ED may encourage men to present for investigation, enabling early detection of CHD.[37]

Clinical Implications of the DEC Syndrome in Older Men

It is certainly possible that older men present with isolated and unrelated symptoms of depression, ED, and CHD. However, epidemiology studies suggest that these three conditions overlap.[27] The existence of one symptom should lead the physician to suspect that the patient may be suffering from the other two as well. However, it may be very difficult to determine which of the three disorders may have caused the others (e.g., did the depression come first and cause ED?). A mutually reinforcing two-way model has been proposed.[1] Health status, demographics, lifestyle, medications, and psychosocial events can adversely influence the DEC syndrome, and may even be responsible for its etiology. Here are some guidelines for clinicians dealing with men over the age of 50:

- Older men presenting with CHD should be screened for symptoms of depression. There are simple and quick geriatric instruments such as the GDS that have been proven to be sensitive and specific for an older population.[41]

- Older men presenting with depression or ED should also be routinely screened for risk factors associated with CHD. It is good practice to do a baseline electrocardiogram, as treatment with some antidepressants, such as the tricyclic antidepressants amitriptyline and nortriptyline may have side effects including dysrhythmias, atrioventricular (AV) blocks, ventricular ectopics, and supraventricular tachycardia. Other side effects can also include dry mouth, exacerbation of glaucoma, postural hypotension, dizziness, and faintness. A thallium stress test can also rule out concomitant coronary problems.

- Depression in older men may sometimes also be associated with hypogonadism, and it is prudent to check testosterone levels, especially if depression is associated with symptoms including fatigue, loss of libido, and memory problems.[4]

- Older men presenting with ED should be asked about symptoms of depression. Sometimes it is the medication for the treatment of depression rather than the depression itself that causes the ED. A complete history of medication use is important to acquire.

Discussion of the Case History

Mr. S. typifies a patient with the DEC syndrome. He may at the same time be hypogonadal because of Leydig cell failure from aging. The root cause of this syndrome is a combination of vasculopathy and hypotestosteronemia. He has had narrowing of his coronary arteries, which required stent placements and as such it will not be surprising to find that his penile vasculature will be compromised, leading to vascular ED. The positive response to Viagra and PPR suggests that he does have vascular problems. Beta-blockers used to treat hypertension (Tenormin) may further compromise his ED, and so the patient was switched to an angiotensin-converting enzyme (ACE) inhibitor (Altace). He was put on a therapeutic trial of androgen replacement, but his response was described as marginal. His depression improved with the androgen replacement. The patient was taught how to self-administer the PPR injection. Key to the management of this patient is control of cardiac risk factors including the control of hypercholesterolemia and blood pressure, weight loss, and nutrition and exercise counseling. The patient was also educated about realistic goals in that he had to play a part in improving his own quality of life by paying attention to diet and exercise and that there was no single magic pill for his problem. He is at the moment considering revascularization surgery for his penis.

Conclusion and Key Points

The prevalence of depression, ED, and CHD increases with age, and the symptoms related to these three illnesses are closely interlinked. When a patient presents with one component of the DEC syndrome, the term used to refer to this triad of comorbid conditions, physicians should also screen for the other two components. Studies have shown that depression may predispose an individual to an increased risk of developing CHD, and older men with CHD are more likely to be depressed. Likewise, patients with ED are more likely to be clinically depressed, and patients with clinical depression often have ED. Furthermore, patients presenting with ED are often hypertensive, and thus have a significantly higher prevalence of cardiovascular complications.

Multifactorial problems require multifactorial approaches, and the care of older men can improve if physicians are aware of this interlinked syndrome. The care of older men may sometimes be fragmented, as patients presenting with depression, for example, may be seen by psychiatrists who may not have the capabilities to screen for CHD. Likewise, it may be possible that urologists attending to issues pertaining to ED may not have the time to deal with treatment options for depression. It is important to remember the acronym DEC, as it can serve to remind the specialist to refer the patient to the primary care physician or to other specialists for treatment of the other two components of the DEC syndrome. Primary care physicians should be attending to all three components and should not neglect the other components

when an older patient presents with a relevant symptom such as depression, ED, or angina. It may sometimes be impossible to address all problems at one sitting, and the primary care physician may want to schedule additional appointments to deal with the other issues. Multifactorial problems require multifactorial approaches, and the care of older men can improve if physicians are aware of this interlinked syndrome.

REFERENCES

1. Goldstein I. The mutually reinforcing triad of depressive symptoms, cardiovascular disease, and erectile dysfunction. Am J Cardiol 2000;86:41F–45F

2. Roose SP, Seidman SN. Sexual activity and cardiac risk: is depression a contributing factor? Am J Cardiol 2000;86:38F–40F

3. NIH Consensus Development Panel on Impotence. Impotence. JAMA 1993;270:83–90

4. Tan RS, Philip PS. Perceptions of and risk factors for andropause. Arch Androl 1999;43:97–103

5. Barrett-Connor E, Von Muhlen DG, Kritz-Silverstein D. Bioavailable testosterone and depressed mood in older men: the Rancho Bernardo Study. J Clin Endocrinol Metab 1999;84: 573–577

6. Ong PJ, Patrizi G, Chong WC, et al. Testosterone enhances flow-mediated brachial artery reactivity in men with coronary artery disease. Am J Cardiol 2000;85:269–272

7. Jeste DV, Alexopoulos GS, Bartels SJ, et al. Consensus statement on the upcoming crisis in geriatric mental health: research agenda for the next 2 decades. Arch Gen Psychiatry 1999;56:848–853

8. Blazer D, Hughes DC, George LK. The epidemiology of depression in an elderly community population. Gerontologist 1987;27:281–287

9. Brown GK, Bruce ML, Pearson JL. High-risk management guidelines for elderly suicidal patients in primary care settings. Int J Geriatr Psychiatry 2001;16:593–601

10. Roeters van Lennep JE, Westerveld HT, Erkelens DW, van der Wall EE. Risk factors for coronary heart disease: implications of gender. Cardiovasc Res 2002;53:538–549

11. Sesso HD, Kawachi I, Vokonas PS, Sparrow D. Depression and the risk of coronary heart disease in the Normative Aging Study. Am J Cardiol 1998;82:851–856

12. Ford DE, Mead LA, Chang PP, et al. Depression is a risk factor for coronary artery disease in men: the precursors study. Arch Intern Med 1998;158:1422–1426

13. Ariyo AA, Haan M, Tangen CM, et al. Depressive symptoms and risks of coronary heart disease and mortality in elderly Americans: Cardiovascular Health Study Collaborative Research Group. Circulation 2000;102:1773–1779

14. Penninx BW, Beekman AT, Honig A, et al. Depression and cardiac mortality: results from a community-based longitudinal study. Arch Gen Psychiatry 2001;58:221–227

15. Glassman AH, Shapiro PA. Depression and the course of coronary artery disease. Am J Psychiatry 1998;155:4–11

16. Pratt LA, Ford DE, Crum RM, et al. Depression, psychotropic medication, and risk of myocardial infarction: Prospective data from the Baltimore ECA Follow-Up. Circulation 1996;94: 3123–3129

17. Barefoot JC, Schroll M. Symptoms of depression, acute myocardial infarction, and total mortality in a community sample. Circulation 1996;93:1976–1980

18. Wassertheil-Smoller S, Applegate WB, Berge K, et al. Change in depression as a precursor of cardiovascular events: SHEP

Cooperative Research Group (Systolic Hypertension in the Elderly). Arch Intern Med 1996;156:553–561

19. Romanelli J, Fauerbach JA, Bush DE, Ziegelstein RC. The significance of depression in older patients after myocardial infarction. J Am Geriatr Soc 2002;50:817–822

20. Lesperance F, Frasure-Smith N, Talajic M, Bourassa MG. Five-year risk of cardiac mortality in relation to initial severity and one-year changes in depression symptoms after myocardial infarction. Circulation 2002;105:1049–1053

21. Bush DE, Ziegelstein RC, Tayback M, et al. Even minimal symptoms of depression increase mortality risk after acute myocardial infarction. Am J Cardiol 2001;88:337–341

22. Chang M, Hahn RA, Teutsch SM, Hutwagner LC. Multiple risk factors and population attributable risk for ischemic heart disease mortality in the United States, 1971–1992. J Clin Epidemiol 2001;54:634–644

23. Musselman DL, Evans DL, Nemeroff CB. The relationship of depression to cardiovascular disease: epidemiology, biology, and treatment. Arch Gen Psychiatry 1998;55:580–592

24. Rosen RC, Lane RM, Menza M. Effects of SSRIs on sexual function: a critical review. J Clin Psychopharmacol 1999; 19(1):67–85

25. Mamdani MM, Parikh SV, Austin PC, Upshur RE. Use of antidepressants among elderly subjects: trends and contributing factors. Am J Psychiatry 2000;157:360–367

26. Kroenke K. A 75-year-old man with depression. JAMA 2002; 287:1568–1576

27. Feldman HA, Goldstein I, Hatzichristou DG, et al. Impotence and its medical and psychosocial correlates: results of the Massachusetts Male Aging Study. J Urol 1994;151:54–61

28. Araujo AB, Durante R, Feldman HA, et al. The relationship between depressive symptoms and male erectile dysfunction: cross-sectional results from the Massachusetts Male Aging Study. Psychosom Med 1998;60:458–465

29. Fagan PJ, Schmidt CW Jr, Wise TN, Derogatis LR. Sexual dysfunction and dual psychiatric diagnoses. Compr Psychiatry 1988;29:278–284

30. Krane RJ, Goldstein I, Saenz de Tejada I. Impotence. N Engl J Med 1989;321:1648–1659

31. Gitlin MJ. Psychotropic medications and their effects on sexual function: diagnosis, biology, and treatment approaches. J Clin Psychiatry 1994;55:406–413

32. Keller Ashton A, Hamer R, Rosen RC. Serotonin reuptake inhibitor-induced sexual dysfunction and its treatment: a large-scale retrospective study of 596 psychiatric outpatients. J Sex Marital Ther 1997;23:165–175

33. Croft H, Settle E Jr, Houser T, et al. A placebo-controlled comparison of antidepressant efficacy and effects on sexual functioning of sustained-release bupropion and sertraline. Clin Ther 1999;21:643–658

34. Muller MJ, Benkert O. Lower self-reported depression in patients with erectile dysfunction after treatment with sildenafil. J Affect Disord 2001;66:255–261

35. Seidman SN, Roose SP, Menza MA, et al. Treatment of erectile dysfunction in men with depressive symptoms: results of a placebo-controlled trial with sildenafil citrate. Am J Psychiatry 2001;158:1623–1630

36. Benet AE, Melman A. The epidemiology of erectile dysfunction. Urol Clin North Am 1995;22:699–709

37. Kirby M, Jackson G, Betteridge J. Friedli K. Is erectile dysfunction a marker for cardiovascular disease? Int J Clin Pract 2001; 55:614–618

38. Burchardt M, Burchardt T, Anastasiadis AG, et al. Erectile dysfunction is a marker for cardiovascular complications and psychological functioning with hypertension. Int J Impot Res 2001;13:276–281

39. Greenstein A, Chen J, Miller H, et al. Does severity of ischemic coronary disease correlate with erectile function? Int J Impot Res 1997;9:123–126

40. Feldman HA, Johannes CB, Derby CA, et al. Erectile dysfunction and coronary risk factors: prospective results from the Massachusetts Male Aging Study. Prev Med 2000;30:328–338

41. Yesavage JA. Geriatric depression scale. Psychopharmacol Bull 1988;24:709–711

16

Premature or Rapid Ejaculation: A Neurobiological Approach to Management

ROBERT S. TAN AND MICHAEL MISTRIC

Case History

Mr. A.B. is a 42-year-old very successful self-employed businessman who owns three companies. He spent most of his earlier life building his career and had just recently married a 20-year-old secretary in his company. He never had sexual intercourse before he got married. Of late, he has been very depressed, as he is concerned that he cannot please his new wife in the bedroom. He consulted the sexual dysfunction clinic, because he did not get very far with his primary care physician. He insisted that he did not have erectile dysfunction but that he "comes too quick" and that his previous physician had prescribed Viagra, to no avail. He admits to being anxious and depressed, and concerned that his young wife will leave him for a younger man. On examination, the patient appeared anxious and was somewhat shy about this complaint. His vitals included a blood pressure of 146/88 mm Hg and heart rate of 90/min. Cardiovascular, respiratory, and abdominal examination was unremarkable. Genitalia examination revealed bilaterally descended testes, phallus length of 4 inches at rest, and no deformities. Prostate examination revealed a normal-size prostate but it was slightly tender. He was assessed for depression using the criteria of the *Diagnostic and Statistical Manual of Mental Disorders*, fourth edition (DSM-IV). The patient reported sleep difficulties, lack of interest in life, guilt feelings, lack of energy, concentration problems, and a loss of appetite. He was not suicidal. Urinalysis showed some leukocytes, and seminal analysis also revealed some leukocytes as well. Testosterone levels were in the normal range.

Debates About Definitions and Root Causes

Although there is no universal definition of premature ejaculation (PE), this condition remains a real problem in men. PE has also been referred to as rapid ejaculation (RE). This condition is generally underdiagnosed as it depends in part on patients' self-reporting. Most studies have indicated that men are less forthcoming in reporting sexually related symptoms because of embarrassment. The DSM definition of PE or RE is: "The persistent or recurrent ejaculation with minimal sexual stimulation that occurs sooner than desired, either before or shortly after penetration, and that causes distress to either one or both parties."

The problem with the definition is that it is subjective. Moreover, there are no large-scale studies to determine normal latency time before ejaculation, and there are no data to determine if the aging process affects this. One author has suggested that the average ejaculatory latency is 3 minutes, and that 30% of women reach orgasm with vaginal intercourse regardless of their partner's ejaculatory control.[1]

Central control of ejaculation is from the hypothalamus and the limbic system. In experiments in rats, destruction of the paraventricular nuclei of the hypothalamus has led to significant decreases in seminal emission during ejaculation.[2] On the other hand, peripheral control of ejaculation has four components: seminal emission, closure of bladder neck, ejaculation, and orgasm. The condition has gradually been moving from being a strictly psychological phenomenon to one with an associated neurobiological basis in some cases. This is based on findings that men with PE sometimes have different nerve conduction/latencies times.[3] Intuitively, one may suspect that PE may be associated with hypertestosteronemia, but studies in men with PE have determined that hormonal differences such as testosterone levels have *not* been significantly different.[4] Suffice it to say that the neuroregulation of sexuality by neurotransmitters has not been well characterized. This is partly because most of the experiments supporting these theories have been based on rat models and are

TABLE 16–1. Classification of Ejaculatory Disorders

Premature or rapid ejaculation
Primary
Secondary
Delayed or absent ejaculation
Anejaculation
Retrograde ejaculation
Failure of emission
Aspermia

difficult to translate to humans. Overall, dopamine and serotonin appear to be the major transmitters involved in sexuality. In general, dopamine enhances sexual functioning and promotes ejaculation. On the other hand, serotonin tends to decrease sexual functioning by inhibiting ejaculation. Based on these models, several therapeutic models have been suggested and are discussed in this chapter.

PE, like erectile dysfunction, is becoming more recognized in mainstream medical circles, but it will probably take a breakthrough medication such as Viagra before medical professionals will truly recognize the condition. Table **16–1** summarizes the different ejaculatory disorders that exist in men.

Epidemiological Characteristics

In general, there have been wide variances in the prevalence of this condition, partly because it is a subjective condition, albeit real. The National Health and Social Life Survey estimated that ~30% of American men suffered from PE. There were *slight* but no significant differences in reporting rates between the younger (<40 years, 31%) and older (≥40 years, 29.5%) men, which suggests that aging does not seem to influence this condition substantially. The study also suggested that PE might be more common among blacks (34%) as compared with whites (29%). Table **16–2** shows the relationship of health and lifestyle to premature ejaculation, erectile function, and low desire.[5]

Classification of Premature Ejaculation

Primary Premature Ejaculation

In primary PE, the condition existed because the patient has been sexually active. Primary PE tends to be associated with psychological and psychiatric problems. Psychological problems include relationship distress, psychological distress, guilt, and psychosexual skill deficits. Psychiatric conditions include generalized anxiety state and depression.

Secondary Premature Ejaculation

This usually occurs after a period of normal and satisfactory ejaculation. Medical conditions can lead to secondary PE, but there is often underlying psychological and psychiatric issues as well. For example, a patient suffering from angina may have anxiety about his performance. He may worry that excessive sexual activity may kill him and as such would want to complete the act as soon as possible. As such, his anxiety may lead to a premature ejaculation, which in turn has medical roots, in this case his angina.

In the literature, it has been reported that prostatitis has been associated with PE.[6] An Italian study evaluated segmented urine specimens by bacteriological localization studies before and after prostatic massage and expressed prostatic secretion specimens from 46 patients with PE and 30 controls. The incidence of PE in the subjects with chronic prostatitis was evaluated and they found prostatic inflammation in 56.5% and chronic bacterial prostatitis in 47.8% of the subjects with PE, respectively. When compared with the controls, these novel findings were statistically significant ($p < .05$). The authors suggested that because the prostate gland is responsible in the mechanism of ejaculation that a role for chronic prostate inflammation should be considered in the pathogenesis of some cases of PE. The authors suggested that because chronic prostatitis has been found with a high frequency in men

TABLE 16–2. Relationship of Health and Lifestyle to Premature Ejaculation, Erectile Function, and Low Desire

Predictors	Adjusted Odd Ratios		
	Premature Ejaculation	Erectile Dysfunction	Low Desire
Daily alcohol consumption	0.79	1.63	2.24
Sexually transmitted disease (ever)	1.10	1.29	1.05
Urinary tract symptom	1.67	3.13	1.68
Poor to fair health	2.35	2.82	3.07
Circumcised	0.87	1.30	1.64
Emotional problems or stress	2.25	3.56	3.20

Data from the National Health and Social Survey.

with PE that a careful examination of the prostate be undertaken before any pharmacological or psychosexual therapy for PE.

Pharmacological Treatment of Premature Ejaculation

At present there are no drugs approved by the Food and Drug Administration (FDA) to treat PE. However, clinical trials have evaluated several topical preparations and oral antidepressants. More research is needed in this area, and it is hoped that newer therapies that are well tested will be available in the future.

Topical Treatment of Premature Ejaculation

Lidocaine-Prilocaine Cream 5%

A Turkish group evaluated the efficacy of lidocaine-prilocaine, which could be obtained in the United States via compounding pharmacists.[7] In that study, the authors evaluated the efficacy and optimum usage of lidocaine-prilocaine cream 5% in preventing PE. A total of 40 patients were examined in the study group and randomized into four groups of 10 patients. Patients in group 1 applied lidocaine-prilocaine cream 5% for 20 minutes before sexual contact, the patients in group 2 applied it for 30 minutes, and the patients in group 3 applied the cream for 45 minutes, with all patients covering the penis with a condom. Patients in the fourth group applied a base cream as placebo. In group 1, the preejaculation period increased to 6.71 ± 2.54 minutes without any adverse effects. In group 2, although the preejaculation period increased in four patients up to 8.70 ± 1.70 minutes, six patients in this group and all patients in group 3 had erection loss because of numbness. In the placebo group, there was no change in their preejaculation period. The conclusion of that study was that lidocaine-prilocaine cream 5% is effective in PE and that 20 minutes of application before sexual contact is the optimum period. A further study suggested that fluoxetine plus the topical anesthetic was more effective than fluoxetine alone in treating PE.[8]

SS Cream

This cream is a combination of nine natural products including bufosteroids and plant extracts and was developed in Asia. A Korean group investigated the clinical efficacy of SS cream in a double-blind, randomized, placebo-controlled, phase III clinical study. One hundred and six patients (mean age 38.7 ± 0.61 years) completed the study. A stopwatch measured the ejaculatory latency. Also, the sexual satisfaction ratio of both partner and patient were recorded twice in the screening period and once after each treatment (one placebo treatment of 0.20 g and five SS-cream treatments of 0.20 g for a total of six treatments). Patients were instructed to apply the cream on the glans penis 1 hour before sexual intercourse in a double-blind randomized fashion.

Clinical efficacy was compared with the prolongation of ejaculatory latency and improvement of the sexual satisfaction ratio before and after each treatment. In the screening period, the mean ejaculatory latency was assessed at 1.37 ± 0.12 minutes, and neither the patients nor their partners were satisfied with their sex lives. After treatment, the mean ejaculatory latency was prolonged to 2.45 ± 0.29 minutes in the placebo group and 10.92 ± 0.95 minutes in the SS-cream group. The clinical efficacy of placebo and SS cream as judged by an ejaculatory latency time prolonged more than 2 minutes was 15.09% and 79.81%, respectively. The improvement of sexual satisfaction to a grade higher than effective was 19.81% and 82.19%, respectively, for placebo and SS cream. Of 530 trials of SS cream, 98 (18.49%) resulted in a sense of mild local burning and mild pain. No adverse effect on sexual function or on the partner and no systemic side effects were observed. It was interesting to note that even placebo improved latency in this study, which suggests a strong psychological component in PE. Also, there was a dose-dependent response to SS cream.

Oral Therapy

Oral therapy consists of antidepressants, typically the selective serotonergic reuptake inhibitors (SSRIs). Tricyclic antidepressants such as clomipramine have also been studied.

Selective Serotonergic Reuptake Inhibitors

Of the SSRIs the drug most studied is paroxetine (Paxil). Altogether, there are about five or 10 clinical trials from different countries evaluating the efficacy of Paxil. Different regimens were used in these trials including paroxetine usage on a continued basis or on demand usage. There is no consensus as to which is better, although one study suggested that continued use was better than on demand use in prolonging ejaculatory latency.[9] Waldinger and his colleagues[10] in 1994 reported one of the earliest studies on the effects of paroxetine on PE. In that small study, 17 male outpatients with PE were randomly assigned to treatment with paroxetine ($n = 8$) or placebo ($n = 9$). After a first-week dose of 20 mg/day, the paroxetine regimen was increased to 40 mg/day for 5 weeks. Patients and their female partners were interviewed separately. Patients treated with paroxetine had significantly greater clinical improvement than the patients given placebo.

In another study performed in Israel, *sildenafil* (Viagra) combined with *paroxetine* and psychological and behavioral counseling alleviated PE in patients in whom other treatments failed.[11] The intravaginal ejaculatory latency time (IVELT) was graded on a scale of 0 to 3 (0 = longer than 5 minutes, 3 = shorter than 1 minute). The 138 men who scored their PE as 4 or greater and IVELT as 2 or greater comprised the study group. Psychological and behavioral counseling was provided during the study. PE was graded using the same scales 3 months after the initiation of each treatment. Topical 5% lidocaine ointment comprised the initial treatment: dissatisfied patients (PE grade 4 or greater, IVELT 2 or greater), took one tablet of paroxetine 20 mg for 30 days and then one tablet 7 hours before intercourse. Sildenafil was added to the treatment of patients dissatisfied with paroxetine alone. The mean initial PE grade was 5.67 ± 0.13 and that for IVELT was 2.9 ± 0.19 for all participants (mean age 28.7 years). Thirty-eight reported improvement (PE grade 2.0 ± 0.8, $p < .01$; IVELT 0.13 ± 0.34, $p < .001$) after local lidocaine application. Of the 100 treated with paroxetine, 42 reported improvement (PE grade 2.5 ± 0.1, $p < .01$; IVELT 0.28 ± 0.46, $p < .001$), and 56 of the remaining 58 who were treated with a combination of paroxetine and sildenafil reported improvement (PE grade 1.78 ± 0.23, $p < .001$; IVELT 0.16 ± 0.37, $p < .001$). Two patients in the trial remained dissatisfied with all treatment modalities. This reality can be transferred to clinical practice as some patients will remain dissatisfied no matter what happens.

The efficacy of sertraline hydrochloride (Zoloft) for the treatment of premature ejaculation has also been evaluated in an Australian study.[1] A total of 37 potent men, 19 to 70 years old (mean age 41), with PE were treated with 50 mg oral sertraline and placebo in a controlled, randomized, single-blind, crossover trial. All men were either married or in a stable relationship, and none of the patients received any formal psychosexual therapy. Chronic open-label treatment with sertraline was continued in 29 patients who had achieved an increase in ejaculatory latency times over pretreatment levels with active drug in the initial crossover study. In an attempt to identify which patients could maintain the improved ejaculatory control after withdrawal of the active drug, serial drug withdrawal was conducted every 4 weeks with drug initiation after a further 2 weeks if improved ejaculatory control was not maintained. The mean pretreatment ejaculatory latency time was 0.3 minute (range 0 to 1). The mean ejaculatory interval after 4 weeks of treatment was 3.2 minutes (range 1 minute to anejaculation) with sertraline and 0.5 minute (range 0 to 1)

with placebo ($p < .001$). Intravaginal ejaculation was achieved for the first time in five patients with primary PE and two patients experienced anejaculation. One patient described minor drowsiness and anorexia, and two patients described mild, transient gastrointestinal upset. Staged drug withdrawal allowed 20 of the 29 patients (69%) on chronic open label treatment with sertraline to discontinue the drug after a mean interval of 7.3 months with a mean ejaculatory latency time of 4.1 minutes (range 1 to 12). This study concluded that sertraline also appears to be a useful agent in the pharmacological treatment of PE.

Tricyclic Antidepressants

Clomipramine (Afranil) is another antidepressant and is in the tricyclic class. It works by inhibiting the membrane pump mechanism responsible for uptake of norepinephrine and serotonin in adrenergic and serotonergic neurons.[12] The inhibition of serotonin is probably how the medication aids in PE. The suggested dosage is 50 mg 2 to 12 hours before sexual activity or 50 mg per day continuously. Clomipramine has been evaluated in different clinical trials for ~20 years. Italian investigators evaluated the effectiveness mechanism of the clomipramine treatment. Investigators measured sacral evoked response (SER) and dorsal nerve somatosensory cortical evoked potential (DN-SEP) testings in 15 patients with true premature ejaculation (TPE). They couldn't demonstrate any significant difference between the values of either latency times or amplitudes of the evoked responses determined just before and at the end of the treatment with clomipramine in these patients. However, the sensory thresholds were $24.4 \pm 4.3\,V$ in the pretreatment term and $30.2 \pm 7.3\,V$ at the end of the treatment. This difference was found to be statistically significant ($p = .0031$). Their results suggest that clomipramine increases the sensory threshold for the stimuli in the genital area.

Role of Nonpharmacological/ Neurobiological Approaches

Behavioral approaches remain important despite recent advances in pharmacology. In some instances, behavioral therapy supersedes medications. The goal of behavioral therapy is to enhance the patient's awareness of his sensory input and learn how to control it. This could be achieved by relaxation training. For example, the patient could be taught how to control and relax his pubococcygeal muscle by Kegel exercises. Cognitive and behavioral techniques could also be used and the start-stop (squeeze) method is probably the most effective. In this method, the patient squeezes the penis near the time of climax to prevent ejaculation. Combination approaches using both the neurobiological and behavioral methods are often needed to treat patients with PE.

Relapses can be common in patients with PE. They can be prevented by managing the patient's stress and anxiety by counseling and medications for anxiety such as benzodiazepines. The patient should also practice the squeeze technique during periods of low intercourse frequency.

Discussion of the Case History

Mr. A.B. somewhat typifies the patient profile with PE/RE. These patients generally have some sort of psychological or psychiatric disorder. It is essential to determine whether the depression led to the PE or whether the PE in itself led to depression. Mr. A.B. probably had a strong psychological component because he experienced performance anxiety. He is married to a young woman and may be afraid that he cannot perform adequately. He had coincidental prostatitis, and studies often show that there is an association of prostatitis with PE, but it may not be causative of PE. In any event, he was given a course of ciprofloxacin, which cleared his prostatitis but not the PE. He returned to the clinic for counseling and was also taught the squeeze technique and to practice the start-stop techniques. In view of his depression, he was started on paroxetine (Paxil). This drug has an "off-label" indication for treatment for PE. Three months later, the patient said there was improvement in his PE, but said he might need a little help with his erections. He was then started on Levitra. This case illustrates that PE is a difficult condition to treat and that doctors sometimes confuse PE with erectile dysfunction and prescribe the wrong treatments. Usually a combined approach using both the neurobiological as well as the psychological models can lead to better outcomes. Relapse is common, and this patient will have to be monitored carefully.

Conclusion and Key Points

- PE is very common, and epidemiological studies indicate a prevalence rate of ~30%. Despite this, very few physicians report encountering this problem, possibly because physicians often do not screen for such problems or are not comfortable in dealing with them.

- There are often overlapping psychiatric or psychological disorders in patients with PE.

- Treatments in the past have been based mainly on a psychological model, but there is increasing evidence for a neurobiological model, wherein drug treatment in combination with psychological methods may be therapeutically useful. Most of the available drugs for treatment are used on an "off-label" basis.

- Patience is required of both the provider and patient in treating this disorder, as it can frequently relapse.

REFERENCES

1. McMahon CG. Treatment of premature ejaculation with sertraline hydrochloride: a single-blind placebo controlled crossover study. J Urol 1998;159:1935–1938
2. Liu YC, Salamone JD, Sachs BD. Impaired sexual response after lesions of the paraventricular nucleus of the hypothalamus in male rats. Behav Neurosci 1997;111:1361–1367
3. Colpi GM, Fanciullacci F, Beretta G. Evoked sacral potentials in subjects with true premature ejaculation. Andrologia 1986; 18:583–586
4. Pirke KM, Kockott G, Aldenhoff J. Pituitary gonadal system function in patients with erectile impotence and premature ejaculation. Arch Sex Behav 1979;8:41–48
5. Laumann EO, Paik A, Rosen RC. Sexual dysfunction in the United States: prevalence and predictors. JAMA 1999;281: 537–544
6. Screponi E. Prevalence of chronic prostatitis in men with premature ejaculation. Urology 2001;58:198–202
7. Atikeler MK, Gecit I, Senol FA. Optimum usage of prilocaine-lidocaine cream in premature ejaculation. Andrologia 2002; 34:356–359
8. Atan A, Basar MM, Aydoganli L. Comparison of the efficacy of fluoxetine alone vs. fluoxetine plus local lidocaine ointment in the treatment of premature ejaculation. Arch Esp Urol 2000; 53:856–858
9. McMahon CG, Touma K. Treatment of premature ejaculation with paroxetine hydrochloride as needed: 2 single-blind placebo controlled crossover studies. J Urol 1999;161:1826–1830
10. Waldinger MD, Hengeveld MW, Zwinderman AH. Paroxetine treatment of premature ejaculation: a double-blind, randomized, placebo-controlled study. Am J Psychiatry 1994;151:1377–1379
11. Chen J, Mabjeesh NJ, Matzkin H. Efficacy of sildenafil as adjuvant therapy to selective serotonin reuptake inhibitor in alleviating premature ejaculation. Urology 2003;61:197–200
12. Colpi GM, Fanciullacci F, Aydos K. Effectiveness mechanism of clomipramine by neurophysiological tests in subjects with true premature ejaculation. Andrologia 1991;23:45–47

17

Prostate Cancer: Controversies and Developments in Screening and Treatments

ROBERT YONG AND ROBERT S. TAN

Case History

A 63-year-old Caucasian man had a routine physical examination by his family physician. He was told that the left lobe of his prostate was slightly firm. A prostate-specific antigen (PSA) and other routine blood work were done. On a return visit to his family physician he was informed that the PSA was 5.5 ng/mL, which is slightly high for his age. He was advised to undergo transrectal ultrasound guided prostate biopsy. Two weeks later he was asked to come to the doctor's office with his wife. He was told that the biopsy was positive for prostate cancer. One of six core biopsies was positive for cancer. The tumor involved 3 mm in length of one core. The Gleason score was 3 + 3 = 6. He was then referred to a urologist for consultation. In the mean time he surfed the Internet and searched for information about prostate cancer. He was overwhelmed by the voluminous amount of information and confusing controversies regarding the treatment of prostate cancer.

Epidemiology

Prostate cancer is the most common male cancer and the second leading cause of cancer death in the United States. The American Cancer Society estimated that 220,900 men would be diagnosed with prostate cancer and 28,900 would die of the disease in 2003.[1] Prostate cancer incidence and mortality vary markedly across geographic regions and populations. The prevalence of histological prostate cancer is similar among ethnic groups, while the incidence of clinically manifest disease differs markedly.[2]

The age-adjusted rates of prostate cancer in 1992 for African-American men and American white men were 249 per 100,000 and 182 per 100,000, respectively. The average age-adjusted incidence of prostate cancer in Kingston, Jamaica, is the highest in the world, 304 per 100,000.[3] Asians, especially the Chinese, have the lowest incidence of prostate cancer, 28 per 100,000 for the Chinese and 39 per 100,000 for the Japanese.[4] It appears that after migrating to the United States, however, the incidence rates for prostate cancer among Asians increase.[5] African-American men have higher stage and grade of prostate cancer when it is diagnosed and the mortality of prostate cancer is higher than for American white men. The introduction of widespread PSA testing contributed to the dramatic increase in the incidence of prostate cancer between 1989 and 1991, and the incidence rates leveled off in 1992 as these prevalent cancers were detected. The mortality of prostate cancer has declined recently. However, because the natural history of prostate cancer is long, this decline in mortality cannot be attributed to widespread PSA testing.

Etiology of Prostate Cancer

The etiology of prostate cancer is multifactorial, and requires the interaction of various factors. Smoking has been associated with prostate cancer and so have the following factors, which should also be considered.

Age Factor

Age is an important factor. The prevalence of prostate cancer increases with age. The probability of developing prostate cancer is less than one in 10,000 in men aged 39 years or younger, one in 103 for men aged 40 to 59 years, and one in 8 for men 60 to 79 years.[6]

Genetic Factors

Ten to 15% of patients with prostate cancer have at least one relative who is affected, and first-degree relatives of patients with prostate cancer have a twofold or threefold increased risk of developing this disease. Men who have a brother who has prostate cancer are more likely to develop this disease than are those who only have a father who is affected, suggesting that the disease is recessive or linked to the X-chromosome.

Dietary Factor

Prostate cancer is positively associated with diet high in fat, meat, and dairy products. The low incidence of prostate cancer in the Asians might be related to consumption of diet with high content of phytoestrogens. Soybean has the highest contents of phytoestrogens and Asians consume soybean in large quantities.

Hormone Factors

Androgen plays an important role in the development and growth of prostate gland. Testosterone in the prostate cell is converted to dihydrotestosterone by 5α-reductase enzyme, which combines with the androgen receptor stimulating the prostate cell growth. Gann et al[7] found that men with higher circulating levels of testosterone had an increased risk of prostate cancer compared with men with lower levels of testosterone. However, this association of prostate cancer with testosterone does not mean that testosterone if administered is causative of prostate cancer. Interestingly, men who were castrated before puberty or have congenital absence of 5α-reductase do not develop prostate cancer. A recent chemoprevention study has shown that 5α-reductase inhibitor reduces prostate cancer prevalence by 25%.[8] This may support the hypothesis that dihydrotestosterone is associated with prostate cancer.

Principles of Screening Based on Epidemiological Principles

Does It Apply to Prostate Cancer?

In general, the purpose of screening is to find persons with risk factors in which preventive interventions could be used to identify individuals with early or asymptomatic treatable disease. This is in a sense an application of the principles of optimal aging, whereby the individual wants to minimize morbidity and mortality through preventive strategies after early detection. However, there are several principles that should guide a clinician as to whether it is worth screening in a population:

- The disease should be sufficiently common in the community because the detection of rare diseases may lead to high cost-benefit ratios. Prostate cancer is in no way a rare disease, and in fact it is the most common cancer in men. Each year ~200,000 new cases of prostate cancer are diagnosed in the United States.[9] Moreover, prostate cancer is a function of age, and prevalence rates increase as men age. Current statistics point toward a larger pool of baby boomers over the next few decades.

- The burden of suffering or the morbidity and mortality should be substantial. This is where screening can sometimes be seen as controversial. There is no doubt that this cancer is very common among aging men, but the morbidity and mortality from prostate cancer seem to vary from individual to individual. There may be also ethnic, dietary, and environmental factors that can determine the aggressiveness of these tumors. Many men die *with* prostate cancer rather than *of* prostate cancer. Often, complications from cardiovascular, pulmonary, infectious, and dementing disease cause the demise of older men, rather than prostate cancer itself. European experience, which includes watchful-waiting, has indicated that an aggressive approach to all patients with early disease would entail substantial overtreatment.[10]

- An effective preventive intervention or treatment should be available. Prevention of prostate cancer is very important. Lifestyle modifiable factors such as smoking, poor diet, and lack of exercise can influence the rates of prostate cancer. There have been some inroads in our knowledge of prostate cancer prevention recently. For instance, it was recently reported in a trial involving 18,882 men that finasteride might reduce the risk of prostate cancer.[11] Finasteride is an inhibitor of 5α-reductase, and it inhibits the conversion of testosterone to dihydrotestosterone, which is the primary androgen in the prostate. In the Prostate Cancer Prevention Trial, it was reported that finasteride delays the appearance of prostate cancer, but the possible benefit and a reduced risk of urinary problems must be weighed against sexual side effects and the increased risk of high-grade prostate cancer. The role of selenium, vitamin E, zinc, and other supplements seem promising in preventing prostate cancer, but they have to be studied in greater detail, such as in the finasteride trial, before the widespread prescription of these supplements can be endorsed. The treatment of prostate cancer is discussed in more detail in subsequent sections, but it can be summarized under three options: radical prostatectomy, radiotherapy, and watchful waiting.

- The screening program should be acceptable and available for routine use in the general population. Here lies one of the biggest dilemmas in screening for prostate cancer. The most common test is the PSA. This test is neither very specific nor sensitive. For example, the PSA test is not very specific as only *one third* of men with an abnormal serum PSA level actually have cancer.[12] Also, conditions such as prostatitis or even benign prostatic hyperplasia may even cause a rise in the PSA levels, sometimes causing false alarms. PSA as a serum marker is only *70 to 80% sensitive* for prostate cancer. This serum marker is a protein made only by prostate cells. Serum PSA levels are proportional to either the total volume of prostate tissue or the amount of irritation in the prostate (such as occurs with carcinoma or inflammation). Either increased volume or irritation causes PSA to spill from the prostate into the bloodstream. Paradoxically, low levels of PSA can even indicate presence of cancers as well.[13] Longitudinal studies in men suggest that PSA velocity of *0.7 ng per mL per year* rise may be suggestive of cancer.[14] Table **17–1** shows the relationship between percent PSA levels and the probability of prostate cancer.

Arguments for Screening

Prostate cancer is the most common cancer in men; as such, screening would help diagnose many of these patients in the early stage of their disease. Moreover, the American Cancer Society and the American Urological Association recommend the use of a PSA-based screening program to detect prostate cancer in men 50 years of age and older.[15] Some evidence also shows that, compared with screening by rectal examination alone, routine screening of asymptomatic patients with PSA testing and digital rectal examinations detects a higher percentage of cancers that are localized to the prostate.[16] Although PSA alone may not be accurate enough, PSA density and PSA velocity may guide a clinician to a higher level of suspicion for this cancer.

Smith et al[17] determined, for the first 4 years of serial PSA-based screening, the trends in compliance, prevalence of abnormal screening test results, cancer detection rates, and stage and grade of cancers detected. In a community-based study of serial screening with PSA measurements, a total of 10,248

male volunteers at least 50 years old were screened at 6-month intervals for a minimum of 48 months. At 48 months, 79% of volunteers returned for screening. During this interval there was a decrease in the proportion of volunteers with serum PSA levels higher than 4.0 ng/mL (from 10% to 6–7%), in cancer detection rates (from 3% to <1%), and in the proportion with clinically advanced cancer (from 6% to 2%). In men who underwent surgery, the proportion with high-grade cancer decreased (from 11% to 6%), and the proportion with pathologically advanced cancer was proportionately reduced but not significantly reduced. This study concluded that with serial PSA-based screening, the proportion of men with abnormal test results *decreased*, and the prostate cancer detection rate decreased to near the reported population-based incidence rate. There was also a shift to detection of cancers at an *earlier clinical stage* and detection of lower-grade cancers.

Arguments Against Screening

Some have argued that because prostate cancer screening does not fulfill all of the requirements for an effective screening program that it should not be performed routinely. Both the American Academy of Family Physicians and the U.S. Preventive Services Task Force recently recommended against the use of routine prostate cancer screening for two reasons: (1) early prostate cancer detection has no proven benefit; and (2) the potential side effects of treatment may outweigh the benefits.[18]

Although an individual PSA test is rather inexpensive, operating costs multiply when a patient with an abnormal PSA test must be evaluated. Transrectal ultrasound examination costs approximately $170 per patient, and random biopsies cost another $230. Pathologic evaluation of the biopsy specimens costs approximately $167 per patient. When compounded by the fact that three patients without cancer must be evaluated for each cancer that is detected, the estimated overall cost of initiating a nationwide prostate cancer screening and treatment program for all eligible men ranges from $8.5 to $25.7 billion per year.[19]

TABLE 17–1. The Relationship Between Percent PSA Levels and the Probability of Prostate Cancer

PSA	Probability of Prostate Cancer (DRE Neg.), %	Free PSA, %	Probability of Prostate Cancer (DRE Neg.), %
0–2	1	0–10	56
2–4	15	10–15	28
4–10	25	15–20	20
>10	>50	20–25	16
		>25	8

In all likelihood, many cancers that have been detected by PSA screening would never have become symptomatic in the patient's lifetime. Many older men die of illnesses other than prostate cancer. Autopsy of elderly men often demonstrates prevalence rates of prostate cancers approaching 80%.[20] However, the cause of death is often pneumonia, heart disease, and neurological disease. The natural history of prostate cancer is also not well defined, and there is great variation as to the aggressiveness of the tumor. A study in Sweden using watchful waiting rather than radical prostatectomy or hormonal therapy illustrates this point. Johnasson et al[10] studied the natural history of initially untreated early-stage prostate cancer with a prospective cohort study; 642 patients with prostate cancer of any stage with a mean age of 72 years were studied. In the entire cohort, prostate cancer accounted for 201 (37%) of all 541 deaths. Among 300 patients with a diagnosis of localized disease (T0-T2), 33 (11%) died of prostate cancer. In this group, the corrected 15-year survival rate was similar in 223 patients with deferred treatment [81%; 95% confidence interval (CI), 72–89%] and in 77 who received initial treatment (81%; 95% CI, 67–95%). The corrected 15-year survival was 57% (95% CI, 45–68%) in 183 patients with locally advanced cancer (T3-T4) and 6% (95% CI, 0–12%) in those 159 who had distant metastases at the time of diagnosis. The Swedish study concluded that patients with localized prostate cancer have a favorable outlook following watchful waiting, and the number of deaths potentially avoidable by radical initial treatment is limited. Without reliable prognostic indicators, an aggressive approach to all patients with early disease would entail substantial overtreatment. In contrast, patients with locally advanced or metastatic disease may need trials of aggressive therapy to improve prognosis.

Controversies on Guidelines for Screening

Various professional organizations have variable screening recommendations:

- The *American Cancer Society* recommends that health care professionals should offer the PSA and digital rectal examination (DRE) yearly, beginning at age 50, to men who have at least a 10-year life expectancy. Men at high risk, such as African-Americans and men who have a first-degree relative diagnosed with prostate cancer at an early age, should begin testing at age 45.[21]

- The *American Urological Association* recommends that men older than 50 with a greater than 10-year life expectancy should be offered prostate cancer screening with PSA and DRE. Men at high risk should begin testing at age 45.[22]

- However, the *U.S. Preventive Services Task Force* (USPSTF) concludes that the evidence is insufficient to recommend for or against routine screening for prostate cancer using PSA testing or DRE.[23]

- The *American College of Preventive Medicine* recommends against routine population screening with DRE and PSA. Men age 50 or older with a life expectancy of greater than 10 years should be given information about the potential benefits and harms of screening and limits of current evidence of screening and should be allowed to make their own choice about screening, in consultation with their physician, based on personal preferences.[24]

- Whether screening for prostate cancer with early detection and treatment will be beneficial and reduce prostate cancer mortality will await the result of the ongoing large randomized control trials conducted by the United States Prostate, Lung, Colorectal and Ovarian Cancer Screening Trial[25] and the European Randomized Study for Screening for Prostate Cancer.[26]

Screening Tools Available to the Clinician

Digital Rectal Examination (DRE)

Historically, DRE is the first line of prostate cancer detection. Abnormal DRE, including in duration, irregular surface, asymmetry, nodule, and hardness, calls for prostate biopsy. DRE is subject to interpersonal and intrapersonal variation. DRE by itself is a poor screening tool. Despite the shortcomings of the DRE, up to 25% of prostate cancers are still detected by DRE in men with normal PSA levels. Therefore, a suspicious DRE should be followed by prostate biopsy.[27] DRE when used with PSA together becomes a most effective tool for detecting prostate cancer in its earliest stages; 50% of men with abnormal DRE and PSA were found to have prostate cancer on biopsy.[28]

Prostate Specific Antigen (PSA)

PSA is a serine protease produced by the epithelial cells of the prostate and the periurethral glands. It has a half-life of 3 days. It is the best tumor marker. When comparing PSA to mammography in terms of cancer detection, an abnormal PSA of a man over the age of 50 is twice as likely to detect cancer than is an abnormal mammogram in a woman over the age of 50. PSA is not prostate cancer specific, however. Normal prostate gland, benign prostate hyperplasia, prostatitis, urinary tract infection, riding a bicycle in the previous 48 hours, and sexual intercourse with ejaculation within 48 hours will increase the serum levels of PSA. The normal range of PSA is 0 to 4.0 ng/mL. The sensitivity, specificity, and positive predictive value of PSA is 79%, 59%, and 40%, respectively.[29] Using PSA 4 ng/mL as a cutoff, 25% of men with benign prostate hyperplasia (BPH) have a PSA >4 ng/mL and 25% of men with prostate cancer have a PSA <4 ng/mL. The low specificity of PSA creates false positivity, resulting in unnecessary prostate biopsy, patient's anxiety, and medical cost. To improve the predictive value of PSA for cancer, it is recommended that the levels of PSA be adjusted for age (age-specific PSA) and for prostate volume (PSA density, PSAD), and be evaluated for the rate of change (PSA velocity). Recently, the molecular forms of PSA in free and bound fractions in the serum were quantified. The measurement of free and bound PSA has been evaluated to differentiate BPH from cancer. The age-specific range of normal PSA includes < 2.5 ng/mL for age 40 to 49; <3.5 ng/mL for age 50 to 59; <4.5 ng/mL for 60 to 69; and <6.5 ng/mL for age 70 to 79. When PSAD is >0.15 and PSA is 4 to 10 ng/mL with a normal DRE, prostate biopsy is recommended. The rate of change of PSA >0.75 ng/mL per year is significant to warrant a prostate biopsy. It has been shown that men with prostate cancer have a greater fraction of PSA bound to α-antichymotrypsin and a lower percentage of PSA that is free. If the ratio of free PSA to total PSA is ≤25% in men with total PSA between 4 and 10 ng/mL and if the prostate is normally palpable, 95% of prostate cancers were detected and 20% of unnecessary biopsies were avoided.[30]

Transrectal Ultrasonography (TRUS)

Studies have confirmed that TRUS cannot localize early prostate cancer.[31] The major role of TRUS is to biopsy the suspicious area and do a systemic biopsy of the prostate gland.

Beyond Prostate Specific Antigen

Since 1998, the "free" or "unbound" PSA test has been Food and Drug Administration (FDA)-approved to augment information available from the total PSA. The "free" PSA test is used as a ratio with the total PSA to help give a more precise determination of the risk of prostate cancer. In essence, the lower the amount of free PSA, the higher the likelihood of cancer. The free PSA test is often used following a nonsuspicious DRE and a total PSA test that shows moderately elevated PSA levels (between 4 and 10 ng/mL) in men aged 50 years and older.

In addition, two novel molecular forms of "free" PSA exist. BPSA and proPSA are distinct molecular forms of

free PSA in serum. *BPSA (benign PSA)* and is more often associated with benign enlargement of the prostate gland. On the other hand, the inactive precursor of PSA, or *truncated proPSA*, is more often associated with cancer. ProPSA is composed of native proPSA as well as two truncated proPSA forms, [−2]pPSA and [−4]pPSA, and both have been shown to be more cancer-associated. Preliminary studies indicate that BPSA is a biomarker for clinical BPH. Early studies also reveal that truncated proPSA significantly increases the specificity for prostate cancer especially in the 2- to 4-ng/mL PSA range.[32] It is estimated that as many as 40% of prostate cancer patients are within the 2- to 4-ng/mL total PSA range. Further studies are currently underway to determine the potential range and application of clinical utility of truncated proPSA in prostate cancer detection and management. These new tests may further enhance the physician's ability to differentiate between prostate cancer and benign disease in men with slightly elevated PSA levels.

Other exciting possibilities for more precise prostate cancer screening include the *TGG β growth factor* blood test and the *IL-6 soluble receptor* blood test that can help distinguish between more and less aggressive forms of prostate cancer. Shariat et al[33] reported that plasma interleukin-6 (IL-6) and IL-6sR levels were dramatically elevated in men with prostate cancer metastatic to bone. In patients with clinically localized prostate cancer, the preoperative plasma IL-6 and IL-6sR levels independently predicted biochemical progression after surgery, presumably because of an association with occult metastatic disease present at the time of radical prostatectomy. Another test that is being explored is the *human glandular kallikrein-2*.[34]

Prostate Biopsy and Grading of Prostate Cancer

The Gleason grading system is the most commonly used histological grading system for prostate cancer. It is based on the architectural pattern, the primary predominant pattern, and the secondary most predominant pattern. Both patterns are assigned a grade from 1 to 5. Pick the two most predominant patterns, for example, 3 and 4. Add the most predominant patterns to arrive at the Gleason score, for example, 3 + 4 = 7. If only one pattern predominates, double this pattern, for example, 3 + 3 = 6. The Gleason scores are grouped according to the following ranges: 2 to 4, 5 to 6, 7, and 8 to 10. The Gleason score correlates with prognosis after radical prostatectomy. A Gleason score 7 tumor behaves significantly worse than Gleason score 5 to 6 tumors but fare better than Gleason score 8 to 9 tumors. The biopsy Gleason score can also be combined with serum PSA values and clinical stage to predict

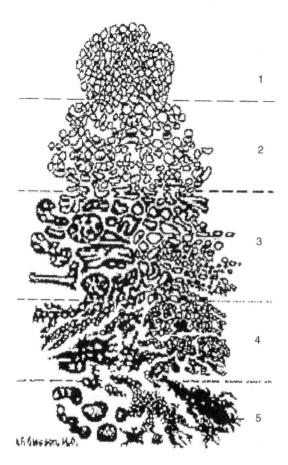

FIGURE 17–1. Histological guide to Gleason score.

organ-confined versus non–organ-confined disease and the risk of progression after radical prostatectomy.[35,36] Like the DRE, there is interobserver variation in assigning Gleason score.

Recently, Lattouf and Saad[37] set out to assess the correlation of the Gleason score on the initial prostate biopsy and the final pathology after radical prostatectomy (RP) for prostate adenocarcinoma. Often there was poor correlation with the prostate biopsy reading and the final pathological report after surgery. As such, the investigators concluded that Gleason grading of the prostate biopsy could be a poor predictor of pathological outcome. Assessment by the same pathologist may reduce the discrepancy. The authors suggest that clinicians should be aware of these limitations when using the biopsy Gleason grade in decision making. Fig. **17–1** summarizes the histological Gleason grading.

Staging of Prostate Cancer

The tumor-node-metastasis (TNM) classification system is the most common system used to evaluate local and distant extent of disease for the staging of prostate cancer. T1 is clinically localized tumor not palpable on DRE. T1a is focal tumor [<5% of resected tissue on transurethral resection of the prostate (TURP)] and low grade. T1b is diffuse tumor >5% of resected tissue on TURP or high grade. T1c is the tumor diagnosed based on elevated PSA and not palpable on DRE. T2 tumor is clinically localized, tumor palpable. T2a tumor involves half a lobe or less; T2b tumor involves more than half a lobe; T2c tumor involves both lobes; T3 is locally invasive beyond the prostatic capsule and tumor is palpable. T3a is unilateral extracapsular extension; T3b is bilateral extracapsular extension; and T3c is seminal vesicle invasion. T4 tumor invades adjacent tissues (e.g., bladder, rectum, levator ani muscles), and N/M indicates the metastatic disease. N1 is microscopic pelvic lymph node metastasis; N2 is gross pelvic lymph node metastasis; and N3 is extrapelvic lymph node metastases. M is distant metastases (lung, bones, liver, and brain). Staging prostate cancer depends on the DRE finding. There is interpersonal variation, in that two examiners examining the same patient may have two different findings.

Treatment of Prostate Cancer

Successful treatment of prostate cancer depends on patient selection. The use of PSA testing combined with DRE has revolutionized our ability to detect prostate cancer at its early and curable stage. However, not all patients with prostate cancer require treatment. According to the study by Albertsen et al,[38] in 4 to 7% of patients with a Gleason score of 2 to 4, 6 to 11% of patients with a Gleason score of 5, 18 to 30% of patients with a Gleason score 6, 42 to 70% of patients with a Gleason score 7, and 60 to 87% of patients with a Gleason score of 8 to 10, prostate cancer progressed to cause death in 15 years. The majority of patients diagnosed with prostate cancer are 65 years and older, when they may have many other comorbidities that may compete with prostate cancer to cause death. The candidate for treatment of clinically localized prostate cancer with curative intent should have a life expectancy of 10 to 15 years. Clinically localized prostate cancer can be treated with radical prostatectomy, external beam radiation therapy, brachytherapy, or conservative therapy (watchful waiting) with acceptable good result. A recent randomized control trial from Sweden has shown that the cancer-specific mortality is significantly reduced in the radical prostatectomy group versus watchful-waiting group, though the overall survival is no different.[39] However, these patients' diagnosis of prostate cancer is clinical based rather than PSA based. *There is no randomized control trial* among radical prostatectomy, external beam radiation therapy, and brachytherapy to compare the superiority of these treatment modalities. Therefore,

TABLE 17–2. Percentage of 1870 Patients with Postoperative Complications Excluding Impotency and Incontinence

	No. of patients	*% (95% Confidence Interval)*
Anastomotic stricture	71	4 (2.9–4.7)
Thromboembolic	39	2 (1.4–2.7)
Inguinal hernia	25	1 (0.8–1.8)
Miscellaneous*	21	1 (0.6–1.6)
Infectious	15	0.8 (0.4–1.2)
Incisional hernia	11	0.6 (0.2–0.9)
Lymphatic	7	0.4 (0.1–0.6)
Neurological	5	0.3 (0.03–0.5)
Myocardial Infarction	2	0.1 (0–0.2)
Total	196	10 (9.0–11.8)
Death	0	NA

* Includes six cases of Peyronie's disease, five unknown hernia, one cholecystitis, one catheter-related complication, one wound hematoma, one wound seroma, one rectal injury, one uretal injury, and two unrelated complications.

Adapted from Catalona WJ, Carvalhal GF, Mager DE, Smith DS. Potency, continence, and complication rates in 1,870 consecutive radical retropubic prostatectomy. J Urol 1999;162:433, with permission.

the choice among radical prostatectomy, external beam radiation, or brachytherapy is hard to make both for physician and patient.

Radical Prostatectomy

Radical prostatectomy is performed through either the retropubic or perineal approach. Most urologists are trained in retropubic prostatectomy. Radical retropubic prostatectomy is an effective modality for treatment of clinically localized prostate cancer. The mortality rate of radical prostatectomy has been low and approaches zero in recent series (Table **17–2**).[40] Excellent long-term results can be obtained with radical retropubic prostatectomy. The Johns Hopkins Hospital group reported that in 2404 men who underwent radical retropubic prostatectomy for clinically localized prostate cancer, the overall actuarial PSA progression-free survival at 5, 10, and 15 years were 84%, 74%, and 66%, respectively. The actuarial metastasis-free survival at 5, 10, and 15 years were 96%, 90%, and 82%, respectively, and the actuarial cancer-specific survival at 5, 10, and 15 years were 99%, 96%, and 90%, respectively.[41] The improvement of knowledge of surgical anatomy with discovery of the dorsal venous complex around the apex of the prostate[42] and the cavernous nerve in the path of the capsular branch of the inferior vesical artery[43] has reduced the intraoperative blood loss, the perioperative morbidity, and increased the rate of continence and potency preservation of radical retropubic prostatectomy. Radical perineal prostatectomy has the advantages of less morbidity, much less blood loss and easier access to the vesicourethral anastomosis, and shorter hospital stay. However, it is believed that radical perineal prostatectomy creates more positive margins of greater extent than does the radical retropubic prostatectomy.[44]

Complications of Radical Retropubic Prostatectomy

INTRAOPERATIVE COMPLICATIONS

Blood loss can average 1000 mL[45] and is replaced by allogenic or autologous blood transfusion. Cell salvage can be used as well. It is a safe method of transfusion during radical retropubic prostatectomy.[46] Rectal injury with an occurrence rate of 0 to 5.3%[47] can be repaired by the two-layer method without colostomy, provided that preoperative bowel preparation has been performed. Ureteral injury rates are 0 to 1.6%[47] and can be repaired with placement of a double J-stent. Obturator nerve injury, very uncommon during pelvic lymphadenectomy, if recognized can be repaired primarily, though the long-term functional status after repair is not available.

PERIOPERATIVE COMPLICATIONS

Like any other major surgery in the older patient, radical retropubic prostatectomy may be associated with medical complications such as pneumonia, atelectasis, pulmonary embolism, deep vein thrombosis, myocardial infarction, and postoperative ileus. Early surgical complications include wound infection, wound dehiscence, delayed bleeding, catheter dislodgment, anastomotic leakage, and lymphocele. Hedican and Walsh[48] reported that 7 out of 1350 patients undergoing radical retropubic prostatectomy (0.5%) developed delayed bleeding requiring acute blood transfusion, and four who were managed by exploration had good outcomes. Catheter dislodgment before 3 or 4 days probably may require replacement of a new catheter using a flexible cystocope with a guide wire. Leaking urine from the urethrovesical anastomosis occurs in from 0.1 to 21.1%, and spontaneous resolution usually results in prolonged catheterization.[47] Lymphocele may occur after radical retropubic prostatectomy with bilateral pelvic lymphadenectomy. It may require percutaneous drainage with ultrasound guidance or open drainage with marsupialization for persistent lymphocele.

LONG-TERM COMPLICATIONS

The long-term complications of radical retropubic prostatectomy can include urinary incontinence, erectile dysfunction, and bladder neck contracture.

Urinary Incontinence

Urinary incontinence is a devastating complication of radical prostatectomy. The definition of incontinence is not universal. This leads to different incidence rates of incontinence after radical retropubic prostatectomy. The Johns Hopkins Hospital group treated 593 consecutive patients who had 92% complete continence, with some degree of stress incontinence in 8%.[49] The American College of Surgeons Committee on Cancer surveyed 1796 preoperatively continent men who had undergone radical retropubic prostatectomy at 484 hospitals and found that 19% had incontinence and required pads daily and 3.6% had total incontinence.[50] Minor incontinence or stress incontinence can be managed by pads or a Cunningham clamp. Severe or total incontinence may require artificial urethral sphincter placement.

Erectile Dysfunction

Erectile dysfunction was universal after radical retropubic prostatectomy before Walsh and Donker[43] in 1982 discovered the autonomic branches of the pelvic plexus innervating the corpora cavernosa and developed techniques to preserve them. Preservation rates of erectile function after radical retropubic prostatectomy vary widely. Quinlan et al[51] reported the potency rate in 600 patients who underwent the nerve-sparing approach. Of 503 patients who were potent preoperatively, 68% were potent postoperatively. Age, stage of the disease, and either unilateral or bilateral nerve-sparing procedure were significant factors for preservation of potency. Potency was preserved in 91% of men aged less than 50 year, whereas it was only 25% of men aged 70 years or older. The potency rates in younger patients were better when bilateral nerve-sparing procedures were performed.

However, when nonsurgical members of the health care team questioned patients about their sexual function, potency rates were preserved in 31.9% and 13.3% of bilateral and unilateral nerve-sparing procedures, respectively.[52] The cavernous nerve innervating the corpora cavernosa is a microscopic structure and the capsular branch of the inferior vesical artery is not readily identified during the radical retropubic prostatectomy. A device, the CaverMap Surgical Aid (UroMed, Boston, MA), was designed to aid the surgeon in identifying and preserving neurovascular bundles (NVBs). However, the size of the CaverMap nerve stimulator may make it difficult to trace the cavernous nerves before the prostate is removed, particularly in obese men or in patients who have a large prostate or a narrow pelvis. In a randomized, controlled study, the use of the CaverMap during radical prostatectomy resulted in improved nocturnal erections, but did not lead to improved overall sexual function.[53] Sural nerve graft during radical retropubic prostatectomy may enhance the recovery of erectile function when the neurovascular bundles were resected.[54] The importance of nerve grafting is uncertain.[55] Despite nerve-sparing procedure recovery of erectile function may take months to 2 years. Therapy may be necessary for some patients to restore the erectile function in the interim. There are various means that can restore erectile function. Zippe et al[56] reported that sildenafil (Viagra) helped men who had undergone bilateral nerve-sparing radical retropubic prostatectomy to obtain erections sufficient for intercourse, whereas no men who had undergone excision or ligation of both neurovascular bundles had a positive response to this agent. Intracavernous injection with vasoactive agents such prostaglandin E_1 is effective to achieve sufficient erection for vaginal intercourse more than 90% of the time.[57] Transurethral application of prostaglandin E_1 may help patients achieve erection 64.9% of the time.[58] Using a vacuum erection device can help men achieve an adequate erection. Penile prostheses, either inflatable or semirigid, are the ultimate solution to erectile dysfunction after radical retropubic prostatectomy.

Bladder Neck Contracture

The incidence of bladder neck contracture ranged from 1.3 to 27% in early studies.[59] The scarring at the site of the urethrovesical anastomosis was thought to result from the extravasation of urine. With eversion of the bladder neck mucosa and watertight urethrovesical anastomosis, the bladder neck contracture rates have reduced. Poon et al[60] reported that there was no statistical difference with regard to bladder neck contracture rates whether the urethrovesical anastomosis is achieved by bladder neck preservation, tennis racket type closure, or anterior bladder tube bladder neck reconstruction. Bladder neck contracture may present clinically as urinary incontinence or obstruction. The treatment is bladder neck dilation, incision, or resection.

External Beam Radiation Therapy (EBRT)

Prostate cancer is a radiosensitive tumor. The higher the radiation dose delivered to a given volume of cancer, the more likely that the cancer will be permanently controlled. External beam radiation therapy is usually administered in once-daily fractions, 5 days per week over a period of 7 to 8 weeks. Each treatment session lasts ~10 to 20 minutes.

Conventional External Beam Radiation Therapy

The 10-year disease-specific survival of 1557 prostate cancer patients treated with conventional external beam radiation therapy alone was 87%, 75%, and 44% for tumors of Gleason score 2 to 5, 6 to 7, and 8 to 10, respectively.[61] PSA failure definitions after EBRT vary from >0.5 to >4 ng/mL among radiation oncologists. The American Society for Therapeutic Radiology and Oncology recommends that PSA failure definition is three consecutive increases in PSA from the nadir value. The date of failure should be the midpoint between the nadir and the first of the three PSA rising values, and PSA testing should be performed every 3 to 4 months.

Zietman et al[62] reported that less than 50% of the T1- or T2NxM0 and less than 20% of the T3- or T4NxM0 prostate cancer patients receiving conventional radiation therapy were biochemical disease free at 10 years. Conventional external beam radiation therapy irradiated large volumes of normal tissues as well as the prostate gland. To avoid late side effects such as rectal bleeding, the radiation dose is limited to 60 to 70 Gy, delivered in 2-Gy daily fractions.

Three-Dimensional Conformal Radiation Therapy (3D-CRT)

With the development of three-dimensional conformal radiation therapy (3D-CRT), a higher radiation dose can be delivered to the tumor target improving treatment efficacy while reducing the radiation to the normal tissues and avoiding late side effects. Study of patients treated with 3D-CRT with tumor target dose increasing from 64.8 to 81 Gy was reported. PSA nadir ≤1.0 ng/mL was achieved in 90% of the patients treated with doses 75.6 or 81 Gy, in 76% of those treated with 70.2 Gy, and in 56% of those treated with 64.8 Gy. The 5-year PSA relapse-free survival rate was

85% for patients with a good prognosis, 65% for intermediate prognosis, and 35% for unfavorable prognosis. A positive biopsy at >2.5 years after 3D-CRT was observed in only 1/15 (7%) of patients receiving 81.0 Gy, compared with 12/25 (48%) after 75.6 Gy, 19/42 (45%) after 70.2 Gy, and 13/23 (57%) after 64.8 Gy (p < .05). This study provides evidence for a significant effect of dose escalation on the response of human prostate cancer to irradiation.[63] The GU Radiation Oncologists of Canada recommends classification of prostate cancer into three groups based on prognostic grouping and risk of recurrence: (1) low-risk group with PSA ≤10 ng/mL, Gleason score ≤6 and stage T2a or less; (2) intermediate-risk group with PSA ≤20 ng/mL, Gleason score <8, and stage T1/T2; and (3) high-risk group with PSA >20 ng/mL, Gleason score ≥8 and stage ≥T3a. Conformal treatment techniques should be offered to patients receiving prostatic radiation therapy. Low-risk patients receive 70 Gy in 2-Gy fractions, intermediate-risk patients receive 75 to 78 Gy in 2-Gy fractions, and the high-risk patients are at high risk of systemic metastases. It is unclear how dose escalation using local therapy would improve outcome. It was felt that conformal doses in the range of 75 to 78 Gy is a reasonable option for high-risk group patients.[64]

Intensity Modulated Radiotherapy

Recently, it has become possible not only to shape the borders of the radiation fields to conform to the target volume, but also to define spatial variation in beam intensity within each field. This technique, known as intensity modulated radiotherapy (IMRT), enables the high-dose radiation envelope to form complex shapes, providing even greater conformality of radiation dose to the target volume. It is therefore expected to further reduce the risk of normal tissue complications, and so enable greater dose-escalation. At New York Memorial Sloan-Kettering Cancer Center, a total of 698 patients (90%) were treated with 81.0 Gy, and 74 patients (10%) were treated with 86.4 Gy. Using the IMRT technique for clinically localized prostate cancer, the 3-year rate of grade ≥2 rectal toxicity is just 4%.[65]

Complications of External Beam Radiation Therapy

The conventional techniques of external-beam radiation therapy are fairly well tolerated, although grade 2 or higher acute rectal morbidity (discomfort, tenesmus, diarrhea) and/or urinary symptoms (frequency, nocturia, urgency, dysuria) requiring medication occur in ~60% of patients. The incidence of late complications is low. An analysis of 1020 patients treated in two large Radiation Therapy Oncology Group (RTOG) trials[66] demonstrated an incidence of chronic urinary sequelae (i.e., cystitis, hematuria, urethral stricture, or bladder contracture) requiring hospitalization in 7.3% of cases, but the incidence of late urinary toxicity requiring major surgical intervention was only 0.5%. Urethral stricture is more common in patients who had undergone a previous TURP. The incidence of chronic intestinal sequelae (chronic diarrhea, proctitis, rectal or anal stricture, rectal bleeding, or ulcer) requiring hospitalization for diagnosis and minor intervention was 3.3%, with 0.6% of patients experiencing bowel obstruction or perforation. Fatal complications were extremely uncommon (0.2%).[66] Most adverse events are observed within the first 3 to 4 years after treatment; the likelihood of complications developing after 5 years is low.[67] The risks of complications are increased when radiation doses exceed 70 Gy.[68] The risk of rectal toxicity has been correlated with the volume of the anterior rectal wall exposed to the higher doses of irradiation.[69] Erectile potency appears to diminish with advancing time after treatment due to vascular disruption from radiation, with half of patients impotent at 7 years after irradiation.[70]

Prostate Brachytherapy

Prostate brachytherapy is a technique whereby small radioactive implants or seeds are placed within the prostate gland, allowing high radiation dose to tumors (Fig. **17–2**). In the 1970s, Whitmore and colleagues[71] developed the technique of open retropubic approach and free-hand implantation of iodine 125 seeds in the

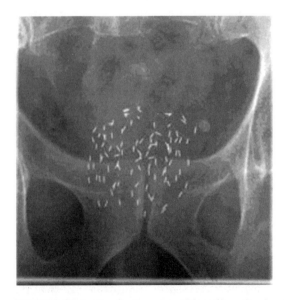

FIGURE 17–2. Brachytherapy with radioactive implants.

treatment of prostate cancer. The poor dose distribution and the inclusion of patients with locally advanced disease led to poor treatment result. The technique was abandoned.[72] Follow-up of Whitmore et al's patients showed a biochemical diseased free rate of only 13% at 15 years.[73] Accurate transperineal placement of radioisotope seeds through a template with real-time transrectal ultrasound guidance developed by Holm et al[74] began the modern era of prostate brachytherapy in 1983. Blasko and colleagues[75] developed and improved the technique with reduced complication and improved outcomes. Iodine 125 (I-125) and palladium 103 (Pd-103) are the commonly used isotopes for prostate brachytherapy. The half-life of I-125 is 60 days and Pd-103 17 days. Based on the consideration that the effective dose of radiation is largely delivered in the first three half-lives, then the duration of treatment can be estimated at 180 days for I-125 and 51 days for Pd-103.[76] The recommended dose for monotherapy is 145 Gy for I-125 and 125 Gy for Pd-103. Pd-103 was preferred for higher grade lesions, though there is no proven data to suggest that Pd-103 is superior to I-125 or vice versa. There are three phases of permanent prostate brachytherapy: the planning phase, the implant procedure, and implant quality evaluation.

During the planning phase the prostate volume is determined by transrectal ultrasound. The ultrasound image of prostate volume is digitalized into the planning computer to generate a treatment plan with the desired dose of radiation delivering to the prostate while assuring that neither the rectum nor the urethra receives excessive radiation. The planning phase can be performed intraoperatively prior to implant procedure. The implant procedure is done through real-time ultrasound-guided transperineal percutaneous implantation on an outpatient basis. The implantation is done in the operating room under spinal or general anesthesia. The procedure usually lasts 45 to 60 minutes. Implant quality evaluation by calculation of the radiation doses can be assessed by radiographs. This method does not provide information relative to the prostate location.[77] Computer tomography (CT)-based dosimetry methods have been recommended to assess implant quality by the American Brachytherapy Society.[78] According to the most recent analysis from the Seattle brachytherapy data, the 10-year PSA progression-free rate following brachytherapy for low-risk prostate cancer was 87%, intermediate-risk 79% and high-risk 51%.[79] Hence, favorable result from prostate brachytherapy is obtainable only in low-risk patients. The American Brachytherapy Society currently recommends prostate monobrachytherapy for clinical T1-T2, Gleason score ≤6 and PSA ≤10 ng/mL disease, that is, low-risk tumor. High-risk patients are recommended to be treated with combination of brachytherapy and EBRT.[78] Ragde[80] reported 12-year observed follow-up on patients treated with brachytherapy alone for the low-risk group and combined EBRT with brachytherapy for the high-risk group. It is interesting that the 12-year PSA progression-free rate was 66% for the low-risk group and 79% for the high-risk group. The rate might have been better than 66% had the low-risk group been treated with a combination of EBRT and brachytherapy. Androgen deprivation therapy to downsize the prostate gland to less than 60 cc may be necessary to avoid pubic arch interference during seed implantation. Luteinizing hormone-releasing hormone (LHRH) agonist with or without antiandrogen is commonly used for androgen deprivation.

Complications of Prostate Brachytherapy

Complications of prostate brachytherapy reported from 13 case series and three cohort studies included acute urinary retention (1–14%), incontinence (5–6%), cystitis/urethritis (14%), urethral stricture (1%), proctitis (1–14%), and impotence (4–50%).[81]

Cryotherapy

The availably of real-time transrectal ultrasound monitoring of the cryoprobe placement, the development of 3-mm cryoprobes that could be placed transperineally into the target areas of the prostate, and the development of urethral warming systems to maintain the periurethral tissues viability have made cryotherapy of prostate cancer feasible. Five or six cryoprobes delivering liquid nitrogen to achieve a temperature of $-40°$ to $-50°C$ at the periphery of the prostate gland are necessary to obtain coagulative necrosis. In addition four or five thermocouple probes are placed to ensure an adequate freeze. The urethral warming catheter systems are necessary to minimize urethral sloughing during cryotherapy. The role of prostate cancer cryotherapy, according to Benoit et al,[79] is principally in the treatment of high-risk clinically localized prostate cancer. It is also the treatment of choice for men with local failure after EBRT. There is no 10- or 15-year result as measured by PSA progression. Five-year actuarial biochemical-free survival post-cryotherapy was 51% and 63% for PSA less than 0.5 ng/mL and PSA less than 1 ng/mL, respectively.[82]

Complications of Cryotherapy

Complications of cryotherapy include dramatic scrotal edema, though resolving spontaneously within 3 weeks, urethral sloughing 15%, urinary incontinence 5.9%, urethral stricture 4.9%, bladder neck contracture 3.2%, perineal pain 0.6%, urethrorectal fistula 0.3%, sepsis 0.8%, urinary retention 1.2%, and stress incontinence 0.3%.[79]

Other Ablative Methods

HIGH-INTENSITY FOCUS ULTRASOUND

High-intensity focus ultrasound (HIFU) delivers intense ultrasound energy, with consequent heat destruction of tissue at a specific focal distance from the probe without damage to tissue in the path of the ultrasound beam. A piezoelectric transducer placed transrectally emits a highly focused convergent ultrasound beam in pulses lasting 3 to 5 seconds. These pulses produce areas of ablation that are ellipsoid in shape and typically measure ~2 cm in height by 2 mm in diameter. Temperatures in the target area range between $85°$ and $100°C$, high enough to produce discrete areas of coagulative necrosis. The transducer is moved sequentially through various areas of the prostate to produce large areas of focal ablation.[83] The procedure is done under general or spinal anesthesia. HIFU can be performed for patients with localized prostate cancer without an incision, with a less severe side-effect profile, and, unlike most other prostate treatments, is repeatable.[84] Long-term PSA result is lacking. Further study with this technology is warranted.

RADIOFREQUENCY INTERSTITIAL TUMOR ABLATION (RITA)

In this procedure, radiofrequency electrodes are placed under sonographic guidance transperineally into target areas of the prostate, and large spherical areas of coagulative necrosis can be produced. Treatment time is short (8 to 12 minutes). Various size lesions can be generated.

Interstitial Thermal Ablation

In this procedure, specially derived cobalt-palladium alloy seeds are placed in the target area (i.e., the prostate) under transrectal sonographic guidance. The seeds are 1 mm in diameter by 14 mm in length and will self-regulate to desired temperature when placed in a magnetic field. Neither RITA nor interstitial thermal ablation has generated enough peer-review clinical information as treatment options for prostate cancer.[82]

Androgen Deprivation and Prostate Cancer

Huggins and Hodges[85] described androgen sensitivity of prostate cancer in 1941. Androgen deprivation induces apoptosis and inhibition of cell proliferation of the prostate cancer tissues.

Neoadjuvant Hormone Therapy

Androgen deprivation used prior to primary treatment of prostate cancer is called neoadjuvant hormone therapy (NHT). NHT 3 to 8 months prior to radical prostatectomy may reduce margin positivity by 50%. It is hoped this will translate into improved disease-free survival. However, studies have shown no difference in disease-free survival with or without NHT up to 5 years of follow-up. Until recurrence rates have been demonstrated to decrease in randomized studies, NHT should be considered investigational and studied in the context of controlled clinical studies.[86]

In high-risk patients, NHT followed by EBRT has been shown to improve biochemical disease-free and metastasis-free survival versus EBRT alone.[87] NHT is used prior to brachytherapy to downsize the prostate gland to less than 60 cc to avoid pubic arch interference. Currently there is no evidence that the addition of androgen deprivation to brachytherapy will improve the biochemical disease-free survival.[77]

Adjuvant Hormone Therapy

Androgen deprivation used following primary treatment of prostate cancer when the disease fails to be completely eradicated is called adjuvant hormone therapy (AHT). The therapy is usually started within 6 months of the primary treatment. A report by Messing et al[88] showed biochemical or clinical progression and overall survival advantage when immediate adjuvant hormone therapy was given to the patients with positive pelvic nodes after radical prostatectomy versus observation alone. The most convincing data favoring the use of androgen deprivation therapy in conjunction with EBRT were reported by Bolla et al.[89] A significant difference in overall survival was noted for patients who received adjuvant hormone therapy with radiation when compared with those who underwent radiation therapy alone. Five-year overall survival for the two groups was 79% and 62%, respectively.[89]

Rising Prostate-Specific Antigen After Primary Treatment

At least 35% of men who received primary therapy for clinically localized prostate cancer with curative intent will experience PSA failure within 10 years.[90] The management of rising PSA after radical prostatectomy or radiotherapy is an increasingly common problem facing patients and physicians. The objectives of treating a rising PSA level are to prevent metastasis, symptoms, or death due to prostate cancer. Pound et al[91] reported the outcomes of 1997 men who underwent a radical prostatectomy over a 15-year period. The median follow-up was 5.3 years. At the time of analysis, 315 patients' disease had recurred as shown by detectable PSA. The importance of this study is that it provides evidence that a *rising PSA after radical prostatectomy does not mean a death sentence for all patents*. It pointed out that for Gleason score 5 to 7 prostate cancer, if the rising PSA recurred in 2 years or more after surgery and PSA doubling time was 10 months or more, the probability of metastatic progression was 5%, 14%, and 18% in 3, 5, and 7 years, respectively. The median time to metastatic progression was 8 years, yet 63% of the patients with rising PSA remained free of metastasis at 5 years. Once the metastatic disease was documented, the median time to death was 5 years.[91] Rising PSA after surgery can be effectively treated with salvage EBRT if there is no distant disease. Androgen deprivation therapy may be considered in patients at high risk of metastatic progression. The incidence of biochemical failure after EBRT or brachytherapy is at least as high as after radical prostatectomy.

In Critz et al's[92] study, in which a combination of brachytherapy and EBRT was administered to 660 men with T1–T2 disease, patients who achieved a PSA nadir of ≤0.5 ng/mL had a 5-year disease-free survival of 93%, whereas only 26% of patients in whom the nadir was ≥0.6 ng/mL were disease-free at 5 years. All men with a nadir of >1 ng/mL eventually failed treatment. The authors concluded that a PSA nadir of ≤0.5 ng/mL should be reached for a good prognosis. However, this value is not accepted by all radiation oncologists. Patients who failed after EBRT can be treated with salvage radical prostatectomy, cryotherapy, or brachytherapy if the site of recurrent disease is considered to be local rather than distant. Salvage surgery is feasible but is associated with significant morbidity. Short-term follow-up of salvage cryotherapy has shown reduction of PSA, though with high urinary obstruction required transurethral surgery.[93] Salvage brachytherapy has significant risk of incontinence. The majority of patients with rising PSA after EBRT are managed by androgen deprivation therapy.

Method of Androgen Deprivation Therapy

ORCHIECTOMY

Androgen deprivation therapy can be surgical or medical. Bilateral orchiectomy is a simple, safe, and low-cost surgical procedure.[94] For decades this has been the most common and effective treatment for metastatic prostate cancer. Serum testosterone levels were reduced by 95% within 3 hours of castration. Therefore, some patients may suffer psychological trauma from an orchiectomy.

ESTROGEN

Estrogen in the form of diethylstilbestrol 3 to 5 mg daily reduces the testosterone secretion from the Leydig cells of the testes to castration levels by *downregulating* the secretion of luteinizing hormone (LH) and follicle-stimulating hormone (FSH) by the pituitary gland. This is achieved at between 3 and 9 weeks.[95] Diethylstilbestrol may cause direct cytotoxic effects on prostate cancer cells by cytoplasmic microtubules disruption.[96] The side effects of diethylstilbestrol include nausea, vomiting, gynecomastia, fluid retention, and cardiovascular events.

ANTIANDROGEN

There are two classes of antiandrogen: steroidal and nonsteroidal. Steroidal antiandrogens such as cyproterone acetate and megestrol acetate inhibit c21–9 decarboxylate, preventing adrenal androgen synthesis and gonadotropin release from the hypothalamus. Nonsteroidal antiandrogens include flutamide, bicalutamide, and nilutamide. They block the androgen receptor from binding to dihydrotestosterone and testosterone. The serum levels of testosterone will increase. Flutamide may induce gastrointestinal and liver toxicity, whereas bicalutamide has fewer of these side effects. Nilutamide may cause flushing, poor visual adaptation to dark, and pulmonary fibrosis. Monotherapy using bicalutamide 150 mg daily compared with orchiectomy or LHRH agonist has been shown to have equivalent survival with an improvement of sexual function and physical capacity for those patients treated with bicalutamide.[97]

LUTEINIZING HORMONE–RELEASING
HORMONE (LHRH) AGONISTS

Medical castration can be effected by inhibition of LH secretion by giving LHRH-agonist injection subcutaneously. Lupron (leuprolide acetate), Zoladex (goserelin acetate), and Suprefact (buserelin) are LHRH agonists, available for injection every 3 months. Initially the LHRH agonists stimulate secretion of LH and FSH from the pituitary gland. The LH promotes testosterone production from the Leydig cells of the testes causing a *testosterone surge*. The testosterone surge is called flare phenomenon. Constant exposure to an LHRH agonist causes downregulation of receptors in the pituitary, inhibiting LH and FSH release and decreasing in testosterone production. LHRH agonists should be given with prior administration of antiandrogen of 2 weeks to prevent flare phenomenon.

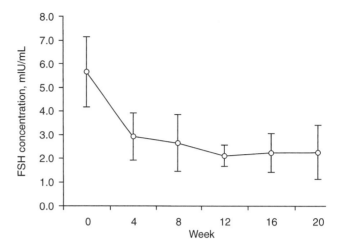

FIGURE 17–3. Effect of Abarelix on serum follicle-stimulating hormone (FSH). (Adapted from Beer TM, Garzotto M, Eilers KM, Lemmon D. Phase II study of abarelix depot for androgen independent prostate cancer progression during gonadotropin-releasing hormone agonist therapy. J Urol 2003;169:1738–1741, with permission.)

Luteinizing Hormone–Releasing Hormone (LHRH) Antagonists

Abarelix, a direct LHRH antagonist, unlike LHRH agonists, avoids the flare phenomenon. A recent phase III randomized trial comparing Abarelix with leuprolide has shown that Abarelix caused rapid medical castration in 24% of men 1 day after treatment and 78% after 7 days compared with 0% of men treated with leuprolide acetate on either day.[98] Fig. 17–3 shows the effect of Abarelix on FSH concentrations. A comparable percentage of men achieved and maintained castration between days 29 and 85 in each group. The PSA had a statistically significant decrease for the first month in patients treated with Abarelix. Dihydrotestosterone, LH, PSA, and FSH showed similar rapid reductions without an initial increase.[98] As this study does not have a mature follow-up, it is not possible to determine if Abarelix and leuprolide will provide identical rates of disease control.[99]

Complete Androgen Blockade (CAB)

Ninety percent of serum testosterone is produced by the testis and 10% by the peripheral conversion of the adrenal androgens including dehydroepiandrosterone, dehydroepiandrosterone sulfate, and androstenedione. Testosterone in the prostatic cells is converted to dihydrotestosterone (DHT) by 5α-reductase enzyme. DHT binds with the androgen receptor, which activates the androgen-responsive gene promoting transcription. Labrie et al[100] demonstrated that ~60% of total intraprostatic DHT originates from the testis, whereas

40% is of adrenal origin. Based on this information CAB using LHRH-agonist or bilateral orchiectomy with antiandrogen became a popular treatment of advanced prostate cancer. There have been many studies on whether patients should be treated with CAB or monotherapy alone, creating a great deal of controversy. The Overview Consensus Statement recommends that complete androgen blockade cannot be considered standard therapy in asymptomatic patients. Meta-analyses of clinical trial data would suggest that perhaps 2% to 3% of patients realize a survival benefit with CAB, compared with LHRH agonist alone. It is not possible to identify a subset of patients who will benefit, so combined androgen blockade should be considered the exception rather than the rule.[101]

Time to Initiate Androgen Deprivation Therapy

When to initiate androgen deprivation therapy (ADT) is another controversial subject in the management of prostate cancer. Based on the results of the 1970s Veterans Administration Cooperative Urological Research Group (VACURG) study[102] in which ADT was administered to patients only after the onset of symptoms, the rate of prostate cancer-specific mortality in patients with locally advanced or metastatic disease did not increase. But reanalysis of the VACURG studies and the results of the Medical Research Council Trial[103] would suggest that there is benefit to initiating therapy early in terms of disease-specific and overall survival. There is a tendency to treat patients with recurrent prostate cancer with ADT at some point when the PSA is rising prior to symptom development.

Continuous or Intermittent Androgen Deprivation Therapy

Continuous androgen deprivation therapy has been the standard treatment for advanced or metastatic prostate cancer. Intermittent therapy consists of starting ADT and stopping treatment when a predefined clinical response is achieved to let the tumor cells repopulate with androgen-sensitive cells and delay evolution of androgen-insensitive cells. In vitro and animal models of prostate cancer studies suggest that intermittent ADT delays progression to the androgen independent state but does not prevent it.[104] This observation has not been confirmed in human clinical trials. The Overview Consensus Statement suggested that given the lack of sufficient long-term data to demonstrate equivalence to conventional ADT, it is mandatory to explain to the patient that intermittent ADT is an investigational rather than a standard approach to ADT, and to document that the patient understands this distinction.[101] Goldenberg et al,[105] in a very heterogeneous group of prostate cancer patients treated with various forms of intermittent combinations of androgen blockage, reported that during the off-treatment period patients had an improved sense of well-being and near-normal sexual function.

Adverse Effects of Androgen Deprivation Therapy

Androgen deprivation therapy, either bilateral orchiectomy or LHRH-agonist injection, results in a 90% reduction of serum testosterone levels, inducing acute and relatively complete hypogonadism. Clinical symptoms of hypogonadism include loss of libido, poor sexual function, hot flashes, irritability, poor concentration, depression, mood change, reduced cognitive function, weight gain, change of body composition with an increase of fat mass and a decrease of muscle mass, easy fatigability, and loss of bone mineral density leading to osteopenia or osteoporosis and anemia.[106–109]

Chemotherapy and Prostate Cancer

Despite adequate ADT for advanced prostate cancer, hormone-refractory prostate cancer (HRPC) develops after a median time of 2 years.[110] Clinical treatment of HRPC is limited and the patient's median survival is short, 10 to 12 months.[111] Single cytotoxic chemotherapeutic agents including doxorubicin, mitoxantrone, cyclophosphamide, vinblastine, paclitaxel, docetaxel, estramustine, and etoposide have ~15 to 20% response. Several combinations of these agents have shown significant palliative benefit and antitumor effects in patients with HRPC. Chemotherapy with mitoxantrone and prednisone provides palliation for some patients with symptomatic hormone-resistant prostate cancer.[112]

Discussion of Case History

In the case at the beginning of this chapter, prostate cancer was diagnosed by early detection rather than by screening. This is a significant cancer, as both the PSA and DRE were abnormal. A healthy 63-year-old man has a life expectancy of 17.06 years.[113] Treatment with curative intent is appropriate. With the parameters of a PSA of 5.5 ng/mL, a Gleason score of $3 + 3 = 6$, and clinical stage T2a, this patient's prostate cancer is in a low-risk category. The predicted pathological finding would be organ-confinement 51%, capsule penetration 44%, seminal vesicle involvement 3%, and lymph node involvement 2%.[35] This prediction is not 100% proven due to a possible sample error in the biopsy and an inaccurate DRE in determining tumor stage. A Gleason score of 5 to 6 on biopsy corresponded to the same grade in the radical prostatectomy specimen in 64% of the cases.[114] The patient does not require a bone scan. The chance of bone metastasis in patients with a PSA less than 10 ng/mL and Gleason score less than 7 is less than 1%.[115] The treatment options include radical prostatectomy, EBRT, brachytherapy, and watchful waiting. Watchful waiting is not appropriate for this patient for two reasons. First, he has more than 15 years of life expectancy. Though low-risk prostate cancer progresses slowly, if not treated it will eventually pose a threat to his life. Second, studies have shown that clinically localized prostate cancer treated with radical prostatectomy has a cancer-specific mortality of 4.6%, local progression rate of 19.3%, and metastasis of 13.4% versus 8.9%, 61%, and 27.3%, respectively, if treated with watchful waiting.[39] There is no prospective randomized control study of sufficient power comparing the efficacy of various local therapies for clinically localized prostate cancer. According to D'amico et al,[116] in low-risk prostate cancer the PSA outcome after RP, EBRT, or brachytherapy showed no difference at 5 years. What about a 10-year or 15-year result? A survey showed that the specialists overwhelmingly recommend the therapy they themselves deliver.[59]

Radical prostatectomy is a good choice for this patient. He has a 90% chance of 15-year disease-specific survival and 74% chance of PSA-progression free in 15 years.[41] NHT is not necessary in view of the low PSA, Gleason score, and clinical stage disease. Surgical mortality approaches zero.[40] No pelvic lymphadenectomy preceding prostatectomy is necessary, because potential pelvic lymph node involvement is ~2%.[35] Postoperatively, he will have a period of urinary incontinence. Urinary continence gradually improves with Kegel's exercise with a potential of 3.6% total incontinence.[50] If bilateral-nerve sparing is successful, his chance of potency preservation is ~32%.[52] Rectal injury during surgery may occur in 0 to 5.3%[47] and bladder contracture in ~1.3 to 27%.[117] Pathological stage and grade of his disease will be available from the prostatectomy specimen. This information is essential to prognosticate his future outcome and therapy. This patient should also see a radiation oncologist for consultation as a potential candidate for radiation therapy. Two radiation modalities are available: EBRT and brachytherapy. Conventional EBRT will provide 75% ten-year disease-free survival[61] and 50% biochemically disease free at 10 years.[62] If this low-risk patient is treated with 3D-CRT, the 5-year PSA relapse-free survival rate could be 85%.[63] Treatment-related toxicity includes bowel complications such as chronic diarrhea, proctitis, rectal or anal stricture, rectal bleeding, or ulcer and urinary tract complications such as cystitis, hematuria, urethral stricture, or bladder contracture.[66]

Erectile dysfunction occurs in half of patients at 7 years after irradiation.[70] Neither NHT nor AHT is necessary for this patient because his disease belongs to a low-risk category. This patient is also a good candidate for brachytherapy, which is probably the least invasive therapy. He will not require NHT because the prostate gland is less than 60 cc. The 10-year PSA progression-free result ranges from 66% to 87%.[78,80] The potential side effects of treatment include acute urinary retention (1–14%), incontinence (5–6%), cystitis/urethritis (14%), urethral stricture (1%), proctitis (1–14%), and impotence (4–50%).[81] By choosing radiation therapy there is no pathological information available to confirm whether the stage and the grade of the cancer are what is predicted. The patient, armed with all the information, has to balance the benefits and risks of each treatment and make the final choice of therapy. It is highly recommended that he attends the local prostate cancer support group where he can meet with other patients who have undergone the treatment.

Summary and Key Points

- Prostate cancer is the most common cancer and the second leading cause of cancer death of North American men. However, most men die *with* prostate cancer rather than die *of* prostate cancer.

- There is a great deal of variation in the international incidence of prostate cancer. The highest incidence is in Jamaicans, and the second highest is in African Americans; the lowest incidence is in the Chinese who live in China.

- The unmodifiable risk factors of prostate cancer include age and genetics, and the modifiable risk factors include diet and lifestyle.

- There is heated debate about whether men should be screened for prostate cancer. There is no evidence that screening and early treatment of prostate cancer will reduce mortality of the disease. On the other hand, there is no evidence that it will not.

- It is reasonable to offer annual PSA and DRE to men 50 years or older who a have life expectancy of greater than 10 years. Men at high risk, such as African Americans and men who have a first-degree relative diagnosed with prostate cancer at an early age, should begin testing at age 45.

- Information about the potential benefits and harms of screening and treatment of prostate cancer and the limits of current evidence of screening and treatment of prostate cancer should be thoroughly discussed with patients before testing.

- Patients diagnosed with clinically localized prostate cancer face a difficult task of choosing a treatment option.

- Patients should take active participation in decision making with regard to choices of treatment modality after full consultation with various specialists.

- Future research should aim at cancer prevention and discovering factors that can help identify significant prostate cancer that requires treatment and insignificant prostate cancer that will remain occult the rest of the patient's lifetime and requires no treatment.

- For prostate cancer that has progressed beyond curability, a method should be available to convert the disease to a controllable chronic illness.

REFERENCES

1. American Cancer Society. Cancer Facts and Figures 2003, using 2000 population standard for age adjustment.

2. Carter HB, Piantadosi S, Isaacs JT. Clinical evidence for and implications of the multistep development of prostate cancer. J Urol 1990;143:742–746

3. Glover FE Jr, Coffey DS, Douglas LL, et al. The epidemiology of prostate cancer in Jamaica. J Urol 1998;159:1984–1986

4. Ruijter E, van de Kaa C, Miller G, et al. Molecular genetics and epidemiology of prostate carcinoma. Endocr Rev 1999; 20:22–45

5. Miller G. Prostate cancer among the Chinese: pathologic, epidemiologic, and nutritional considerations. In: Resnick MI, Thompson IM, eds. *Advanced Therapy of Prostate Disease*. New York: BC Decker; 2000:18–27

6. Wingo PA, Tong T, Bolden S. Cancer statistics, 1995. CA Cancer J Clin 1995;45:8–31

7. Gann PH, Hennekens CH, Ma J, Longcope C, Stampfer MJ. Prospective study of sex hormone levels and risk of prostate cancer. J Natl Cancer Inst 1996;88:1118–1126

8. Thompson IM, Goodman PJ, Tangen CM, et al. The influence of finasteride on the development of prostate cancer. N Engl J Med 2003;349:215–224

9. Von Eschenbach A, Ho R, Murphy GP, et al. American Cancer Society guideline for the early detection of prostate cancer: update 1997. CA Cancer J Clin 1997;47:261–264

10. Johansson JE, Holmberg L, Johansson S, Bergstrom R, Adami HO. Fifteen-year survival in prostate cancer: a prospective, population-based study in Sweden. JAMA 1997;277:467–471. Erratum in JAMA 1997;278:206

11. Thompson IM, Goodman PJ, Tangen CM. The influence of finasteride on the development of prostate cancer. N Engl J Med 2003;349:215–224

12. Arcangeli CG, Ornstein DK, Keetch DW, Andriole GL. Prostate-specific antigen as a screening test for prostate cancer. The United States experience. Urol Clin North Am 1997; 24:299–306

13. Fang J, Metter EJ, Landis P, Chan DW, Morrell CH, Carter HB. Low levels of prostate-specific antigen predict long-term risk of prostate cancer: results from the Baltimore Longitudinal Study of Aging. Urology 2001;58:411–416

14. Riffenburgh RH, Amling CL. Use of early PSA velocity to predict eventual abnormal PSA values in men at risk for prostate cancer. Prostate Cancer Prostatic Dis 2003;6:39–44

15. Partin AW, Carter HB. The use of prostate-specific antigen and free/total prostate-specific antigen in the diagnosis of localized prostate cancer. Urol Clin North Am 1996;23:531–540

16. Naitoh J, Zeiner RL, Dekernion JB. Diagnosis and treatment of prostate cancer. Am Fam Physician 1998;57:1531–1539

17. Smith DS, Catalona WJ, Herschman JD. Longitudinal screening for prostate cancer with prostate-specific antigen. JAMA 1996;276:1309–1315

18. Wahid ZU. Task force finds evidence lacking on whether routine screening for prostate cancer improves health outcomes. J Natl Med Assoc 2003;95:A20

19. Lubke WL, Optenberg SA, Thompson IM. Analysis of the first-year cost of a prostate cancer screening and treatment program in the United States. J Natl Cancer Inst 1994;86: 1790–1792

20. Sanchez-Chapado M, Olmedilla G, Cabeza M, Donat E, Ruiz A. Prevalence of prostate cancer and prostatic intraepithelial neoplasia in Caucasian Mediterranean males: an autopsy study. Prostate 2003;54:238–247

21. American Cancer Society Guidelines for the Early Detection of Cancer, 2002

22. Carroll P, Coley C, Mcleod D, et al. Early detection and diagnosis of prostate cancer. Urology 2001;57:217–224

23. Harris R, Lohr KN. Screening for prostate cancer: an update of the evidence for the U.S. Preventative Services Task Force. Ann Intern Med 2002;137:917–929

24. Ferrini RL, Woolf SH. American College of Preventive Medicine practice policy: screening for prostate cancer in American men. Am J Prev Med 1998;15:81–84

25. Gohagan JK, Prorok PC, Kramer BS, et al. Prostate Cancer screening in the prostate, lung, colorectal and ovarian cancer screening trial of the National Cancer Institute. J Urol 1994; 152:1903–1904

26. Standaert B, Louis Denis L. The European randomized study of screening for prostate cancer: an update. Cancer 1997;80: 1830–1834

27. Basler JW, Thompsen IM. Lest we abandon digital rectal examination as a screening test for prostate cancer. J Natl Cancer Inst 1998;90:1761–1763

28. Bretton PR. Prostate-specific antigen and digital rectal examination in screening for prostate cancer: a community-based study. South Med J 1994;87:720–723

29. Catalona WJ, Smith DS, Ratliff TL, et al. Measurement of prostate-specific antigen in serum as a screening test for prostate cancer. N Engl J Med 1991;324:1156–1161

30. Catalona WJ, Partin AW, Slawin KM, et al. Southwick use of the percentage of free prostate-specific antigen to enhance differentiation of prostate cancer from benign prostatic disease: a prospective multicenter clinical trial. JAMA 1998;279: 1542–1547

31. Ellis WJ, Chetner MP, Preston SD, Brawer MK. Diagnosis of prostate carcinoma: the yield of serum prostate specific antigen, digital rectal examination, and transrectal ultrasonography. J Urol 1994;152:1520–1525

32. Mikolajczyk SD, Marker KM, Millar LS, et al. A truncated precursor form of prostate-specific antigen is a more specific serum marker of prostate cancer. Cancer Res 2001;61: 6958–6963

33. Shariat SF, Andrews B, Kattan MW, Kim J, Wheeler TM, Slawin KM. Plasma levels of interleukin-6 and its soluble receptor are associated with prostate cancer progression and metastasis. Urology 2001;58:1008–1015

34. Nam RK, Diamandis EP, Toi A, et al. Serum human glandular kallikrein-2 protease levels predict the presence of prostate cancer among men with elevated prostate-specific antigen. J Clin Oncol 2000;18:1036–1042

35. Partin AW, Kattan MW, Subong EN, et al. Combination of prostate-specific antigen, clinical stage, and Gleason score to predict pathological stage of localized prostate cancer: a multi-institutional update. JAMA 1997;277:1445–1451

36. Kattan MW, Eastham JA, Stapleton AM, Wheeler TM, Scardino PT. A preoperative nomogram for disease recurrence following radical prostatectomy for prostate cancer J Natl Cancer Inst 1998;90:766–771

37. Lattouf JB, Saad F. Gleason score on biopsy: is it reliable for predicting the final grade on pathology? BJU Int 2002;90:694–698

38. Albertsen PC, Hanley JA, Gleason DF, Barry MJ. Competing risk analysis of men aged 55 to 74 years at diagnosis managed conservatively for clinically localized prostate cancer. JAMA 1998;280:975–980

39. Holmberg L, Bill-Axelson A, Helgesen F, et al. A randomized trial comparing radical prostatectomy with watchful waiting in early prostate cancer. N Engl J Med 2002;347:781–789

40. Catalona WJ, Carvalhal GF, Mager DE, Smith DS. Potency, continence, and complication rates in 1,870 consecutive radical retropubic prostatectomies. J Urol 1999;162: 433–438

41. Han M, Partin AW, Pound CR, Epstein JI, Walsh PC. Long-term biochemical disease-free and cancer-specific survival following anatomic radical retropubic prostatectomy: the 15 year Johns Hopkins experience. Urol Clin North Am 2001;28: 555–565

42. Reiner WG, Walsh PC. An anatomical approach to the surgical management of the dorsal veins and Santorini's plexus during radical retropubic surgery. J Urol 1979;121:198–200

43. Walsh PC, Donker PJ. Impotence following radical prostatectomy: insight into etiology and prevention. J Urol 1982;128: 492–497

44. Stamey TA. Techniques for avoiding positive surgical margins during radical prostatectomy. Atlas of the Urol Clin North Am 1994; 2:37–51

45. Walsh PC. *Anatomic Radical Retropubic Prostatectomy; Campbell Urology.* 7th ed. Philadelphia: WB Saunders; 1998

46. Davis M, Sofer M, Gomez-Marin O, Bruck D, Soloway MS. The use of cell salvage during radical retropubic prostatectomy: does it influence cancer recurrence? BJU Int 2003;91: 474–476

47. Shekarriz B, Upadhyay J, Wood DP. Intraoperative, perioperative, and long-term complications of radical prostatectomy. Urol Clin North Am 2001;28:639–653

48. Hedican SP, Walsh PC. Postoperative bleeding following radical prostatectomy. J Urol 1994;152:1181–1183

49. Steiner MS, Morton RA, Walsh PC. Impact of anatomical radical prostatectomy on urinary continence. J Urol 1991;145: 512–515

50. Murphy GP, Mettlin C, Menck H, et al. National patterns of prostate cancer treatment by radical prostatectomy: results of a survey by the American College of Surgeons Committee on Cancer. J Urol 1994;152:1817–1819

51. Quinlan DM, Epstein JI, Carter BS, Walsh PC. Sexual function following radical prostatectomy: influence of preservation of neurovascular bundle. J Urol 1991;145:998–1002

52. Geary ES, Dendinger TE, Freiha FS, Stamey TA. Nerve sparing radical prostatectomy: a different view. J Urol 1995;154: 158–159

53. Kim HL, Mhoon DA, Brendler CB. Does the CaverMap device help preserve potency? Curr Urol Rep 2001;2:214–217

54. Kim ED, Scardino PT, Hample O, Mills NL, Wheeler TM, Nah RK. Interposition of sural nerve restore function of cavernous nerves resected during radical prostatectomy. J Urol 1999;161: 188–192

55. Meuleman EJH, Mulders PFA. Erectile function after radical prostatectomy: a review. Eur Urol 2003;43:95–102

56. Zippe CD, Kedia AW, Kedia K, Nelson DR, Agarwal A. Treatment of erectile dysfunction after radical prostatectomy with sildenafil citrate (Viagra). Urology 1998;52:963–966

57. Linet OI, Ogrinc FG. Efficacy and safety of intracavernosal alprostadil in men with erectile dysfunction: the Alprostadil Study Group. N Engl J Med 1996;334:873–877

58. Padma-Aathan H, et al. Treatment of men with erectile dysfunction with transurethral alprostadil. N Engl J Med 1997; 336:1–7

59. Fowler FJ Jr, Collins M, Albertsen PC, Zietman A, Elliott DB, Barry MJ. Comparison of recommendations by urologists and radiation oncologists for treatment of clinically localized prostate cancer. JAMA 2000;283:3217–3222

60. Poon M, Ruckle H, Bamshad BR, Tsai C, Webster R, Lui P. Radical retropubic prostatectomy bladder neck preservation versus reconstruction. J Urol 2000;163:194–198

61. Roach M III, Lu J, Pilepich MV, et al. Long-term survival after radiotherapy alone: Radiation Therapy Oncology Group Prostate Cancer Trials. J Urol 1999;161:864–868

62. Zietman AL, Coen JJ, Dallow KC. Shipley WU. The treatment of prostate cancer by conventional radiation therapy: an analysis of long-term outcome. Int J Radiat Oncol Biol Phys 1995;32:287–292

63. Zelefsky MJ, Leibel SA, Gaudin PB, et al. Dose escalation with three-dimensional conformal radiation therapy affects the outcome in prostate cancer. Int J Radiat Oncol Biol Phys 1998;41:491–500

64. Lukka H, Warde P, Pickles T, Morton G, Brundage M, Souhami L. Controversies in prostate cancer radiotherapy: consensus development. Can J Urol 2001;8:1314–1322

65. Zelefsky MJ, Fuks Z, Hunt M, et al. High-dose intensity modulated radiation therapy for prostate cancer: early toxicity and biochemical outcome in 772 patients. Int J Radiat Oncol Biol Phys 2002;53:1111–1116

66. Lawton CA, Won M, Pilepich MV, et al. Long-term sequelae following external beam irradiation for adenocarcinoma of the prostate: analysis of RTOG studies 7506 and 7706. Int J Radiat Oncol Biol Phys 1991;21:935–939

67. Shipley WU, Zeitman AL, Hanks GE, et al. Treatment related sequelae following external beam radiation for prostate cancer: a review with an update of patients with stages T1 and T2 tumor. J Urol 1994;152:1799–1803

68. Leibel SA, Hanks GE, Kramer S. Patterns of care outcome studies: results of the national practice in adenocarcinoma of the prostate. Int J Radiat Oncol Biol Phys 1984;10:401–409

69. Smith WG, Helle PA, Putten WL, et al. Late radiation damage in prostate cancer patients treated by high dose external radiotherapy in relation to rectal dose. Int J Radiat Oncol Biol Phys 1990;18:23–29

70. Asbell SO, Krall JM, Pilepich MV, et al. Elective pelvic irradiation in stage A2, B carcinoma of the prostate: analysis of RTOG 77–06. Int J Radiat Oncol Biol Phys 1988;15: 1307–1316

71. Whitmore WF Jr, Hilaris B, Grabstald H. Retropubic implantation of iodine-125 in the treatment of prostate cancer. J Urol 1972;108:918–920

72. Schellhammer PF, L-el Mahdi AE, Ladaga LE, Schutheiss T. Iodine implantation for carcinoma of the prostate: 5-year survival free of disease and incidence of local failure. J Urol 1985;134:1140–1145

73. Zelefsky MJ, Whitmore WF Jr. Long-term results of retropubic permanent 125-iodine implantation of the prostate for clinically localized prostatic cancer. J Urol 1997;158: 23–29

74. Holm HH, Juul N, Pedersen JF, Hansen H, Stroyer I. Transperineal iodine seed implantation in prostatic cancer guided by transrectal ultrasonography. J Urol 1983;130:283–286

75. Blasko JC, Ragde H, Grimm PD. Transperineal ultrasound guided implantation of the prostate: morbidity and complications. Scand J Urol Nephrol Suppl 1991;137:113–118

76. Ellis WJ. Prostate brachytherapy. Cancer Metastasis Rev 2002; 21:125–129

77. Blasko JC, Mate T, Sylvester JE, Grimm PD, Cavanagh W. Brachytherapy for carcinoma of the prostate: techniques, patient selection, and clinical outcomes. Semin Radiat Oncol 2002;12:81–94

78. Nag S, Beyer D, Friedland J, Grimm P, Nath R. American brachytherapy society (ABS) recommendations for transperineal permanent brachytherapy of prostate cancer. Int J Radiat Oncol Biol Phys 1999;44:789–799

79. Benoit RM, Cohen JK, Miller RJ Jr, Merlotti L. *Role of Cryosurgery in the Treatment of Prostate Cancer: Advanced Therapy of Prostate Disease.* London: BC Decker Hamilton; 2000

80. Ragde H, Korb LJ, Elgamal A, Grado GL, Nadir BS. Modern prostate brachytherapy prostate specific antigen results in 219 patients with up to 12 years of observed follow-up. Cancer 2000;89:135–141

81. Crook J, Lukka H, Klotz L, et al. Systematic overview of the evidence for brachytherapy in clinically localized prostate cancer. CMAJ 2001;164:975–981

82. Long JP, Bahn D, Lee F, Shinnohara K, Chinn DO, Macaluso JN Jr. Five-year retrospective, multi-institutional pooled analysis of cancer-related outcomes after cryosurgical ablation of the prostate. Urology 2001;57:518–523

83. Long JP. New methods of focal ablation of the prostate. In: Kantoff PW, Carroll PR, D'Amico AV. *Prostate Cancer, Principles and Practice.* Philadelphia: Lippincott Williams & Wilkins; 2002;358–367

84. Uchida T, Sanchvi NT, Gardner TA, et al. Transrectal high-intensity focused ultrasound for treatment of patients with stage T1b-2N0M0 localized prostate cancer: a preliminary report. Urology 2002;59:394–399

85. Huggins C, Hodges CV. Studies on prostate cancer, I: the effect of castration, of estrogen, and of androgen injection on serum phosphatase in metastatic carcinoma of the prostate. Cancer Res 1941;1:293

86. Gleave ME, La Bianca S, Goldenberg SL. Review of neoadjuvant hormonal therapy prior to radical prostatectomy: promises and pitfalls. Prostate Cancer Prostatic Dis 2000;3:136–144

87. Pilepich MV, Krall JM, Al-Sarraf M, et al. Androgen deprivation with radiation therapy compared with radiation therapy alone for locally advanced prostatic carcinoma: a randomized comparative trial of the Radiation Therapy Oncology Group. Urology 1995;45:616–623

88. Messing EM, Manola J, Sarosdy M, Wilding G, Crawford ED, Trump D. Immediate hormonal therapy compared with observation after radical prostatectomy and pelvic lymphadenectomy in men with node-positive prostate cancer. N Engl J Med 1999;341:1781–1788

89. Bolla M, Gonzalez D, Warde P, et al. Improved survival in patients with locally advanced prostate cancer treated with radiotherapy and goserelin. N Engl J Med 1997;337:295–300

90. Schulman CC, Altwein JE, Zlotta AR. Treatment options after failure of local curative treatments in prostate cancer: a controversial issue. BJU Int 2000;86:1014–1022

91. Pound CR, Partin AW, Eisenberger MA, Chan DW, Pearson JD, Walsh PC. Natural history of progression after PSA elevation following radical prostatectomy. JAMA 1999;281:1591–1597

92. Critz FA, Levinson AK, Williams WH, Holladay DA, Holladay CT. The PSA nadir that indicates potential cure after radiotherapy for prostate cancer. Urology 1997;49:322–326

93. Pisters LL, von Eschenbach AC, Scott SM, et al. The efficacy and complications of salvage cryotherapy of the prostate. J Urol 1997;157:921–925

94. Chon JK, Jacobs SC, Naslund MJ. The cost value of medical versus surgical hormonal therapy for metastatic prostate cancer. J Urol 2000;164:735–737

95. Cox LE, Crawford ED. Estrogens in the treatment of prostate cancer. J Urol 1995;154:1991–1998

96. Robertson CN, Roberson KM, Padilla GM, et al. Induction of apoptosis by diethylstilbestrol in hormone-insensitive prostate cancer cells. J Natl Cancer Inst 1996;88:908–917

97. Iversen P, Tyrrell CJ, Kalsary AV, et al. Bicalutamide monotherapy compared with castration in patients with nonmetastatic locally advanced prostate cancer: 6.3 years of follow-up. J Urol 2000;164:1579–1582

98. McLeod D, Zinner N, Tomera K, et al. A phase 3, multicenter, open-label, randomized study of abarelix versus leuprolide acetate in men with prostate cancer. Urology 2001;58:756–761

99. Hellerstedt BA, Pienta KJ. The current state of hormonal therapy for prostate cancer. CA Cancer J Clin 2002;52:154–179

100. Labrie F, Dupont A, Belanger A. Complete androgen blockade for the treatment of prostate cancer. In: DeVica VP Jr, Hellman S, Rosenberg SA, eds. *Important Advances in Oncology.* Philadelphia: JB Lippincott; 1985:193–271

101. Carroll PR, Fair WR, Grossfeld GD, et al. Overview consensus statement. Urology 2001;58:1–4

102. Byar DP. Proceedings: the Veterans Administration Cooperative Urological Research Group's studies of cancer of the prostate. Cancer 1973;32:1126–1130

103. The Medical Council Prostate Cancer Working Party Investigators Group. Immediate versus deferred treatment for advanced prostatic cancer: initial results of the Medical Research Council Trial. Br J Urol 1997;79:235–246

104. Akakkura K, Bruchovsky N, Goldenberg SL, et al. Effects of intermittent androgen suppression on androgen-dependent tumors. Cancer 1993;71:2782–2789

105. Goldenberg SL, Bruchovsky N, Gleave ME, et al. Intermittent androgen suppression in the treatment of prostate cancer: a preliminary report. Urology 1995;45:839–845

106. Stone P, Hardy J, Huddart R, et al. Fatigue in patents with prostate cancer receiving hormone therapy. Eur J Cancer 2000;36:1134–1141

107. Tayek JA, Heber D, Byerley LO, et al. Nutritional and metabolic effects of gonadotropin-releasing hormone agonist treatment for prostate cancer. Metabolism 1990;39:1314–1319

108. Diamond T, Campbell J, Bryant C, et al. The effect of combined androgen blockage on bone turnover and bone mineral densities in men treated for prostate carcinoma: longitudinal evaluation and response to intermittent cyclic etidronate therapy. Cancer 1998;83:1561–1566

109. Atala A, Amin M, Garty JI. Diethylstilbestrol in treatment of post-orchiectomy vasomotor symptoms and its relationship with serum follicle-stimulating hormone, luteinizing hormone, and testosterone. Urology 1992;39:108–110

110. Scher HI, Steineck G, Kelly WK. Hormone-refractory prostate cancer: refining the concept. Urology 1995;46:142–148

111. Mahler C, Denis CJ. Hormone-refractory disease. Semin Surg Oncol 1995;11:77–83

112. Tannock IE, Osolba D, Stockler MR, et al. Chemotherapy with mitoxantrone plus prednisone or prednisone alone for symptomatic hormone-resistant prostate cancer: a Canadian randomized trial with palliative end points. J Clin Oncol 1996;14:1756–1764

113. Period Life Table. Updated February 2003, Social Security Administration (*www.ssa.gov/*)

114. Steinberg DM, Sauvageot J, Piantadosi S, et al. Correlation of prostate needle biopsy and radical prostatectomy Gleason grade in academic and community settings. Am J Surg Pathol 1997;21:566–576

115. Gleave ME, Coupland D, Drachenberg D, Cohen L, Kwong S, Goldenberg SL. Ability of serum prostate-specific antigen levels to predict normal bone scans in patients with newly diagnosed prostate cancer. Urology 1996;47:708–712

116. D'Amico AV, Whittington R, Malkowicz SB, et al. Biochemical outcome after prostatectomy, external beam radiation therapy, or interstitial radiation therapy for clinically localized prostate cancer. JAMA 1998;280:969–974

117. Surya BV, Provet J, Johanson KE, et al. Anastomotic strictures following radical prostatectomy: risk factors and management. J Urol 1990;143:755–758

18

Growth Hormone Replacement in Aging Men

SENG-HIN TEOH AND ROBERT S. TAN

Case History

Mr. J.M. is a 50-year-old white man who was referred by a colleague because of his interest in optimal aging and antiaging issues. The patient has been around the world seeking management of his health issue. In reality, he is healthy and is just keen on preserving his functionality and maintaining his health. He realizes that there is no real "fountain of youth," but he expressed interest in growth hormone (GH) replacement. In fact, he is seeking a second opinion, as his previous physician had already started him on growth hormone after testing him and determining he had low insulin-like growth factor-I (IGF-I) levels. He is currently on 16 IU of growth hormone injections per week. On direct questioning, he admitted to feeling tired and slow prior to the therapy. He also said that the therapy has helped maintain his body fat at ~15%, which was previously in the 20% range. He exercises daily with weights for ~20 minutes and does aerobic exercises for another 20 minutes. He does not smoke, but he drinks socially. He loves life and travels the world for ballooning adventures. He is also a successful businessman and is semiretired. His only significant past history is gastroesophageal reflux, and he finds relief with Nexium. Examination reveals a healthy person, weighing 170 pounds and with a height of 5 feet 10 inches. Skin turgor was normal and within the range for his age. Body fat by caliper was 13%. Right hand grip strength was 110 pounds, with a grip differential of 10 pounds. Short-term visual memory was above average for his age. Neurological and cardiovascular examinations were within normal limits. Blood pressure was normal and he did not have any evidence of carpal tunnel syndrome. There was also no swelling noted.

Introduction

Many of the physical changes associated with aging mimic those seen in the adult GH-deficiency syndrome, such as muscle atrophy, osteopenia, obesity, cardiovascular deterioration, exercise intolerance, decreased metabolic rate, dyslipidemia, thinning of the skin, and low quality of life in terms of energy and social life.[1–4] The successful treatment of adult GH deficiency with recombinant human GH has raised the possibility that GH supplementation could reverse or slow down some of the deleterious features of aging. This chapter discusses GH, specifically its biosynthesis, secretion, effects, relation to aging, use, the secretagogues, and the future uses of this hormone.

Structure and Biosynthesis of Growth Hormone

Human GH is a 191-amino-acid single-chain polypeptide with a molecular weight of 22 kd. GH is made as a prehormone by somatotrophs, which are acidophilic cells in the pars tuberalis of the anterior lobe of the pituitary gland.[5] GH circulates in the plasma bound to a specific GH binding protein (GHBP)—GHBPI.[6] The circulating half-life of GH is ~20 to 25 minutes.[7]

Regulation of Synthesis and Secretion of Growth Hormone

Growth hormone secretion is controlled by two factors secreted in the hypothalamus. Hypothalamic growth hormone–releasing hormone (GHRH) is the major stimulator of GH synthesis and secretion. GHRH is a 44-amino-acid peptide. The major inhibitor of GH synthesis and secretion is somatostatin (SS), another hypothalamic 14-amino-acid peptide. Both are secreted

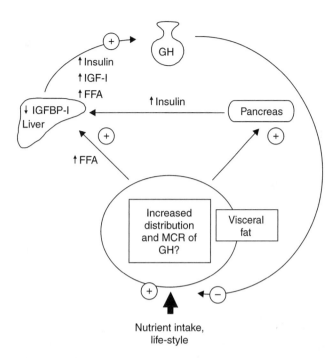

FIGURE 18–1. Summary of regulation and secretion of growth hormone (GH). GH secretion has both direct and indirect effects. The direct effects are the result of GH binding its receptor on target cells. Adipocytes have GH receptors and GH stimulate them to break down triglyceride to free fatty acids. The indirect effects are mediated by insulin-like growth factor-I (IGF-I), which is secreted by the liver in response to GH. The majority of growth promoting effects of GH is due to IGF-I, a chip on target cells.

into the capillaries of the median eminence and then pass via the portal veins to the pituitary gland.[5] The interplay of stimulation and inhibition is further modulated by a local negative feedback by GH itself on the pituitary. Fig. **18–1** summarizes the regulation and secretion of growth hormone.

The secretion of GH is characteristically pulsatile. In humans there are at least five to seven pulses lasting 1 to 2 hours.[8] About two thirds of daily GH secretion occurs at night, with pulses occurring especially during slow-wave sleep. This pulsatility is in turn modulated by pulsatile GHRH stimulation.[9] There are other factors that can stimulate GH synthesis and secretion, and these include hypoglycemia; α-adrenergic, dopaminergic, and serotonergic agents; and amino acids such as arginine.[7] Besides SS, hyperglycemia, free fatty acids, and β-adrenergic agonists can inhibit synthesis and secretion of GH.[7]

Physical exercise is a physiological stimulus to GH secretion. Repeated bouts of aerobic exercise in a 24-hour period has been shown to result in increased 24-hour integrated GH concentrations. However, the effect of chronic resistance training on 24-hour GH release is still

not well established.[10] Exercise not only stimulates GH secretion but also causes a direct increase in circulating IGF-I independent of GH.[11,12]

GH secretion is also under the control of a stimulatory pathway that may be activated by some synthetic compounds, the GH secretagogues (GHSs). Examples of these secretagogues are the GH-releasing peptides MK-0677 and hexarelin. These secretagogues bind to the GHS receptor (GHS-R), a family of G-protein–coupled receptors found in the hypothalamus and the pituitary. They act via the phosphatidyl inositol triphosphate pathway to stimulate and amplify GH release.[13]

Much interest has surrounded a 28-amino-acid peptide, Ghrelin, which is a natural ligand for the GHS-R first identified by Kojima and colleagues.[14] Rat studies have revealed that it is produced mainly in the stomach but also in the duodenum, ileum, cecum, aorta, thyroid, lung, and testes. Ghrelin and GHS exert many similar effects by stimulating and augmenting the release of GH, and they also stimulate the release of prolactin, adrenocorticotropic hormone (ACTH), and cortisol. Ghrelin is involved in energy homeostasis. In humans, Ghrelin was shown to stimulate food intake and also reduces cardiac afterload and increases cardiac output.[13] Figure **18–2** summarizes the regulation of Ghrelin.

The Actions of Growth Hormone and Insulin-Like Growth Factors

Besides being essential for normal growth in childhood, GH has many direct and indirect effects on metabolism, body composition, and immunity in adults. The direct metabolic effects of GH include increased protein synthesis, increased mobilization of fatty acids from adipose tissue, and decreased rate of glucose utilization throughout the body, resulting in increased blood sugar.[5,15] The indirect effects are through the stimulation of the production of insulin-like growth factors IGF-I and IGF-II. Of the two, IGF-I is the more important.

GH stimulates IGF-I production in the liver[16] and in several other tissues.[17] The major source of circulating IGF-I comes from the liver. In other tissues, IGF-I acts on other cells in a paracrine manner.[18] Circulating IGF-I levels depend mainly on GH. IGFs, including IGF-I, are bound to circulating proteins, six in all, known as IGF binding proteins (IGFBPs). IGFBP-3 is the most abundant and accounts for 95% of the total circulating IGFBPs in adults. It binds 80 to 90% of IGF-I and increases the half-life of IGF-I by 100-fold.[19] Compared with the half-life of GH, that of IGF-I is much longer, 12 to 15 hours.[17]

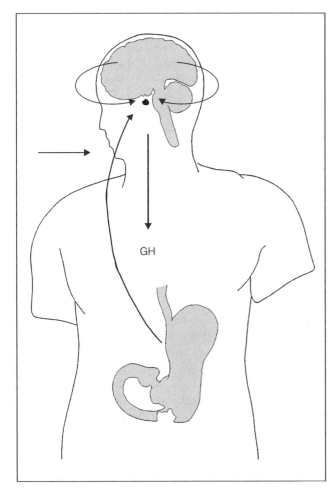

FIGURE 18–2. Regulation of Ghrelin, a natural ligand for GHS-R that exerts effects similar to GHS by stimulating and augmentating the release of GH via "the pituitary-stomach axis." GH is produced and secreted from the anterior pituitary gland and exerts its effects by binding to GH receptor. Ghrelin on the other hand, is found in the stomach and has been discovered to regulate GH, suggesting the presence of a "pituitary-stomach" axis. (Adapted from Kojima M, Hosoda H, Matsuo H, et al. Ghrelin: discovery of the natural endogenous ligand for the growth hormone secretagogue receptor. Trends Endocrinol Metab 2001;12:118–126, with permission.)

Growth Hormone and Insulin-Like Growth Factor-I in Aging

GH secretion is reduced in older adults, arising from a gradual decline of GH with aging.[20] The rate of decline is ~14% per decade of adult life.[21] The decline probably starts in the third decade and approaches a plateau from about age 60 years.[22,23] It must also be noted that GH levels in age-related decline of GH are higher than in adult GH deficiency secondary to disease.[24,25] Levels of IGF-I also fall with age.[26] The age-related reduction of IGF-I is also observed as early as

TABLE 18–1. Symptoms and Signs of Growth Hormone Deficiency

Symptoms	Signs and Physiological Effects
Psychological problems	↓ Lean body mass
Poor general health	↓ Extracellular volume
Impaired self-control	↓ Bone mineral density
Lack of positive well-being	Increased body fat
Mood changes, including depression, anxiety	Increased waist-hip ratio
Reduced vitality	↓ HDL, ↑ LDL
Sexual dysfunction	↓ GFR ↑ Renal plasma flow
Tiredness	↓ BMR
Emotional problems	↓ Muscle bulk and strength
	↓ Exercise performance
	↓ Anaerobic threshold

BMR, basal metabolism rate; GFR, glomerular filtration rate; HDL, high-density lipoprotein; LDL, low-density lipoprotein.

the third decade.[27] The decline in older age takes the IGF-I levels to ~50% of the levels in healthy young adults.[15] Table **18–1** summarizes the symptoms and signs of growth hormone deficiency.

The age-related decline of GH is due to a decrease in the stimulation of pituitary GH secretion by the hypothalamic GHRH and the increased inhibition of pituitary GH secretion by SS.[25] The pulse amplitude in particular declines with age.[4,22] The pituitary somatotrophs appear to retain adequate or even unchanged secretory capacity in older adults.

It has been questioned whether the increase in body fat and decrease in lean body mass is a consequence of age-dependent decline of GH or is the accumulation of fat mass secondary to factors such as diet and lifestyle, which has led to a feedback inhibition of GH release.[28,29]

The Effect of Growth Hormone Administration in Age-Related Decline of Growth Hormone

There is no controversy about the indications for the use of GH for childhood growth failure, adult growth hormone deficiency (e.g., secondary to pituitary irradiation), and AIDS-associated wasting.[30] But GH replacement in the elderly for age-related decline still remains controversial. Rudman et al's[31] study in 1990 on the effect of GH replacement in the elderly is now considered a landmark. They recruited 21 healthy men between the ages of 61 to 81 years of age. These men had IGF-I levels below 350 U/L during a 6-month baseline period. They received 0.03 mg of biosynthetic human growth hormone per kilogram weight subcutaneously three times a week. A control group of nine men received no treatment. At the end of 6 months, the mean plasma IGF-I level rose to 500 to 1500 U/L in the treatment group. The levels remained at less than 350 U/L in the control group. The members of the treatment group had increased lean body mass and skin thickness, decreased adipose tissue mass, and an increased average lumbar vertebral bone density.

Following this there were other studies.[32,33] All have shown that treatment with GH increased lean body mass, decreased adipose tissue mass, and increases skin thickness. Groups elsewhere have found evidence of improvement in body composition but no functional improvement. Papadakis et al,[34] who studied the effect of GH or placebo administered to 52 healthy elderly men with low IGF-I levels over a period of 6 months, also reported no improvement in functional status in those who had received GH. Recently a double-blind placebo controlled study by Blackman et al[35] on 27 women and 25 men reported similar findings.

Goh and colleagues[36] studied 23 healthy elderly Chinese men between the ages 60 and 69. They were given subcutaneous injections of 0.08 U/kg recombinant GH for 6 months. Three months after starting the study, it was found that these men showed a significant increase in lean body mass and a decrease in body fat. He also found that there were GH-induced increases in thyroid function and serum dehydroepiandrosterone sulfate (DHEAS) levels. The effects of thyroid hormone on metabolism are well known. It has been suggested that DHEAS may have an antiobesity effect, and may enhance memory by improving rapid eye movement (REM) sleep, increase muscle mass, activate immune function, and enhance quality of life in aging men.

The effect of GH on bone mass is less clear. In a review, Geusens and Boonen[37] found that most studies have shown that administration of GH in healthy elderly individuals resulted in stimulation of bone formation and resorption. In a placebo-controlled study, 12 months of GH administration had a beneficial effect on bone density, preventing the age-related reduction of bone mass seen in the placebo group but not increasing bone density in absolute terms.[38] Studies have shown that GH administration did not affect bone density but did prevent the reduction of bone mass.[39,40]

Problems in the Study of Age-Related Decline of Growth Hormone

Although observations on metabolic changes were repeatable in all studies, study design of the effect of GH therapy for age-related decline in otherwise healthy individuals remains a problem. More robust evidence would be obtained if large, randomized, double-blind studies were done. However, one can foresee problems in the conduct of such studies. To begin with, what should be the entry criteria to such a study? Is it going to be based on clinical parameters or on measurements of GH and/or IGF-I? Also, it is known that asymptomatic patients may have low levels of IGF-I, whereas symptomatic patients may not have low levels of IGF-I.[17] So initially there must be studies establishing the relationship between signs of aging on the one hand and the decline of GH and IGF-I on the other. This will not be easy because aging is a complex process involving physiological, psychological, social, and endocrine functions. Another question is how and when shall GH be measured, knowing its secretion is pulsatile, that the majority of its secretion is at night, and that the half-life of GH is short.

Current methods to assess the integrity of GH secretion are mainly the GH stimulation tests, which utilize the factors that trigger the release of GH, namely insulin, arginine, and GHRH. The insulin tolerance test (ITT) induces hypoglycemia and challenges the body's ability to respond by producing GH. However, 10% of normal individuals fail to respond to hypoglycemia, so other tests may be necessary. There are contraindications to such tests, especially in diseases where hypoglycemia is a risk factor, for example, epilepsy. In addition, the test itself is not without risk. Hypoglycemia may be so significant and severe as to cause ketosis acidosis and shock.[41] To further complicate matters, a study on the use of GH in aging should include patients 65 years of age and over (i.e., the elderly) and the American Academy of Clinical Endocrinologists does not recommend the ITT for patients older than 65 years of age.[30]

GH-releasing peptide is also used to stimulate the production of GH by the pituitary. Arginine infusion and GHRH injection have been combined, and the results are apparently comparable with those of the ITT.[7] Arginine alone has also been used to challenge pituitary GH production, and its infusion has been known to raise GH by 70% in healthy individuals.[7]

Considering the problems and complexity associated with challenge tests of GH production, for clinical purposes of management it should be enough to assess serum IGF-I as an indicator of GH levels. Serum levels are fairly stable and do not exhibit a diurnal rhythm; they reflect a composite of daily GH secretion rates. Therefore, a single IGF-I level is a reasonable though indirect measure of the 24-hour GH secretion.[16] However, it has been argued that IGF-I production decreases in the aging liver and that this fact would confound the level measured.[42] Furthermore, IGF-I levels are influenced by many other variables. Malnutrition, severe chronic illness, severe liver disease, and hypothyroidism reduce plasma IGF-I levels.

Studies have taken a predetermined lower percentile of the IGF-I range for young adults as reflective of decline in GH production in the elderly.[31,36]

Bearing in mind the limitations of IGF-I as a biological parameter, thought could be given to establishing a clinical score, much in the same way as is done for osteoporosis, where an assessed low value of serum IGF-I is evaluated as the number of standard deviations below the mean serum IGF-I of healthy young adults. Such a score, although likely to be indicative of GH secretion, will have to be associated with defined clinical end points before it can be widely applicable.

Therefore, for any two studies to be comparable, the methods used to assess the GH–IGF-I axis and the criteria used to define age-related decline must be the same. Another problem in any trial on GH supplementation and aging would be the definition of end points. These shall have to be grouped into physical and psychological end points if studies were to have a holistic impact.

For physical end points, results can be objective; a percentage change in a physical parameter is unequivocal. But how much of a change shall be deemed an appropriate response? Psychological end points are harder to define. Papadakis and colleagues[34] used the Mini–Mental State Examination to evaluate improvement in cognitive function when they looked at a 6-month replacement of GH in 52 healthy elderly men. Although standardized psychological tests are useful, they are to some extent affected by the subject's level of education, attitudes, religious beliefs, and life experiences before and during the period of the test. The psychological and social status of the subject may have changed over the time of the test such as death in the family or onset of morbidity, such as development of cataracts and strokes.

Moving on to the subjects themselves, one must be realistic and accept that, given the demographics of the study population, a high proportion of individuals will be excluded from the study over time. Morbidity and mortality is increased from illnesses to which aging adults are prone, such as strokes, hypertension, diabetes, and cancer. Life situations for the subjects may change, precluding them from suitability during the study, which may affect the assessment of psychological end points. These include the death of a spouse or loss of financial support and change in home situations, for example, moving from home to an institution. In earlier studies, the high incidence of side effects in the use of GH led to a high rate of dropouts.[32,38]

The Approach to the Patient

Given the widespread publicity surrounding GH, the clinician is not uncommonly faced with the dilemma of the approach to an otherwise healthy patient who presents with signs and symptoms attributable to the age-related decline of GH. The dilemma arises for the following reasons and questions:

1. There is a lack of robust evidence or a guide from well-designed clinical trials with meaningful end points.

2. The laboratory diagnosis is problematic and may be controversial.

3. There is a known higher incidence of prostate and breast cancer in subjects with high IGF-I.[8] There are reports suggesting that acromegalic patients have an increased risk of colon cancer, colonic polyps, and malignancies of the lymphoid system.[1,43] Rosen et al,[1] however, pointed out that the production rate of GH in a GH producing adenoma far exceeds the usual recombinant GH (rGH) substitution dose.

4. Is an age-related decline of GH a normal protective mechanism, assuming that administration of GH would result in the survival of aging cells, which have a higher tendency to abnormally proliferate, perhaps leading to cancer?

5. How long should treatment be given?

6. The cost of treatment is high. GH replacement in patients with GH deficiency secondary to disease could cost between $7,000 and $10,000 annually.[44]

7. There are side effects from GH administration, although these are usually dose related. Side effects include edema, arthralgia, myalgia, carpal tunnel syndrome, glucose intolerance, and hypertension.[33,43]

8. The exclusion of its use in patients with coexistent illness such as diabetes mellitus.

9. The possibility of development of tolerance by patients to administered rGH. In Thompson et al's[32] study although nitrogen retention remains significantly higher than baseline values throughout the study, it had declined by the final week of treatment.

A reasonable approach to a patient who is potentially suffering from symptoms and signs associated with age-related decline of GH is one of exclusion of all other causes, for example, hypothyroidism, depression, anemia, and others. The clinician should start with a thorough and complete history including a review of systems. This should then be followed by a meticulous clinical examination. Laboratory tests should include a full blood count, tests of liver and renal functions, serum lipids, cancer markers, and a full endocrine evaluation.

It is not suggested that GH should be offered to the patient as a panacea for obvious concerns addressed earlier. Because signs and symptoms associated with GH decline are not life-threatening, it is only appropriate to consider conservative measures. These are a healthy lifestyle including proper nutrition, appropriate exercise, avoidance of smoking, avoidance of drug and alcohol abuse, and social interactions to maintain good mental health and promote a healthy environment.

The salutary effect of exercise on GH production has been alluded to earlier. Taaffe and colleagues[45] performed a double-blind, placebo-controlled exercise trial involving 18 healthy elderly men who underwent progressive weight training for 14 weeks. Subjects then either received 0.02 mg/kg body weight daily GH subcutaneously or received a placebo while undergoing a further 10 weeks of strength training. They found that although lean body mass increased and fat mass decreased, there was no significant difference in muscle strength observed between the two groups. Thus exercise alone could achieve the increase of lean body mass and decrease of fat mass seen with GH administration. Hurel et al[46] studied GH production in 10 male subjects running over 40 miles per week as compared with 10 healthy age-matched sedentary males. They found that GH was higher in the runners.

Given the cost, complexity, and potential risks of GH use, attention is now turning to GH-releasing peptides, nonpeptidyl secretagogues, and GHRH.[43] As mentioned earlier, it has been shown that the pituitary gland retains its capacity to secrete GH even in the aging population. Thus it is logical to administer agents, ideally by the oral route, to induce release of GH by the pituitary. The benefit is that GH itself has negative feedback on the pituitary, and the risk of excess GH levels in the individual is thus avoided. In other words the release of GH by secretagogues is autoregulated.

There are biosynthetic peptides that could be administered nasally and orally that act on receptors in the hypothalamus and pituitary gland to stimulate the release of GH. These secretagogues have been shown to require the presence of GHRH and are synergistic with GHRH in promoting GH secretion. However, this has not been extensively studied.

Smith et al[47] have described the secretagogue MK-677, a spiropiperidine that can be administered orally. It increases pulsatile GH release and serum IGF-I levels when administered by the oral route. This product fulfills the expectation of an agent mimicking as close as possible the in vivo effect of endogenous agents. Chapman et al[48] investigated the use of this secretagogue in a randomized double-blind placebo-controlled trial involving 32 healthy elderly men and women. They found that once-daily treatment of oral MK-677 for 4 weeks enhanced pulsatile GH release, significantly increased serum GH and IGF-I concentrations, and, at a high dose of 25 mg daily, increased IGF-I levels to that of a young adult. The use of MK-677 alone or in combination with alendronate in the management of osteoporosis was studied in a small sample of 18 postmenopausal women with osteoporosis. The authors found that MK-677 enhanced the effect of alendronate on bone mineral density in the femoral neck but not lumbar spine and total hip.[49] Side effects included fluid retention, increased serum transaminase, and serum glucose in some of the patients.[43]

GHRH has been studied in elderly men and women. IGF-I levels increased without any significant side effects. The metabolic effects were similar to that following GH administration. However, the drawback of this modality is again cost and the need for injections.

Ghrelin is known to be an even more potent stimulator of GH release than GHRH.[1] Clinically this offers another therapeutic possibility in managing the aging decline of GH. Ghrelin may be administered as are the other GHSs to enhance the GH–IGF-I axis in healthy aging patients.[13,14]

What of the Future?

In this era of molecular medicine the prospects of research are exciting. In addition to the 191-amino-acid native peptide, the pituitary secretes other forms of GH. It would be ideal for an analog of GH that stimulates growth but lacks other actions of GH to be developed. It is tempting to speculate that the molecule of GH can be manipulated biochemically, and the altered GH molecule can be used to treat specific age-related problems provided that relevant end points have been defined.

Discussion of the Case History

Mr. J.M. is a somewhat typical person seeking growth hormone replacement. By and large, these individuals are healthy and have the resources, as the treatment is expensive, especially in the United States. It is not covered by insurance or Medicare. These patients tend to be well informed and rely much on the Internet for the latest developments in this area. Although the patient tested low on his IGF-I level, there was no dynamic testing done by the prior physician. The patient was advised that the exercise that he was doing was probably the best stimulator of growth hormone, and that it was wise to taper off the injectable growth hormone. On repeat testing, it was found that his IGF-I level was 402 ng/mL (normal: 90–360) and was in the supraphysiological level range. He was advised about the side effects of growth hormone therapy and also that the long-term scientific literature at this point was scant. However, he was adamant on continuing the therapy, as he has noticed the loss of his body fat and improved quality of life. Informed consent was discussed at length. Counseling was provided and the patient was also advised about other options including plastic surgery. A compromise was reached, and the dose of growth hormone was reduced to 8 IU per week. The patient requested a discussion on GH-releasing peptides, nonpeptidyl secretagogues, and GHRH, and he continues to be followed up.

Conclusion and Key Points

The worldwide population is aging rapidly. Between 2000 and 2050, the population of people over 65 years of age is expected to double from 6.9 to 16.4%.[50] There exists now an altered population structure with a longer life expectancy unmatched by health expectancy. This phenomenon has been largely brought about by increased life expectancy and lower fertility rates. For the individual, to live longer without quality of life is of little benefit, as advancing age brings with it increasing frailty and increasing susceptibility to illnesses. This has led to a search for agents and solutions to mitigate the continuous, universal, progressive, intrinsic, and deleterious process, which is aging. Growth hormone is one agent that has generated much excitement, passion, and controversy.

There is no doubt that GH is very important in body composition and metabolism. From a conservative viewpoint, low levels of GH associated with aging can be taken as a natural phenomenon and accepted as it is. On the other hand, if one takes health as a state of complete physical, social, and mental well-being, then a case can be made for restoring what has declined, to improve the quality of life of the aging individual. "Primum non nocere": as physicians we must first do no harm. We must have no delusions that GH is the panacea for the aging individual. A circumspect and holistic approach is required. Lunenfeld's[51] advice to "prevent the preventable and delay the inevitable" does not start in the autumn of one's life but in the spring. Although we cannot change our genetic destiny, we can at least be wise in the way we live, thereby reducing the negative impact of lifestyle-related diseases.

- It is established that there is a decline of GH with aging.

- The apparent benefits associated with GH therapy in healthy aging individuals seen in existing studies cannot yet be used to justify large-scale interventions among a healthy aging population.

- Large randomized, placebo-controlled studies with defined relevant end points will increase existing knowledge.

- GH secretagogues hold promise in the management of patients with age-related decline of GH, but more work has yet to be done.

- A thorough discussion of the pros and cons of GH replacement is essential, and informed consent is compulsory.

REFERENCES

1. Rosen T, Johannsson G, Johansson J, Bengtsson B. Consequences of growth hormone deficiency in adults and the benefits and risks of recombinant growth hormone treatment. Horm Res 1995;43:93–99

2. Savine R, Sönksen P. Growth hormone—hormone replacement for the somatopause? Horm Res 2000;53:37–41

3. Rudman D, Feller AG, Nagraj HS, et al. Effects of human growth hormone in men over 60 years old. N Engl J Med 1990;323:1–6

4. Khan AS, Sane DC, Wannenburg T, Sonntag WE. Growth hormone, insulin-like growth factor-1 and the aging cardiovascular system. Cardiovasc Res 2002;54:25–35

5. Guyton AC, Hall JE. *Textbook of Medical Physiology.* Philadelphia: WB Saunders; 2000

6. Baumann G, Stalor MW, Amburn K, et al. A specific growth hormone binding protein in human plasma: initial characterization. J Clin Endocrinol Metab 1986;62:134–141

7. Greenspan FS, Gardner DG. *Basic and Clinical Endocrinology.* New York: Lange Medical Books/McGraw-Hill; 2001

8. Ho KY, Evans WS, Blizzard RM, et al. Effects of sex and age on the 24-hour profile of growth hormone secretion in man: importance of endogenous estradiol concentrations. J Clin Endocrinol Metab 1987;64:51–58

9. Holl RW, Hartman ML, Veldhuis JD, et al. Thirty-second sampling of plasma growth hormone (GH) in man: correlation with sleep stages. J Clin Endocrinol Metab 1991;72:854–861

10. Wideman L, Weltman JY, Hartman ML, et al. Growth hormone release during acute and chronic aerobic and resistance exercise: recent findings. Sports Med 2002;32:987–1004

11. Ehrnborg C, Lange KHW, Dall R, et al. The growth hormone/insulin-like growth factor-1 axis hormones and bone markers in elite athletes in response to a maximum exercise test. J Clin Endocrinol Metab 2003;88:394–401

12. Cappon J, Brasel JA, Mohan B, et al. Effect of brief exercise on circulating insulin-like growth factor-1. J Appl Physiol 1994;76:2490–2496

13. Petersenn S. Growth hormone secretagogues and Ghrelin: an update on physiology and clinical relevance. Horm Res 2002;58:56–61

14. Kojima M, Hosoda H, Matsuo H, et al. Ghrelin: discovery of the natural endogenous ligand for the growth hormone secretagogue receptor. Trends Endocrinol Metab 2001;12:118–126

15. Martin FC, Yeo A, Sonksen PH. Growth hormone secretion in the elderly: aging and the somatopause. Baillieres Clin Endocrinol Metab 1997;11:223–250

16. Clemmons DR, Van Wyk JJ. Factors controlling blood concentration of somatomedin C. J Clin Endocrinol Metab 1984;13: 113–143

17. Gooren L. Age-related decline of growth factor hormone (somatopause). In: Gooren L, Lim PHC, eds. *The Aging Male.* Singapore: MediTech Media Asia Pacific; 2001:33–41

18. Holly JMP, Wass JAH. Insulin-like growth factors; autocrine, paracrine or endocrine? new perspectives on the somatomedin

hypothesis in the light of recent developments. J Endocrinol 1989;122:611–618

19. Guler HP, Zapf J, Schmid C, et al. Insulin-like growth factors I and II in healthy men: estimations of half-lives and production rates. Acta Endocrinol (Copenh) 1989;121:753–758

20. Rudman D, Kutner MH, Rogers CM, et al. Impaired growth hormone secretion in the adult population. J Clin Invest 1981; 67:1361–1369

21. Iranmanesh A, Lizarralde G, Veldhuis JD. Age and relative obesity are specific negative determinants of the frequency and amplitude of growth hormone (GH) secretory bursts and the half-life of endogenous GH in healthy men. J Clin Endocrinol Metab 1991;73:1081–1088

22. Zadik Z, Calew SA, McCarter RJ, et al. The influence of age on the 24-hour integrated growth hormone concentration in normal individuals. J Clin Endocrinol Metab 1985;60:513–516

23. Corpas E, Hartman SM, Blackman MR. Human growth hormone and human aging. Endocr Rev 1993;14:20–39

24. Reutens AT, Veldhuis JD, Hoffman D, et al. A highly sensitive growth hormone (GH) enzyme-linked immunosorbent assay uncovers increased contribution of a tonic mode of GH secretion in adults with organic GH deficiency. J Clin Endocrinol Metab 1996;81:1591–1597

25. Toogood AA, O'Neill PA, Shalet SM. Beyond the somatopause: growth hormone deficiency in adults over the age of 60 years. J Clin Endocrinol Metab 1996;81:460–465

26. Landin-Wilhelmsen K, Wilhelmsen L, Lappas G, et al. Serum insulin-like growth factor 1 in a random population of men and women: relation to age, sex, smoking habits, coffee consumption and physical activity, blood pressure and concentrations of plasma lipids, fibrinogen, parathyroid hormone and osteocalcin. Clin Endocrinol (Oxf) 1994;41:351–357

27. Rudman D, Mattson DE. Serum insulin-like growth factor-I in healthy older men in relation to physical activity. J Am Geriatr Soc 1994;42:522–527

28. Jørgensen JOL, Troels KH, Flavia LC, et al. Somatopause and body composition. In: Lunenfeld B, Gooren L, eds. Textbook of Men's Health. London: Parthenon Publishing Group; 2002

29. Vahl N, Jorgensen JOL, Jurik AG, et al. Abdominal adiposity and physical fitness are major determinants of the age associated decline in stimulated GH secretion in healthy adults. J Clin Endocrinol Metab 1996;81:2209–2215

30. Consensus Guidelines for the Diagnosis and Treatment of Adults with Growth Hormone Deficiency: Summary Statement of the Growth Hormone Research Society Workshop on Adult Growth Hormone Deficiency. J Clin Endocrinol Metab 1998;83:379–381

31. Rudman D, Feller AG, Nagraj HS, et al. Effects of human growth hormone in men over 60 years old. N Engl J Med 1990; 323:1–6

32. Thompson JL, Butterfield GE, Marcus R, et al. The effects of recombinant human insulin-like growth ractor-I and growth hormone on body composition in elderly women. J Clin Endocrinol Metab 1995;80:1845–1852

33. Cohn L, Feller AG, Draper MW, et al. Carpal tunnel syndrome and gynecomastia during growth hormone treatment of

elderly men with low circulating IGF-1 concentrations. Clin Endocrinol (Oxf) 1993;39:417–425

34. Papadakis MA, Grady D, Black D, et al. Growth hormone replacement in older men improves body composition but not functional ability. Ann Intern Med 1996;124:708–716

35. Blackman MR, Sorkin JD, Munzar T, et al. Growth hormone and sex steroid administration in healthy aged women and men: a randomized controlled trial. JAMA 2002;288:2282–2292

36. Goh VHH, Mu SC, Gao F, et al. Changes in body composition and endocrine and metabolic functions in healthy elderly Chinese men following growth hormone therapy. The Aging Male 1998; 1:264–269

37. Geusens PP, Boonen S. Osteoporosis and the growth hormone-insulin-like growth factor axis. Horm Res 2002;58:49–55

38. Holloway L, Butterfield G, Hintz R, et al. Effects of recombinant human growth hormone on metabolic indices, body composition and bone turnover in healthy elderly women. J Clin Endocrinol Metab 1994;79:470–479

39. Aloia JF, Zanzi I, Ellis K, et al. Effects of growth hormone in osteoporosis. J Clin Endocrinol Metab 1976;43:992–999

40. Yarasheski KE, Zachwieja JJ, Campbell JA, et al. Effect of growth hormone and resistance exercise on muscle growth and strength in older men. Am J Physiol 1995;268:E268–E276

41. Pagana KD, Pagana TJ. Mosby's Manual of Diagnostic and Laboratory Tests. St. Louis: Mosby; 1998

42. Blackman MR. Growth hormone and aging in men. In: Lunenfeld B, Gooren L, eds. Textbook of Men's Health. London: Parthenon Publishing Group; 2002

43. Khorram O. Use of growth hormone and growth hormone secretagogues in aging: help or harm. Clin Obstet Gynecol 2001; 44:893–901

44. Vance ML. Can growth hormone prevent aging? N Engl J Med 2003;348:779–780

45. Taaffe DR, Pruitt L, Reim J, et al. Effect of recombinant human growth hormone on the muscle strength response to resistance exercise in elderly men. J Clin Endocrinol Metab 1994; 79:1361–1366

46. Hurel SJ, Koppiker N, Newkirk J, et al. Relationship of physical exercise and aging to growth hormone production. Clin Endocrinol (Oxf) 1999;51:687–691

47. Smith RG, Cheng K, Schoen WR, et al. A nonpeptidyl growth hormone secretagogue. Science 1993;260:1640–1643

48. Chapman IM, Bach MA, Van Cauter E, et al. Stimulation of the growth hormone (GH)-insulin-like growth factor I axis by daily oral administration of a GH secretagogue (MK-677) in healthy elderly subjects. J Clin Endocrinol Metab 1996; 81:4249–4256

49. Murphy MG, Weiss S, McClung M, et al. Effect of alendronate and MK-677 (a growth hormone secretagogue), individually and in combination, on markers of bone turnover and bone mineral density in postmenopausal osteoporotic women. J Clin Endocrinol Metab 2001;86:1116–1125

50. Department of International Economic and Social Affairs. Periodical on Aging. New York: United Nations; 1985;1:1–61

51. Lunenfeld B. Healthy aging for men. Climacteric 1999;2:9

19

Bioidentical Hormone Replacement with Testosterone in Men

CHRISTOPHER B. CUTTER AND ROBERT S. TAN

Case History

Mr. J.J. is a 49-year-old man who has been referred for management of hypogonadism. Another physician told him that his levels of total testosterone were below 300 ng/dL. His wife, who is convinced that his symptoms were attributed to his low testosterone state, accompanied him to the appointment. On direct questioning, he admitted to tiredness, loss of libido, mood changes, and memory loss. He admitted to high stress in his job, as his business was failing. He was not particularly happy to be in the doctor's office, but said that his wife had insisted that something be done about him. He gave a past medical history of coronary artery disease, hypercholesterolemia, and a transient ischemic attack (TIA). He drinks up to 24 alcoholic beverages a week. His current medications included Cozaar, Lipitor, and aspirin. Physical examination revealed a facial blush and rhinophyma suggestive of alcoholic abuse. The liver was not enlarged, but he had an increased waist-hip ratio. Blood pressure (BP) was 136/78 mm Hg, weight was 178 lbs, and body fat content by caliper was 26.5%. A repeat morning sample of total testosterone revealed a level of 267 ng/dL. His prolactin, ferritin, estradiol, luteinizing hormone (LH), and prostate-specific antigen (PSA) were in the normal range.

His wife said that they wanted to explore various options for testosterone replacement therapy and mentioned that cost may be an issue because they have only calamity insurance. It was stressed to the patient that chronic alcoholism certainly can suppress testosterone levels and that he needed to stop drinking completely. Various options were discussed with them, and they felt they wanted to try a commercial topical version of 50 mg testosterone (Androgel) priced at approximately $120 per month. After a few weeks, the dose was escalated to 100 mg but the patient had

felt not much change in his mood or memory, although he did acknowledge that his libido was improved. At this point, the wife asked if he could try compounded testosterone ($50 per month), and a similar dose of the generic gel testosterone was prescribed. Laboratory testing revealed poor absorption of this particular version of compounded testosterone. It was explained to the patient that it could have been due to skin absorption characteristics, as a small minority of patients do not absorb well. The patient said that the pharmacist had told him that an injection might work better. The patient obtained a 10-dose, 200 mg each dose, vial for $89.95, not including the cost of the needles and syringes. His wife, who was an ex-nurse, was comfortable with injecting him every 2 weeks. After a year of injections, the patient said that he did not find any major improvements in symptoms as compared with the gels, and at most had a slight improvement in symptoms in only the first week of the injection. At this point, he wanted to consider a testosterone implant, and was asked to return to the clinic for further counseling.

History of Testosterone Replacement in Men

The association of the testicles with sexual and reproductive function dates back to thousands of years B.C., with accounts of the Babylonians castrating men as punishment for crimes such as adultery, and even castrating young choir boys in the early Christian Church to preserve their soprano voices. Animal husbandry depended on the ability to preserve the potency of desired studs, while removing other males from the gene pool and domesticating them via castration.

The premodern approach to androgen enhancements consisted of a variety of untested, possibly harmful, and dubiously effective concoctions that doctors and charlatans alike devised and sold to the public without scientific testing. These included oral preparations of the testicles of various animals that purportedly enhanced virility and sexual potency. The famous French physician Brown-Sequard may have inadvertently started the field of reproductive endocrinology in 1889, when he began performing uncontrolled experiments whereby he injected dogs (and even himself) with aqueous extracts of animal testes. Brown-Sequard published exuberant accounts of the improvement in their general health, muscular strength, and even cognitive functions. These and other reports led to the fad of using the testicles of every animal available, including monkeys, bulls, and pigs, to prepare both aqueous and glycerol extracts that were then injected or ingested into well-to-do patients who believed they were onto a veritable fountain of youth. This came to an abrupt, and appropriate, halt when an international committee of pragmatic physicians made a proclamation that these claims of rejuvenation were unfounded and probably unsafe.

Thus, androgen research was thrown back into the hands of the appropriate people: scientists. By the end of the 1920s, several researchers developed bioassays of androgen potency using the regeneration of sex organs of castrated mice and rats, or, similarly, the growth of the rooster comb and waddle in capons. Combined with the advancing techniques of modern organic chemistry, these researchers eventually isolated various "male hormones" from the urine of human and animal males and laid the groundwork for the eventual identification of testosterone.

In May 1935, a chemist in Amsterdam named David, working with the famous group of Laqueur and Freud, isolated 10 mg of a pure crystalline compound from 100 kg of bull testes. He named it "testosterone." It turned out to be identical to a compound that had been synthesized by Butenandt and Ruzicka from cholesterol. Thus, not only did scientists now have the most active male hormone in a purified state, they knew its chemical structure and how to synthesize it from cholesterol. Within a few months Ruzicka and his group reported the synthesis of an orally active form of testosterone made by substituting a methyl group at the 17 position of the original molecule. It was called methyltestosterone.

So as early as 1935, scientists were studying an oral form of testosterone and at the same time synthesizing hundreds of new compounds based on the original steroid nucleus of the testosterone molecule. Because it is a three-dimensional molecule, the number of stereoisomers, α and β positions of the side groups, and the potential to add various other side groups and oxidize or reduce the A ring are important factors, there are *thousands* of possible cogeners that could be produced for

experimentation. Fortunately, for the purpose of our discussion we need only mention a handful of them.

Oral Androgens

Characteristics of Oral Agents for Androgen Replacement

The perfect agent for androgen replacement therapy should have the following characteristics:

1. Deliver androgen in a physiological manner

2. Have a reproducible and predictable pharmacological profile

3. Be amenable to routine laboratory assay so that levels can be monitored

4. Completely reverse the signs and symptoms of androgen deficiency

5. Be unattractive to those who would abuse such compounds

6. Be affordable

7. Be easily administered in a patient-friendly manner

It is helpful to look at the classification of testosterone modifications (Figs. **19–1** and **19–2** [pp. 171, 172, respectively]). As research progressed with both the oral and injectable forms of testosterone, scientists became aware of the two somewhat distinct actions of the various androgens they were using. *Anabolic* effects were those features that caused a positive nitrogen balance and increased the mass of various tissues and organs such as muscle, bone, and blood. *Androgenic or virilizing* effects were those features that caused the development of facial and body hair, increased sebum production (and acne), deepening of the voice, and enlargement of the prostate gland and other male reproductive organs.

Researchers throughout the 1940s, 1950s, and 1960s used various animal biological assays to assess the anabolic versus the androgenic actions of their newly developed compounds. Hypertrophy of the castrated rat kidney or levator ani muscle was used as an indicator of anabolic activity, whereas the same effect on the rodent prostate or seminal vesicles was used to assess the androgenic properties of a newly developed steroid. The search for a pure anabolic agent without virilizing effects was finally found to be futile, as all of the tested molecules would eventually have virilizing effects in humans. This was particularly unacceptable in women and children for whom agents for breast cancer or aplastic anemia were being sought.

What those researchers did not know is that there is a single androgen receptor in humans, and that unlike

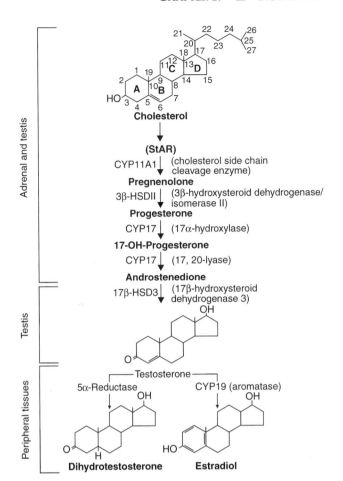

FIGURE 19–1. Pathway of testosterone formation in the testis and the conversion of testosterone to active metabolites in peripheral tissues (sTAR, steroidogenic acute regulatory protein). (Adapted from Griffin JE, Wilson JD. Disorders of the testes. In: Braunwald E, Fauci AS, Kasper DL, et al, eds. *Harrison's Principles of Internal Medicine.* 15th ed. New York: McGraw-Hill; 2001, with permission.)

the estrogen receptor, which has α and β forms that are tissue specific, they would not be successful at separating the effects on individual tissues.

Interestingly, we now can explain the effects of the various synthetic agents on humans by identifying how these molecules compare with the endogenous natural steroids that are metabolized in vivo. It is now apparent that the net effects of testosterone (T) in the human male are the sum of the effects of the two major metabolites of T [E_2 and dihydrotestosterone (DHT)] as well as the effects of T itself. Many tissues, such as adipose tissue, have high concentrations of an aromatase enzyme that converts T into estradiol (E_2). Other tissues such as the skin (especially in the perineal region and scrotum) have high concentrations of a 5α–reductase enzyme that converts T into

5α-dihydrotestosterone (DHT). E_2 is responsible for many important effects and side effects of T therapy including raising the high-density lipoprotein (HDL), closure of bone epiphyses, and the complete formation of strong bones. DHT, while binding the same receptor as T, is responsible for the development and hypertrophy of the prostate and also for many of the skin effects that we see from T therapy.

Certain modifications of the A ring of the steroid nucleus makes the molecule behave like DHT. DHT, which we now know binds five to 10 times more tightly to the androgen receptor than does T. DHT is also nonaromatizable, thus it cannot be turned into E_2, and therefore will not have the benefits of an estrogen (such as raising HDL cholesterol or closing bone epiphyses), nor will it have the undesirable side effects such as causing gynecomastia. In contrast, the esterified cogeners of T are slowly released into the circulation and then serum esterases hydrolyze the fatty acid off the 17 position and convert it into pure testosterone, thus producing a molecule that is capable of being 5α reduced to form DHT, or aromatized at the A ring to produce E_2.

The *Physicians' Desk Reference* (2003 edition) lists only three orally active anabolic-androgenic steroids for current use. These three, methyltestosterone (methyl-T), oxandrolone, and oxymethenolone, all have specific Food and Drug Administration (FDA) indications, even though their differences are more quantitative than qualitative. Oxandrolone is indicated as an adjunctive therapy to promote weight gain following extensive surgery, chronic infections (AIDS), or severe trauma. It is given orally as a 2.5-mg tablet, one to two tablets up to four times a day. Its use is not restricted to males, but obvious limitations and precautions must be observed in females. Oxymethenolone (Anadrol-50) is only FDA approved for the adjunctive treatment of aplastic anemias. It is dosed in children and adult at 1 to 5 mg/kg/day. It comes as a 50-mg tablet. Methyltestosterone is available as a 10-mg tablet by itself (Testred) or in two different dosage combinations with esterified estrogens (in years past, it was also available as a 10-mg orally dissolving tablet that absorbed directly through the venous and lymphatic beds under the tongue). Testred is the only oral FDA approved agent for male hypogonadism, whereas the latter two are used for hormone replacement therapy (HRT) in postmenopausal women in situations where the treating physician believes the T will aid in libido, bone density, or muscle acquisition. For the reasons that will be listed later, methyl-T is seldom used for male HRT, but when it is, a dose of ~50-mg/day, divided q.i.d., is required. As an addition to female HRT, methyl-T has enjoyed some definite, but limited, popularity. It is felt by most experts that the low doses

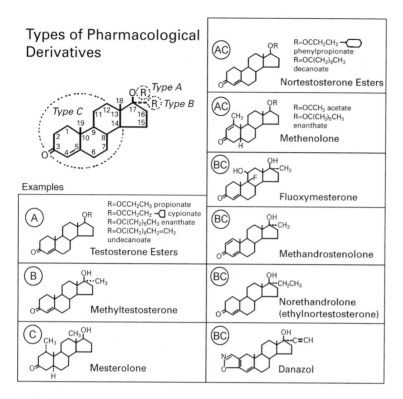

FIGURE 19–2. Types of androgen preparations available for clinical use. Type A derivatives are esterified in the 17β position. Type B derivatives have alkyl substitutions in the 17α position. Type C derivatives involve a variety of alterations of ring structure that enhance activity, impede catabolism, or influence both functions. Most androgen preparations involve combinations of type AC or type BC changes.

used by females are not a health threat or likely to cause liver dysfunction.

The oral route of administration for these agents exposes the liver to very high doses. This has been associated with the rare, but concerning, development of two liver problems. The most common is cholestatic jaundice. This reversible event is generally not accompanied by active liver damage or significant rise in the transaminases. It is considered to be completely reversible upon cessation of the oral steroid. In contrast to this, the long-term use of high doses of some of these agents has led to the much more serious development of *peliosis hepatis*. This formerly rare disorder causes the formation of blood-filled tumors in the liver that can hemorrhage or otherwise lead to serious liver damage. Friedl,[1] in a review of this subject, found 90 cases of reported peliosis in oral and injected steroid users. Other series also found an alarming incidence of this condition in oral steroid users being treated with high doses for aplastic anemia (seven of 19 versus one of 28 in patients not treated with oral steroids). It has also been linked to the development of hepatomas and adenomas of the liver. In his review, Friedl noted 91

reported cases as of 1991 in oral steroid users. Many of these events occurred in patients with Fanconi's syndrome. Some of them actually seemed to regress after withdrawal of the androgenic steroids.

So how do the oral steroid agents hold up to the seven requirements mentioned at the beginning of this chapter for the ideal agent for the treatment of hypogonadism? Well, (1) they don't represent a physiological delivery system; (2) their results are not always reproducible in heterogeneous populations; (3) we cannot easily measure their levels in the bloodstream; (4) because they are not aromatizable they do not reverse all of the signs and symptoms of hypogonadism; (5) they are extensively abused; (6) despite their having been around for more that 50 years they are still extremely expensive (>$600/month for some); (7) and because of multiple doses required per day, they are not convenient. Thus, they satisfy none of the seven requirements, and they have the potential to cause great harm. It is one of the great ironies of medicine, that after 60 years of research by countless scientists, the preferred agent for the reversal of hypogonadism is still testosterone itself.

TESTOSTERONE UNDECANOATE

The undecanoate form of testosterone, marketed as Andriol in Europe and Asia, has yet to be approved by the FDA at the time of this writing. This form of oral testosterone is safer than those mentioned previously, but requires dosing three times daily. Surprisingly, testosterone undecanoate is the most widely used androgen worldwide, but still not available in the United States. Testosterone undecanoate is an oral androgen that provides the hypogonadal patient with the unmodified testosterone molecule. It was introduced in the mid-1970s. In a study by Gooren over 10 years, a small sample of men had biochemical parameters of liver function followed, and the levels remained constant during the study period, indicating that there is no increased hepatic enzymatic breakdown of the androgen over time. Over the 10-year period, some patients on testosterone undecanoate developed mild obstruction of urine flow. Digital examination of the prostate did not reveal signs of prostate tumors.[2] Testosterone undecanoate appears to be a safe oral androgen, but experience in the United States is lacking.

Injectable Androgens

Intramuscular injections of testosterone, usually as an *enanthate* or *cypionate* ester, do not have to be given daily, but are instead given every 1–2 weeks. After injections, blood levels peak about 2 to 3 days after dosing and slowly decline during the next 1 to 2 weeks. The injections are mildly painful, and rise and fall in serum levels of testosterone over time may be accompanied by changes in mood and the sense of well-being. This is called the "roller-coaster" effect, whereby the patient feels best for a week or so after the injection, then loses this effect over the next several days. Arguably, it may also cause a higher rise in PSA as compared with more physiological preparations of testosterone. Injectable therapy usually is the least expensive way to provide testosterone replacement, and it requires the least patient motivation and compliance. Sih and his colleagues[3] validated this form of testosterone replacement. Their study was undertaken to examine the year-long effects of intramuscular testosterone administration in older men. In that study, 15 hypogonadal men (mean age 68 ± 6 year) were randomly assigned to receive a placebo, and 17 hypogonadal men (mean age 65 ± 7 year) were randomly assigned to receive testosterone. Hypogonadism was defined as a bioavailable testosterone $<60 \, ng/dL$. The men received injections of placebo or $200 \, mg$ testosterone cypionate biweekly for 12 months. The main outcomes measured included grip strength, hemoglobin, PSA, leptin, and memory. The men in the testosterone had greater *improvement in bilateral grip strength* and *decreases in leptin* than did those assigned to the control group. There were no significant changes in PSA or memory. Several study subjects were withdrawn because of an abnormal elevation in hematocrit. Intramuscular testosterone supplementation improved strength, increased hemoglobin, and lowered leptin levels in older hypogonadal men. Intramuscular testosterone may have a role in the treatment of frailty in men with hypogonadism and perhaps in controlling obesity through its effects on leptin.[2]

Injectable Testosterone Undecanoate

This form of injectable testosterone is yet to be available in the United States. However, German investigators studied the suitability of intramuscular testosterone undecanoate injections for substitution therapy in hypogonadal men,[4] and it has also been studied for birth control.

The study was small in size and descriptive, and investigators found that testosterone serum levels were never below the lower limit of normal and only briefly after the third and fourth injection above the upper limit of normal, whereas peak and trough values increased over the 24-week observation period. Estradiol and dihydrotestosterone followed this pattern, not exceeding the normal limits. No serious side effects were noted in this small study. Slight increases in body weight, hemoglobin, hematocrit, prostate volume, and PSA, and suppression of gonadotrophins as well as increased ejaculation frequency occurred as signs of adequate testosterone substitution. In the future, injectable testosterone undecanoate may be well suited for long-term substitution therapy in hypogonadism and perhaps even hormonal male contraception.

Injectable Nandrolone

This form of androgen substitution is commonly abused by athletes because of its effects on skeletal muscle and strength. Its effects on sexuality are not pronounced. Medical therapeutic indications for nandrolone in some circumstances may include the HIV wasting syndrome and end-stage renal failure. An Australian study determined the efficacy of nandrolone in HIV wasting syndrome.[5] The changes in weight and body composition (lean body mass, total body water, and nitrogen index) were measured by anthropometry, bioelectrical impedance, and in vivo neutron activation. Subjects who failed to gain weight (10.9%) were treated with nandrolone decanoate (100 mg/mL) by deep intramuscular injection every 2 weeks for 16 weeks. Changes in quality of life were assessed by a short questionnaire. Changes in biochemistry, hematology, and immunology were also measured. The investigators found significant increases in weight (mean, 0.14 kg per week; $p < .05$) and lean body mass (mean, 3 kg by anthropometry; $p < .05$). The change in lean body mass was of similar magnitude across all measurement modalities. Quality of life parameters, especially functionality, increased significantly during the trial. No subject experienced toxicity.

Sublingual Testosterone

A few published reports have defined the characteristics and pharmacokinetics of sublingual T. This form of androgen replacement has several advantages, but ultimately is unsuitable for chronic replacement therapy because of its large peak and trough levels and the very short half-life.

T cyclodextrin consists of unadulterated T passively carried inside a sphere of $\alpha 1$–4 linked glucopyranose molecules. These oligosaccharide macro rings are quite hydrophilic on their exterior, and the T molecule is held in the relatively apolar center. When placed into the mouth, the lipophilic T passively traverses the buccal mucosa and rapidly enters the bloodstream. The remaining carbohydrate moiety is passed into the digestive tract and broken down into harmless, inactive metabolites.

In the late 1990s, the research team at Harbor-UCLA published three excellent studies of a sublingual product that was being investigated and developed by Biotechnology General Corporation. Both 2.5- and 5.0-mg sublingual tablets were compared with T enanthate injectable depot.[6] The results showed that peak serum values were reached in 20 minutes and were near baseline by 2 or 3 hours. Peak levels with the 2.5-mg dose ranged from 800 to 1100 ng/dL. The peak levels of the 5.0-mg dose went as high as 1500 ng/dL. These supraphysiological peaks increased the E_2 levels from baseline, but were small compared with the large increase with the T enanthate injections. Suppression of LH and lowering of sex hormone–binding globulin (SHBG)—both functions of increased and continuous serum T levels—were barely affected by sublingual T, but markedly lowered by the T enanthate.

Sublingual T was further studied to ascertain its effects when given over a longer period of time. A total of 63 hypogonadal men were divided into three groups and randomized to receiving either T enanthate, 200 mg intramuscularly (IM) every 20 days, or 2.5 or 5.0 mg of sublingual T three times per day. Mood changes were apparent in all three groups at the first follow-up visit (day 21) and persisted for the duration of the study (60 days). This included increased energy, sense of well-being, and friendliness. Negative mood parameters such as irritability and nervousness, were generally lowered in all groups, but the 2.5-mg group had the highest level of these feelings.

A separate publication by the same researchers showed that sublingual T can improve lower body muscle mass and strength (but not upper body), increase the calciotropic hormones and the bone turnover markers, and improve several aspects of sexual function. The short duration of the study (6 months) did not allow enough time for dual-energy x-ray absorptiometry (DEXA) measurements to improve. No increases in hemoglobin concentration were found, and the serum PSA went up only slightly. Maximum urine flow did decrease over the time period of the study, and one subject had to quit the study for this reason.

Despite generally positive reports with the sublingual T method, the FDA ultimately did not approve it for long-term hormone replacement in men, largely because of its pharmacokinetic profile, leading to large peak and trough variations and a very short half-life. Thus there is no commercially available sublingual T in the United States. Compounded preparations can be prepared by specially trained pharmacists, and these products have been promoted in some circles for replacement therapy, despite the previously-mentioned drawbacks. It is interesting to consider whether the sublingual T might be useful as an aid for libido and sexual function in both men and women. Its short duration of action and failure to suppress LH might actually be its strengths in this regard. Consistent with this concept, some researchers have found increases in "genital arousal" in women in as little as 45 minutes after administration. Further investigation into this possibility is needed.

Subcutaneous Testosterone Implants

Reports of T implants have been published since the late 1930s. Many different types of implants have been developed and a nearly equal number have been discarded. More recently, however, the use of Silastic capsules and crystalline T pellets has been showing promise as a method of long-term T administration. The bulk of information currently available on T implants comes from researchers and clinicians in the United Kingdom and Australia. Handelsman's group[7] in Victoria, Australia, has had extensive experience using T implants that are made by melting and molding crystalline T into 100- or 200-mg rods and inserting them using aseptic technique into the lateral

FIGURE 19–3. The testosterone pellets, crystalline T in 100 or 200 mg rods, for long-term administration of T to hypogonadal men. When inserted into the abdomen, they provide physiologic levels of T within 1 to 2 days, T level peak at 1 month and then decline to below optimal levels at 4 to 5 months. (Courtesy of the Professional Compounding Association of America.)

abdomen at the level of the umbilicus. The group has reported over 13 years of experience and has found that it is a widely and favorably accepted method for hypogonadal men. Although extrusion of one or more of the implants from the insertion site has become the number one complication of the procedure (~10% of procedures), more serious complications are very rare. Suppurative and/or nonsuppurative cellulitis at the site can occur in anywhere between 3 and 8% of procedures, and can usually be treated with removal of the implants. Interestingly, the likelihood of complications is almost entirely predictable based on the occupation of the patient, with more active patients having more complications.

The insertion of four 200-mg pellets into two or four tracks provides physiological levels of T within a day or two. Fig. **19–3** shows the testosterone pellets. The levels peak at 1 month and then gradually decrease over time so that optimal levels are not present by 4 to 5 months, requiring repeat insertion on the contralateral side. All of the usual benefits of sustained-release testosterone therapy have been demonstrated in the patients treated with this method, and no unusual problems have surfaced. The procedure has also been used in prepubertal hypogonadal boys with success. In the United States, a recent investigation into the use

of the T crystalline pellets for male contraceptive has been completed and will be published in the near future (Dr. Christina Wang, Harbor-UCLA Medical Center, personal communication).

Percutaneous or Topical Testosterone

Arguably, no development in the field of androgen replacement has had the same kind of dramatic effect on clinicians and patients as the advent of topically applied T gels and creams. Although much excitement was generated with the launch of the Testoderm scrotal patch and its successor the Androderm patch, these products soon became known more for their drawbacks than their strengths. Moderate to severe rash formation in as many as 50% of its users soon put a chill on the excitement regarding transdermal therapy for hypogonadal men. Likewise, low serum T, supraphysiological levels of DHT, and the need to shave the scrotum led to low levels of acceptance of the scrotal patches.

If one thing can be said for these products, however, it is that their huge financial backing by their respective pharmaceutical companies led to some of the better and more important research as to the benefit and limitations of androgen replacement in hypogonadal men.

Thus, when Solvay released its product called Androgel in 1999, it was with a bit of skepticism that most clinicians watched to see where its Achilles heel might be. Fortunately for the population of hypogonadal men and the clinicians who treat them, Androgel has turned out to be generally all that it promised, as long as the user and prescriber look to see exactly what was promised.

Principles of Percutaneous Pharmacokinetics

By dissolving T into a lipophilic gel, one can reach concentrations up to ~8 to 10% (wt/vol). This makes it relatively easy to apply as much as 300 mg to the skin of a potential patient. If we keep in mind that the average man needs ~7 to 10 mg of T per day, then one can see that absorption efficiency is key with any topically applied product. By using ethanol as an enhancer and with the careful choice of other ingredients, Solvay (or the original developers, Unimed) created a product in which 50 mg applied to the skin in a 1% solution would bring physiological levels to ~80 to 90% of its users. This equates with an absorption efficiency of ~7/50 X 100 = 14%. By comparison, some of the earlier formulations that this author tried in various compounded gels or creams could only reach a 3% efficiency.

In a study published by Cutter[8] in the *Journal of the American Board of Family Practice*, 10 hypogonadal men were selected with ages ranging from 44 to 77 years. Four of these men had newly diagnosed and six had preexisting hypogonadism. Patients were withdrawn from their previous hormone therapy and baseline laboratory studies were obtained for total testosterone, free testosterone, dihydrotestosterone, estradiol, luteinizing hormone, follicle-stimulating hormone, complete blood counts, lipid panels, and chemistry panels. The patients then started taking increasing dosages of the testosterone gel until physiological levels of testosterone were attained or until the study period of 6 weeks ended. There was no blinding, and each patient served as his own control. Testosterone and free testosterone levels were monitored weekly, and estradiol and dihydrotestosterone less frequently. At the conclusion of this study, all the baseline laboratory tests were repeated. A questionnaire evaluating the psychosexual well-being of the patients was administered before and after the treatment period. The average total testosterone level rose from 136 ng/dL to 442.9 ng/dL ($p < .001$). Average free testosterone levels rose from 34.2 pg/mL to 120.3 pg/mL ($p < .001$). Average dihydrotestosterone levels rose from 20.5 to 199.2 ng/dL ($p = .006$). Average estradiol levels rose only slightly from 34.1 pg/mL to 40.0 pg/mL ($p = .191$). Average total androgens (testosterone plus dihydrotestosterone) rose in all patients to therapeutic levels, from 149.3 ng/dL to 642.1 ng/dL ($p = .001$). The ratio of total androgen to estradiol rose from 5.1 to 17.1 ($p < .002$). Luteinizing hormone was suppressed in the six patients for whom meaningful data were available, and decreased on average from 5.66 to 1.10 mIU/mL ($p = .005$) Lipid effects were measured, and a 15% drop in all cholesterol fractions was noted ($p < .005$). Evaluation of the questionnaire showed considerable improvements in sexual function and overall well-being in all but one patient. No adverse effects or nuisance problems were detected during the duration of the study. The author found that topically applied *compounded* testosterone gels are an effective and convenient means of hormone replacement in hypogonadal men.

Controversy of Effects of Testosterone Replacement and Prostate Cancer

A major interest today involves the effects of androgen supplementation on the prostate gland. Although T administration to hypogonadal men is usually beneficial, there is much concern regarding the potential for serious side effects resulting from T administration. Many physicians remain skeptical of the benefits of testosterone replacement therapy (TRT). It has been well established that androgen administration does *not* stimulate DNA synthesis (and consequently proliferation of stro-

mal prostate cells) in *normal prostates*. However, the administration of T is believed to enhance any *preexisting prostatic malignancy*, such as prostatic carcinoma (CaP). This is based upon the following:

1. Evidence that placing rodents in hypertestosteronemic states results in the appearance of prostatic cancer.

2. The usually dramatic responses of most human prostatic cancers to surgical or medical castration.

3. The controversial yet widely held impression that there exists a positive relationship between serum androgen levels and prostate cancer in humans.

4. The fact that androgens promote prostate cancer development in laboratory animals, which suggests that decreasing androgenic stimulation can lower prostate cancer risk.

In contrast, the following findings, particularly from human studies, refute the notion that TRT may be harmful, that hypotestosteronemia may indeed be a harbinger for CaP, and that TRT is by and large safe in the short term. Long-term data are not available, and the Institute of Medicine is in the process of commissioning one large-scale study.

1. In the literature, four studies have found a positive correlation between high serum T and risk of CaP. In contrast, six studies found that high T levels were actually associated with reduced risk and 15 studies found no difference either way.[9]

2. At the experimental level, a prostate cancer cell line requires initial stimulation by androgens to grow but it is eventually suppressed by androgens.[10] The observation in most studies that *low testosterone* rather than high testosterone is associated with CaP and the subsequent suppressive effects of testosterone have led to the hypothesis that low testosterone rather than high testosterone is harmful.

3. A recent study suggested that CaP *suppresses* testosterone production, and may as such account for the observation that CaP is often detected in men with hypotestosteronemia.[11]

Discussion of the Case History

This case illustrates the difficulty that practitioners sometimes encounter with bioidentical hormone therapy. It also suggests that the practitioner be patient, and that the success of treatment can be variable. Expectations by patients should also be realistic, and a careful explanation of what might be expected should be given to them. Mr. J.J. had a large confounder, which was alcoholism. Secondary causes of hypogonadism were excluded and he was screened for a prolactinoma as well as hemochromatosis. He started off with Androgel, which was effective. However, the patient's demands could sometimes be difficult. He wanted to try a compounded version partly out of curiosity, and also because of cost reasons. Unfortunately, the preparation produced erratic absorption. As such, it is very important to choose well-trained pharmacists such as through the Professional Compounding Association of America (PCCA). The patient subsequently went on injections, producing a "roller coaster" effect in which supraphysiological levels are reached in the initial phase, followed by subphysiological levels. Testosterone implants require a minor surgical process every 6 months, and the equipment and pellets can be obtained through PCCA.

Conclusion and Key Points

- Bioidentical hormonal therapy should be individualized. It takes time and experience to treat hypogonadal symptomatic men.

- There are presently several choices, including topical gels, injections, sublinguals, and sometimes implants.

- Oral therapy in the United States is limited, but testosterone undecanoate may prove promising when available.

- The commonly held belief that testosterone replacement in normal hypogonadal men can cause prostate cancer is untrue, but monitoring is essential.

- Before starting therapy, it is essential that the patient understand the benefits and limitations of therapy.

REFERENCES

1. Friedl KE. Effects of anabolic steroids on physical health. In: Yesalis CE, ed. *Anabolic Steroids in Sports and Exercise.* Champaign, IL: Human Kinetics; 1993:107–150

2. Gooren LJ. A ten-year safety study of the oral androgen testosterone undecanoate. J Androl 1994;15:212–215

3. Sih R, Morley JE, Kaiser FE, et al. Testosterone replacement in older hypogonadal men: a 12-month randomized controlled trial. J Clin Endocrinol Metab 1997;82:1661–1667

4. Nieschlag E, Buchter D, Von Eckardstein S, et al. Repeated intramuscular injections of testosterone undecanoate for substitution therapy in hypogonadal men. Clin Endocrinol (Oxf) 1999;51:757–763

5. Gold J, High HA, Li Y, et al. Safety and efficacy of nandrolone decanoate for treatment of wasting in patients with HIV infection. AIDS 1996;10:745–752

6. Salehian B, Wang C, Alexander G, et al. Pharmacokinetics, bioefficacy, and safety of sublingual testosterone cyclodextrin in hypogonadal men: comparison to testosterone enanthate—a clinical research center study. J Clin Endocrinol Metab 1995;80: 3567–3575

7. Handelsman DJ, Mackey MA, Howe C, et al. An analysis of testosterone implants for androgen replacement therapy. Clin Endocrinol (Oxf) 1997;47:311–316

8. Cutter CB. Compounded percutaneous testosterone gel: use and effects in hypogonadal men. J Am Board Fam Pract 2001; 14:22–32

9. Slater S, Oliver RTD. Testosterone: its role in the development of prostate cancer and the potential risks from use as hormone replacement therapy. Drugs Aging 2000;17:431–439

10. Carter HB, Pearson JD, Metter EJ, et al. Longitudinal evaluation of serum androgen levels in men with and without prostate cancer. Prostate 1995;27:25–30

11. Zhang PL, Rosen S, Veeramachaneni R, et al. Association between prostate cancer and serum testosterone levels. Prostate 2002;53:179–182

20

Nutraceuticals as Preventive Medicine in Aging Men's Health

BRUCE BIUNDO AND ROBERT S. TAN

Case History

Mr. L.K. is a 52-year-old man who was referred to the Optimal Aging Clinic. An internist had previously managed him, but the patient wanted to have a more in-depth evaluation of his health. His concerns were that he felt that he was aging prematurely, and he wanted some directions as to nutrition. He feels that although he eats well and is very watchful over his diet, he may be "nutrient deplete." Since undergoing plastic surgery, including liposuction and an abdoplasty, he has been watching his weight very carefully. He does report some loss of libido and fatigue. He does not complain of symptoms relating to the lower urinary tract (LUT), but wants to have his prostate checked. Currently, he is on several over-the-counter medications and nutraceuticals including aspirin, vitamin E, vitamin D, omega 3, Osteobiflex, selenium, dehydroepiandrosterone (DHEA), and zinc. He also supplements his diet with protein powder, creatine, and glycolide. He asked if he should be on saw palmetto "to protect my prostate from aging." His father died of complications from Alzheimer's disease, but also had prostate cancer that was not aggressive. The patient exercises every other day, and does mainly resistance training and each session lasts ~45 minutes. His diet consists mainly of vegetables, fruits, meat, and very low amounts of processed carbohydrates. Clinical examination reveals a well-sculptured man, blood pressure (BP) 132/80 mm Hg, and height 5 feet 11 inches. There was a long scar on his abdominal flank. Body fat as measured by electrical impedance was 23.8%, and body fat was 45 pounds. Rectal examination revealed a normal-size prostate with no nodules, and a soft consistency. The patient requested his blood to be sent to a specialized laboratory in Arizona. He wanted assessment for oxidative stress levels (e.g., carotene, peroxides, tocopherols, etc.), age-related hormones (cortisol, estradiol, testosterone etc.), trace metals, and cancer markers including prostate-specific antigen (PSA).

The Rationale for Nutraceuticals

In the United States alone, the business of nutritional supplements increases by millions of dollars each year, with expected sales of $8 billion to $10 billion in 2003.[1] The number of products on the shelves of health food stores and pharmacies includes nutraceuticals that were unknown just a few years ago. It is quite evident that many people are taking a more active role in their personal health. Purchases of nutritional supplements reflect that this is taking place. In the past century, there have been major medical victories won over disease states and illnesses that previously were major factors in determining length of life. As a result, the average life span for men born in the 20th century is considerably higher than for those born in the 19th century. As people live longer and causes of death have shifted to more chronic conditions such as coronary artery disease and cancer, the focus of health care has also shifted. It has become more important to consider what is commonly referred to as wellness care, as the public searches for ways to live longer, more active lives.

Chronic conditions that affect the aging population such as diabetes, osteoporosis, and Alzheimer's disease are being targeted as conditions that people who are knowledgeable, aware, and actively investing in their health can avoid, or, at the very least, minimize their degenerative effects. By the time a man or woman has been diagnosed with osteoporosis, the opportunity for preventive medicine is gone, at least for that condition. And when a man is identified as having prostate cancer, it is often at a late stage of the condition, when therapeutic options are limited. We are learning that there may be hope for many in avoiding Alzheimer's disease through, at least partially, proper nutrition and supplements. Also in recent years there have been studies suggested that inflammatory processes are instrumental in many of the conditions associated with aging, including cardiovascular diseases. It now appears that many people do not have to face debilitating illness as natural consequences of aging, but, in fact, can enjoy more of their years in a healthy state. We may not necessarily be able to add years to our life, but quite possibly add life to our years. It is in this context that we address the issue of nutritional supplements for men's health. The advertising that accompanies many products on the market today often overstates or oversimplifies the way a particular product works. We are faced with a vast array of nutraceuticals, products that promise a happier sex life, stronger bones, loss of obesity, and greater physical fitness. Do we need them? Are there real benefits to be gained from them? Or do our normal diets provide all of the nutrition that we need?

Regrettably, for various reasons, our diets in fact do not fully provide all that we need. To mention one factor, we are a people frequently in a hurry, eating on the run or depending on fast-food restaurants to nourish us. So, many of us likely will benefit from taking supplements that will offer us more of the good health that we desire now and for the future. As the interest in nutritional supplements has grown, so has scientific inquiry; although it is sometimes difficult to discern real value from product claims and marketing hype, there are genuine, controlled, objective studies being performed, often with definitive results. The vastness of this topic is such that we choose to be somewhat selective in this chapter, focusing on agents that are available and well known. A major criterion for inclusion in this chapter is that there is good research information available in a scientific publication. For some agents, the data support their use, whereas for others, the conclusion is less positive. We focus on nutritional supplements that address prostate cancer, LUT symptoms, cardiovascular disease, and sex hormone formation.

Prostate Cancer

With the introduction of PSA testing in the 1980s, the diagnosis of prostate cancer sharply increased. In the United States, it is one of the most common cancers, and is the leading cause of cancer deaths among American men. According to studies, ~180,000 men in the U.S. develop prostate cancer each year, and 37,000 men die from it each year.[2] According to a study published in 2002, prostate cancer is not one disease but a family of diseases that may also include a strong contribution from multiple genetic susceptibility genes.[3] Thus it is likely that just as there are various contributing factors, there will be multiple factors in prevention and treatment of prostate cancer. Although genetics and dietary and environmental factors all are important, the role of nutritional supplements has been and continues to be investigated. The agents that have generated the greatest interest include vitamins A, D, and E, the minerals selenium and zinc, lycopene and other carotenoids, and omega-3 fatty acids.

VITAMIN E

Vitamin E is the most commonly used supplement among men with prostate cancer, and it seems to hold promise in both delaying the progression of and in preventing prostate cancer.[4] Vitamin E is not a single entity but a group of related compounds including four tocopherols and four tocotrienols. Although the benefits of γ-tocopherol have been studied, the most recently published research indicates that α-tocopherol is the most natural form of vitamin E, and it has the highest bioactivity. In future studies, it likely to be useful to study these two forms of vitamin E in combination, as both are believed to be significantly active in prostate cancer prevention. There are various properties shown that make vitamin E a possible anti–prostate cancer agent. Perhaps the most important is that, as a potent antioxidant, vitamin E *reduces oxidative stress* in limiting the formation of reactive oxygen species (ROS), which are known to increase oncogene expression. In addition, vitamin E acts as an *antiprostaglandin*, while increasing the release of arachidonic acid. These two pathways are under greater scrutiny in prostate carcinogenesis.

Vitamin E has been studied in combination with other agents, such as beta carotene and selenium. In a study in which vitamin E alone, beta carotene alone, a combination of the two, and a placebo were given to male smokers, there was no reduction of the incidence of lung cancer.[5] But the administration of vitamin E alone did result in fewer cases of prostate cancer than in the other groups.

To date, the largest study on cancer prevention is the Selenium and Vitamin E Cancer Prevention Trial (SELECT).[6] Begun in 2001, this study will enroll 32,500 men who will be given daily supplements of selenium and vitamin E in a randomized, placebo-controlled, double-blind study. In this study, which will entail 12 years of research and analysis, patients will take 200 μg of selenium, vitamin E 400 mg/day, a combination of 200 mg selenium and 400 mg vitamin E, or a placebo. The results of this study should provide definitive answers as to the prevention of and/or the halting of progression of prostate cancer.

TABLE 20–1. Dietary Sources Rich in Vitamin E

Food	mg ATE*
Almonds, 1 oz (24 nuts)	7.4
Hazelnuts, 1 oz (20 nuts)	4.3
Canola oil, 1 tbsp	2.9
Broccoli, 1 cup cooked	2.6
Peanuts, 1 oz (28 nuts)	2.2
Olive oil, 1 tbsp	1.7
Wheat germ, 1 tbsp	1.3
Red bell pepper, 1 cup	1.0
Kiwifruit, 1 medium	0.9
Olives, 5 large	0.7
Spinach, 1 cup raw	0.6
Avocado, 1 oz	0.4
Brown rice, 1 cup	0.4
Apple, 1 medium	0.4
Banana, 1 medium	0.3
Sesame seeds, 1 tbsp	0.2
Romaine lettuce, 1 cup	0.2
Beef, ground, 3 oz	0.2

* Currently, the vitamin E nutrient information on reference databases is measured in ATE (α-tocopherol equivalence). Most food companies and the government have not yet updated their information to measure vitamin E in α-tocopherol.

Data from USDA Nutrient Database for Standard Reference, Release 15.

In a recently published study, vitamin E was shown to interfere with the manufacture of both PSA and the androgen receptor.[7] Although PSA levels in the study dropped by 80% to 90%, there was also a 25% to 50% reduction in the numbers of cancer cells. The authors report that different types of vitamin E produced different effects, with α-tocopherol succinate being the most effective in halting prostate cancer growth. Table 20–1 lists foods that are rich in vitamin E.

VITAMIN D

In recent years, there have been indications that vitamin D may be useful in both the treatment and prevention of prostate cancer. As far back as 1990, vitamin D deficiency has been suggested as a risk factor for prostate cancer.[8,9] Epidemiological studies report that *sunlight exposure*, which is necessary for the synthesis of vitamin D, is inversely proportional to prostate cancer mortality and prostate cancer is greater in men with lower levels of vitamin D.[10] The active form of vitamin D in the body is calcitrol, 1,25-dhydroxy vitamin D. One of the consistent side effects of calcitrol treatment is hypercalcemia.[11] More than 200 analogs of vitamin D with modified side chains have been developed to avoid this side effect. Several analogs have been identified as being more effective than the parent compound with little or no hypercalcemia. Several clinical trials are underway with less hypercalcemic analogs along with a study using dexamethasone to offset the toxicity of

calcitrol and to allow higher doses (oral doses of 0.5 to 1.5 μg daily).

VITAMIN A

The value of vitamin A in prostate cancer prevention and treatment is not clear at this time. Levels of vitamin A, chemically known as retinol, have been found to be reduced by five to eight times in prostate cancer tissue compared with normal prostate tissue. A study by Zhang[12] reported that vitamin A and its natural and synthetic analogs (retinoids) induce apoptosis in vitro and in animal studies. Newer synthetic retinoids are being studied and may show promise for the future in inducing apoptosis in both hormone-dependent and hormone-independent prostate cancer cell levels. The role of vitamin A, or retinol, is not clearly distinguished from retinoids and carotenoids, although many studies are focusing on the latter group.

LYCOPENE

Lycopene is a carotenoid that is found in tomatoes and other foods and is responsible for the red coloring of these foods. Lycopene is the most prevalent carotenoid in the Western diet, and as a potent antioxidant it is the most significant free radical scavenger in the carotenoid family (which includes beta carotene). One significant difference between beta carotene and lycopene is that beta carotene can be converted to vitamin A, whereas lycopene cannot. In 1995, Giovannucci et al[13] conducted a prospective study including 47,894 men that showed an association between lycopene intake and lowered risk of prostate cancer. More recently, it was concluded that the benefit of lycopene might be most pronounced in the protection against more advanced or aggressive prostate cancer.[14] The bioavailability of lycopene appears to be dependent on how it is prepared, or altered from the raw tomato. Processed tomato products, such as tomato paste or sauce, have superior lycopene absorption when compared with tomato juice or even raw tomatoes. Absorption appears to be enhanced by processing the tomato with oil, along with the fact that lycopene is released from the fibrous food matrix in preparation. One explanation for the anti–prostate cancer value of lycopene is its ability to inhibit insulin-like growth factor-I (IGF-I), which has been shown to be associated with a greater risk of prostate cancer. Prophylactic doses of lycopene are not yet established, but the usual carotenoid intake is ~5 mg per day. At this time, the ideal intake is not known. One dietary suggestion offered is for men to have at least approximately five servings of tomato products per week. Table 20–2 shows lycopene content in the diet.

TABLE 20–2. The Relative Content of Lycopene in Natural Foods

Food	Lycopene Content, mg/100 g
Tomato, raw	0.9–4.2
Tomato, cooked	3.7–4.4
Tomato sauce	7.3–18.0
Tomato paste	5.4–55.5
Tomato soup (condensed)	8.0–10.9
Tomato juice	5.0–11.6
Ketchup	9.9–13.4
Watermelon, fresh	2.3–7.2
Papaya, fresh	2.0–5.3
Grapefruit, pink/red	0.2–3.4

TABLE 20–3. The Selenium Content of Selected Foods

Food	Selenium, μg
Brazil nuts, ¼ cup	1036
Oysters, 3.5 oz	115
Chicken liver 3.5 oz	71
Raw oyster, 3.5 oz	70
Steamed clams, 3.5 oz	64
Beef liver, 3.5 oz	57
Sardines, 3.5 oz	46
Crab, 3.5 oz	40
Whole wheat pasta, 1 cup	36
White pasta, 1 cup	30
Wheat germ, ¼ cup	28
Molasses, blackstrap, 2 tbsp	25
Sunflower seeds, ¼ cup	26
Cooked oatmeal, 1 cup	19
Soy nuts, ½ cup	17
Freshwater fish, 3.5 oz	15
Egg, boiled, one	13
Tofu, ½ cup	11

Courtesy of the Northwestern University Feinburg School of Medicine nutrition web page.

SELENIUM

Along with vitamin E, selenium is one of the most popular dietary supplements for those seeking to reduce the risk of prostate cancer. In the SELECT trial cited in the section on vitamin E, selenium will be given at a daily dose of 200 mg by itself or in combination with vitamin E, and compared with a placebo group. Studies have shown an inverse relationship between serum selenium and risk of prostate cancer.[15] Selenium is believed to exert its preventive effect through the action of *glutathione peroxidase*, an enzyme that protects the cells from oxidative damage, and is selenium dependent. At this time, results of studies linking selenium to lower incidence of prostate cancer are compelling, and future research will be studied closely. Table **20–3** lists the selenium content of selected foods, and Table **20–4** lists the natural sources of lycopene, vitamin E, selenium, vitamin D, and omega-3 fatty acids.

SOY

The interest in soy as a preventative for prostate cancer comes from epidemiological data showing a lowered incidence among Asian men compared with men in the United States. In Japan, soy is widely consumed, and is believed to possess anticancer properties. The active agents believed responsible for these benefits are the *isoflavones genistein and diadzein*. American men, by comparison, consume little soy; thus, it has been theorized that their dietary difference is significant in the relative incidence of prostate cancer. However, studies in rats do show that dietary genistein from soy can suppress chemically induced prostate cancer.[16] Clearly, the basis for future testing has been established, without an expectation of determining an effective daily or weekly intake of soy protein. Although genistein and diadzein have theorized modes of action, exact mechanisms have not yet been determined.

OMEGA-3 FATTY ACID

Fat is the dietary component most frequently associated with prostate cancer.[17] Because the association of low-fat diets with a low incidence of prostate cancer is well established, the constituent fatty acids have been the subject of very interesting studies. Particularly under investigation are the essential fatty acids that are *unsaturated* and are of two types: omega-6 fatty acids from linoleic acid and omega-3 fatty acids, derived from linolenic acid. Linolenic acid is mainly found in fish oils, whereas linoleic acid comes mainly from vegetable seed oils. This interest in the essential fatty acids coincides with research linking cancer to inflammatory events. Although the similarity of their names can be confusing, these two groups have very different activities relative to prostate cancer. For example, although omega-3 fatty acids, docahexanoic acid, and eicosapentaenoic acid (EPA) inhibit the growth of androgen-unresponsive prostate cancer line, linoleic acid stimulates its growth. The differing effects are believed to be linked, in part, to the manner in which they affect prostaglandin synthesis. Linoleic acid contributes to the formation of the inflammatory prostaglandins, whereas linolenic acid metabolizes the series-3 antiinflammatory prostaglandins (PGE_3).

In addition, fatty acids appear to influence sex hormone levels. It is well known that testosterone is converted to the active metabolite dihydrotestosterone (DHT) by the 5α-reductase enzyme. Dihydrotestosterone is believed to be the androgen primarily involved in the progression of prostate cancer. Linolenic acid has been shown by Liang and Liao[18] to *reduce the activity of 5α-reductase*, thus showing another possible

TABLE 20–4. Natural Sources of Lycopene, Vitamin E, Selenium, Vitamin D, and Omega-3 Fatty Acids

Nutrient	Best Sources	How Much?	Considerations
Lycopene	Tomato products (sauces, ketchup, soup, juice) watermelon, papaya	10 servings per week or 6.5–15 mg of lycopene	Lycopene from cooked or processed tomato products is better absorbed than that from raw tomatoes
Vitamin E	Vegetable fats and oils, nuts, seeds, grain products, vegetables	50 IU per day	It is difficult to get this much from food alone
Selenium	Grains, fish, meats, broccoli from soil	200 µg per day	Selenium content of foods varies widely because of difference in soil conditions
Vitamin D	Sunshine, milk, butter, cheese, fish, oysters, fortified cereals	Unknown	Vitamin D is made in the skin during exposure to UV light, status is determined by both diet and UV exposure
Omega-3 fatty acids (fish oils)	Fatty fish (salmon, tuna)	2 fish meals per week or 0.9–1.0 g omega-3 fatty acids	

mechanism in prostate cancer prevention. There are unanswered questions at this time regarding how to modify the diet and optimize the effects of omega-3 fatty acids. It seems feasible that, at a minimum, men should be encouraged to regularly include fish in their diet, particularly cold-water fish rich in EPA.

ZINC

Zinc is an important mineral in men. Fascinatingly, the concentration of zinc in the prostate is about ten times more than any other body tissue. It has been reported that men with prostate cancer are reported to have *low zinc levels*. As such, some researchers believe that zinc is essential to prostate function and may even help prevent cancer. Most of the studies have been inferred from in vitro models. Liang et al[19] reported in 1999 that zinc inhibits human prostatic carcinoma cell growth, possibly due to induction of cell cycle arrest and apoptosis. There are no large longitudinal human studies demonstrating the protective role of zinc. However, lack of these studies does not prove that it is ineffective. Zinc in modest doses is safe, but it may cause gastrointestinal problems if taken orally in high doses above 50 to 100 mg. To offset this, zinc can be compounded into a gel and applied topically. Some clinicians not only think that zinc is preventive of prostrate cancer, but also advocate treating benign prostate hyperplasia (BPH) and chronic prostatitis with this mineral. Zinc is found in abundance in pumpkin seeds, oysters, lean meats, and wheat germ.

Sex Hormone Formation

Various nutritional supplements are involved in the steroid cascade of sex hormone formation, such as *prohormones* (e.g., DHEA), and agents that alter the metabolic pathways (e.g., 5α-reductase inhibitors).

DEHYDROEPIANDROSTERONE

Prohormones are hormones that have some intrinsic value in themselves, but primarily serve as precursors to other, more potent hormones. Perhaps the most commonly used prohormone in the United States is dehydroepiandrosterone (DHEA), which, at the time of this writing, is available without a prescription. Many countries in the world require a physician's prescription. Charts of the sex hormone pathways show DHEA as a direct precursor to androstenedione. DHEA is recommended for various purposes, but our major concern here is to show that, as it serves to form androstenedione, it is a vital link in the sex hormone progression.

ANDROSTENEDIONE

Androstenedione is a well-known prohormone, coming into the spotlight through its use by popular professional athletes. Commonly referred to as "andro," this agent possesses anabolic and androgenic properties of its own; it is mainly considered for use because it is a direct precursor to testosterone. Testosterone is a controlled substance in the U.S., requiring a prescription, but at the time of this writing, androstenedione is available without a prescription. However, tighter restrictions have been suggested for several of these prohormones that would move them to prescription-only status. A major consideration for those taking this supplement is that it not only serves as a direct precursor to testosterone, but also acts in the very same way for estrone, an estrogen with very different properties from testosterone. Therefore, it is likely (and has been reported) that supplements of androstenedione will increase levels of both testosterone and estrone. Oral bioavailability is quite variable, with dosing suggestions ranging

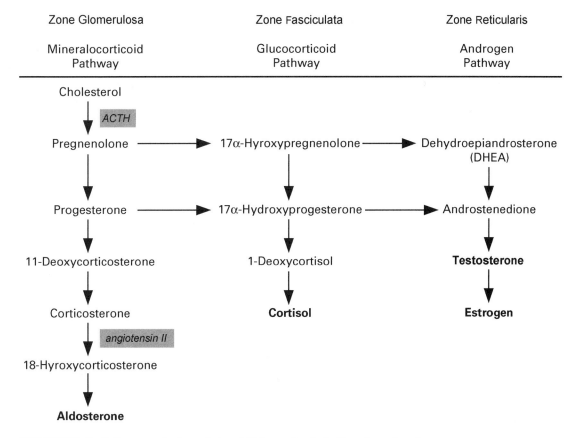

FIGURE 20–1. Pathways of adrenal steroid biosynthesis in the adrenal cortex.

from 25 to 500 mg per day. It also has been sold in topical formulations because of perceived greater bioavailability from that dosage form. Studies on androstenedione do not offer conclusive evidence that it is an efficient way to boost testosterone in men. Fig. **20–1** summarizes androgen and pro hormones.

ANDROSTENEDIOL

Androstenediol is chemically the alcohol version of the ketone androstenedione, and possesses virtually the same actions. It serves in the same position in the steroid pathway, that is, it is also a direct precursor to testosterone. Although touted as not having an estrone pathway, several studies have indicated elevated estrone levels after its use. It has also been recommended for use topically, for the same reason as androstenedione. Perhaps in the future, randomized controlled studies will be performed that will offer more definitive answers as to the value of these popular agents.

SAW PALMETTO

Saw palmetto, a naturally occurring plant product from the berries of *Serenoa repens*, is a very widely used supplement. Pharmacies and health food stores sell millions of dollars of this product each year to men, in the belief that it contributes to their prostate health. One mode of action is that of a *5α-reductase inhibition*, thus lowering the amount of testosterone that is converted to dihydrotestosterone. The usual dose is 160 mg of a standardized extract taken twice a day.

NETTLEROOT

Nettle root is also called stinging nettle. The actions of nettle root are believed to be similar to those of saw palmetto, in that it also acts as a *5α-reductase inhibitor*. It may have another role in testosterone metabolism, as it is believed to displace testosterone from sex hormone–binding globulin (SHBG). In that action, it does not increase the level of total

Chrysin

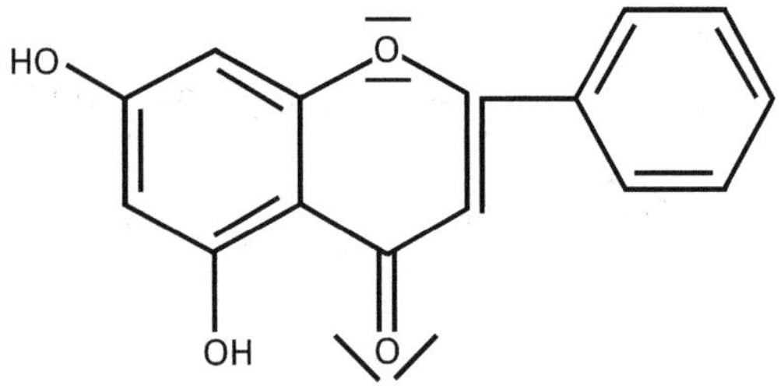

FIGURE 20–2. The molecular structure of Chrysin, an inhibitor in the aromatase enzyme system, where testosterone is converted to estradiol. Zone glomerulosa, zone fasciculate, and zone reticularis are different layers of the adrenal cortex. The pathways occur at the different sites.

testosterone, but can increase the unbound, free hormone.

CHRYSIN

In some men, the conversion of testosterone to estradiol occurs at a greater rate than desired, resulting in estradiol levels that are elevated. Because that process occurs through the aromatase enzyme system, agents have been studied that inhibit that process.[19] Chrysin is one such aromatase inhibitor, a naturally occurring flavonoid that has undergone some study with encouraging results (Fig. 20–2). An ideal aromatase inhibitor would both lower estradiol and elevate testosterone levels without toxic side effects. Although the side-effect profile of Chrysin is favorable, so far as is known at this time, the results of studies have not shown conclusive benefits. Oral bioavailability is known to be poor, with daily doses reported from 500 mg to 3 g per day. Some physicians have prescribed Chrysin in a topical application with the expectation of higher blood levels with a lower dose than that of the oral form. At this time, reports are very limited as to the efficacy of Chrysin from either topical or oral administration. Randomized controlled studies are needed to determine the safety, efficacy, and proper dosing for Chrysin.

Saw Palmetto: A Preventive and Therapeutic Nutraceutical for Benign Prostate Hyperplasia and Lower Urinary Tract Symptoms

One of the most common ailments in aging men involves urinary symptoms such as increased frequency of urination, increased urge, decreased flow rate, and nocturia. Commonly referred to as LUTS (lower urinary

tract symptoms), this condition is very similar in effect on men to benign prostatic hyperplasia (BPH), sometimes referred to as benign prostatic obstruction. Although many men report their urinary problems to their physicians as their first response, many others seek out help with nonprescription supplements. By far, the most common supplement that men take for their BPH/LUTS condition is saw palmetto. This herbal plant, known officially as *Serenoa repens*, is derived from the berry of the American dwarf palm tree. It is, in fact, the most widely used product for voiding problems. Saw palmetto contains various ingredients, such as B-sitosterol, that contribute to its effects. Most studies consider all of the plant's beneficial effects to be those of saw palmetto, not its component parts.

The primary goal for treating men with BPH is to reduce LUTS and increase quality of life.[20] Thus, at this time, clinicians are focused not on eliminating this chronic condition, but to reduce symptoms that lessen quality of life.

What kinds of measures are available to determine whether saw palmetto, or any other agent, can reduce the LUTS incidence in men? There are several, notably the International Prostate Symptoms Score (IPSS), whereby the patients evaluate the occurrence of symptoms such as frequency, urge, incomplete bladder emptying, and nocturia. In addition, urinary flow rate can be measured, along with postvoid residual volume and PSA levels. With these measures in mind, researchers have real parameters by which therapeutic agents can be evaluated and/or compared with others.

Saw palmetto for BPH and LUTS has been the subject of many studies, evaluated separately and also in comparison with finasteride.[21,22] In a randomized, double-blind, placebo-controlled study, saw palmetto led to a statistically significant improvement in urinary symptoms in men with LUTS compared with placebo.[23] Another study concluded that saw palmetto caused contraction of prostate epithelia cells and suppression of dihydrotestosterone levels in men with BPH.[24] The studies that compared saw palmetto with finasteride have also provided significant evidence of the efficacy of the herbal product. In a widely quoted study by Carraro et al,[25] a group of 1089 men were randomized to receive either saw palmetto or finasteride; the results showed almost equal benefit, although *less sexual dysfunction* was experienced with the herbal product.

The exact mechanisms for all of the benefits of saw palmetto have not been determined. As one action, saw palmetto supplements reduce the tissue levels of dihydrotestosterone. This has led to the belief that there is an inhibition of 5α-reductase involved (5α-reductase is the enzyme system that converts testosterone to dihydrotestosterone). Other mechanisms are less clear; although saw palmetto was shown to act as an α_1 blocker in in vitro studies,[26] it was not shown in in vivo studies.[27] From a diagnostic standpoint, it is interesting to note that saw palmetto does not reduce PSA levels, as does finasteride. Because of its reputed 5α-reductase inhibition, some have looked at saw palmetto as a novel agent for treatment of androgenic alopecia. In time, trials may be conducted that can validate this effect. At this time, clinical and/or observational data are scant. In summary, saw palmetto is considered a safe and effective agent for treatment of LUTS as an alternative to finasteride. It is usually dosed orally at 160 mg twice a day.

The Role of Compounding in Nutritional Supplements

From all available information, it is quite clear that there are substantial benefits to be obtained from nutritional supplements. In fact, the evidence is so compelling that it begs the questions: How does one take these agents? Furthermore, how do you find the right doses of the specific nutraceuticals that you are interested in? The authors of this chapter have heard presentations that extol the virtues of so many individual products that it would require a very large inventory as well as a check-off sheet and schedule to be sure that all desired agents are consumed. To be sure, there are combination commercial products on store shelves for sale now. For example, numerous products are sold for prostate health, perhaps containing selenium, vitamin E, and saw palmetto. These combination commercial products will never fill the needs of all; there has to be a way for an individual, in consultation with his physician and pharmacist, to obtain a preparation that fits his own personal needs. That means that the particular nutraceuticals desired for that person, in the specific optimal dose for each agent, in a dosage form that works well and is compliance-friendly, could be made available. It sounds like a lot to ask, but it is a real and present option.

Fortunately, because of the skill and knowledge of a compounding pharmacist, patients can get exactly what they need in a way that works best for them. There are many pharmacists today who have training and expertise in an up-to-date, modernized version of a traditional pharmacy practice. With the availability of top-quality pure powder, granules, or liquid form of the various nutraceuticals in their individual form, knowledge of the art of preparation, and the use of modern equipment and technology, these pharmacists can prepare a product uniquely designed for individual needs. Many of these compounded products can be put in capsules, the most common dosage form. Others can be made into pleasant-tasting, stable oral liquids when that better fits the patient's needs. And many pharmacists have the expertise and capability required for sterile injections, such as those that are occasionally used for water-soluble vitamins. The future holds the promise of even more convenient dosage forms such as transdermal creams and chewable wafers or treats, as compounding pharmacists seek to fill a void for those with special needs. Although many pharmacies do not have the option of compounding personalized supplements, the number of those that can is steadily growing. In 2003, there were approximately 3000 such pharmacies in the United States.

Discussion of the Case History

Mr. L.K. has requested a consultation to assess his physiological age and to embark on a preventive medicine strategy. He has altered his body fat content substantially through plastic surgery. A DEXA bone densitometry reading was performed to assess his osteoporosis risk. He also requested to be referred for electron beam tomography (EBT) to assess cardiac risk as well. In the clinic, he also underwent a stress electrocardiogram and several blood tests for his vitamin levels, hormonal profile, as well as oxidative stress profiles. His PSA was normal.

Oxidative stress is the harmful state that occurs when there is an overload of free radicals or a decrease in antioxidant levels. Scientific studies have revealed that free radicals are indeed culprits of many diseases of aging, including neurodegenerative disorders like Alzheimer's and Parkinson's disease. Oxidative stress can also affect conditions affecting vision such as cataracts and macular degeneration. Strategies aimed at limiting and repairing the damage attributed to oxidative stress may slow the advance of numerous age-related diseases. After the assessments, it was discovered that the patient's cardiac health was good. He further consulted an exercise physiologist and nutritionist to complete his optimal aging screening program. He continued with his current nutritional supplements and compounded testosterone was added, as he was found to be hypogonadal.

Conclusion and Key Points

Most physicians receive little education about nutrition. Although nutrition may influence health and be useful for prevention, the trials that verify the use of nutraceuticals are generally smaller than those verifying medications. There is sometimes a fine line between medications and over-the-counter nutraceuticals, and physicians should be cognizant of what their patients are consuming. Some of these nutraceuticals can interact with medications, and it is prudent for physicians to monitor for side effects. As nutraceuticals is such a large industry, and compounded by multilayer marketing approaches, it impacts the health of millions of people. Some of the nutraceuticals have been evaluated more thoroughly than others. Policy making also decides which nutraceutical may be available to the consumer directly. For instance, DHEA is available over the counter in the United Sates but not in many other countries.

- Prostate cancer seems to be more common in certain races and regions, and scientists have postulated links and possible preventive and even therapeutic interventions.

- So far, vitamin E, vitamin D, vitamin A, lycopene, selenium, soy, omega-3 fatty acid, and zinc have been implicated in preventive roles, but the evidence is not conclusive.

- Certain nutraceuticals may influence androgen metabolism, and of these DHEA, androstenedione, androstenedione, saw palmetto, nettle root, Chrysin, and zinc have been implicated.

- Saw palmetto has been studied quite extensively in the treatment of BPH and LUTS.

- Nutraceuticals can be compounded to alter bioavailability and be specifically useful for some patients.

REFERENCES

1. Business Communications Company, March 11, 2003. *www.bccresearch.com/editors/RGA*

2. Yip I, Heber D, Aronson W. Nutrition and prostate cancer. Urol Clin North Am 1999;26:403–411

3. Miller EC, Giovannucci E, Erdman JW, Bahnson R, Schwartz SJ, Clinton SK. Tomato products, lycopene, and prostate cancer risk. Urol Clin North Am 2002;29:83–93

4. Fleshner NE. Vitamin E and prostate cancer. Urol Clin North Am 2002;29:107–113

5. The Alpha-Tocopherol, Beta Carotene Cancer Prevention Study Group. The effect of vitamin E and beta carotene on the incidence of lung cancer and other cancers in male smokers. N Engl J Med 1994;330:1029–1035

6. Moyad MA. Selenium and vitamin E supplements for prostate cancer: evidence or embellishment? Urology 2002;59:9–19

7. Zhang Y, Ni J, Messing EM, Chang E, Yang C, Yeh S. Vitamin E succinate inhibits the function of androgen receptor and the expression of prostate-specific antigen in prostate cancer cells. Proc Natl Acad Sci USA 2002;99:7408–7413

8. Schwartz GC, Hulka BS. Is Vitamin D deficiency a risk factor for prostate cancer? Anticancer Res 1990;10:1307–1311

9. Polek TC. Wiegel Nl. Vitamin D and prostate cancer. J Androl 2002;23:9–17

10. Bodiwala D, Luscombe CJ, Liu S, et al. Prostate cancer risk and exposure to ultraviolet radiation: further support for the protective effect of sunlight. Cancer Lett 2003;192:145–149

11. Konety BR, Getzenberg RH. Vitamin D and prostate cancer. Urol Clin North Am 2002;29:95–106

12. Zhang XK. Vitamin A and apoptosis in prostate cancer. Endocr Relat Cancer 2002;9:87–102

13. Giovannucci E, Aschero A, Rim EB, et al. Intake of carotenoids and retinal in relation to prostate cancer. J Natl Cancer Inst 1995;87:1767–1776

14. Miller EC, Giovannucci E, et al. Tomato products, lycopene, and prostate cancer risk. Urol Clin North Am 2002;29:83–93.

15. Vogt TM, Ziegler RG, Granbard BI, et al. Serum selenium and risk of prostate cancer in U.S. blacks and whites. Int J Cancer 2003;103:664–670

16. Wang J, Eltoum JE, Lamartiniere CA. Dietary genisten suppresses chemically induced prostate cancer in Lobound-Wistar rats. Cancer Lett 2002;186:11–18

17. Kushi L, Giovannucci E. Dietary fat and cancer. Am J Med 2002;113:63S–70S

18. Liang T, Liao S. Inhibition of steroid 5 alpha reductase by specific aliphate unsaturated fatty acids. Biochem J 1992;285:557–562

19. Liang JY, Liu YY, Zou J, Franklin RB, Costello LC, Feng P. Inhibitory effect of zinc on human prostatic carcinoma cell growth. Prostate 1999;40:200–207

20. Jeong HJ, Shin YG, Kim IH, et al. Inhibition of aromatase activity by flavinoids. Arch Pharm Res 1999;22:309–312

21. Coleman CI, Hebert JH, Reddy P. The effect of phytosterols on quality of life in the treatment of benign prostatic hyperplasia. Pharmacotherapy 2002;22:1426–1432

22. Berges RR, Kassen A, Senge T. Treatment of symptomatic benign prostatic hyperplasia with beta-sitosterol: an 18-month follow-up. BJU Int 2000;85:842–846

23. Gerber GS, Kuznetsov D, Johnson BC, Burstein JD. Randomized, double-blind, placebo-controlled trial of saw palmetto in men with lower urinary tract symptoms. Urology 2001;58:960–964

24. Veltri RW, Marks LS, Miller C, et al. Saw Palmetto alters nuclear measurements reflecting DNA content in men with symptomatic BPH: evidence for a possible molecular mechanism. Urology 2002;60:617–622

25. Carraro JC, Raynaud JP, Koch G, et al. Comparison of phytotherapy (Permixon) with finasteride in the treatment of benign prostate hyperplasia: a randomized international study of 1,098 patients. Prostate 1996;29:231–240

26. Goepel M, Hecker U, Krege S, et al. Saw palmetto extracts potently and noncompetitively inhibit human alpha1-adrenoceptors in vitro. Prostate 1999;38:208–215

27. Goepel M, Dinh L, Mitchell A, et al. Do saw palmetto extracts block human alpha1-adrenoceptor subtypes in vivo? Prostate 2001;46:226–232

Index

The letter "t" or "f" following a page number indicates the entry on that page is a table or figure, respectively.